Progress in Green Energy

Xianguo Li
Editor

Green Energy

Basic Concepts and Fundamentals

Springer

Prof. Xianguo Li
University of Waterloo
Department of Mechanical Engineering
200 University Avenue West
Waterloo
Ontario
Canada N2L 3G1
x6li@uwaterloo.ca

ISSN 2191-561X
ISBN 978-1-4471-2698-0
DOI 10.1007/978-1-84882-647-2
Springer London Dordrecht Heidelberg New York

e-ISSN 2191-5628
ISBN 978-1-84882-647-2 (eBook)

British Library Cataloguing in Publication Data
A catalogue record for this book is available from the British Library

Softcover reprint of the hardcover 1st edition 2011

Cover design: eStudioCalamar, Girona/Berlin

Printed on acid-free paper

Springer is part of Springer Science+Business Media (www.springer.com)

Preface

Energy has become the prime commodity in modern civilization, and the amount of energy consumption has become the indicator for the standard of living and the degree of industrialization. It has long been recognized that associated with excessive energy use is the adverse impact on the environment, economy, and society, from local air and water pollution to the threat of global warming (the mean temperature increase around the globe) and climate variability (the temperature fluctuations around the mean); and from the economic difficulties arising from the rapid increase and swings in energy prices to the international tensions and geopolitics arising from access to and distribution of energy resources. The sustainable development of humanity and the economy with security of energy has topped national agendas around the world. It is imperative to develop energy strategies, policies, and technologies to achieve this objective through an energy system(s) that have no observable (or net) negative impact on environment, economy, and society – such energy systems are being referred to as green energy (systems). By extension, energy systems that have minimal or reduced negative impacts are being referred to as greener energy. Green energy systems thus include many essential elements that affect the impact of energy use, ranging from the green alternative and renewable energy sources, energy carriers and related energy conversion technologies – the traditional energy sectors to enforcible energy policies and instruments that promote sustainability; feasible energy conservation and management programs for efficient and effective energy use, including changes in energy user behaviors; system integration and optimization for reduced emissions of chemical, thermal, and greenhouse gas pollutants; monitoring and assessment of environmental impact and mitigation methods. Above all, green energy technologies and systems must be economical in order to compete and win in the marketplace.

The objective of this book series is therefore to provide a summary of the progress that has been made in the field of green energy, the current state-of-the-art knowledge and technology, and future developments and directions, as well as the significant technical and non-technical challenges that must be tackled. This book series can serve as a valuable resource on green energy and related issues in the development of sustainability and energy security for the beginners as well as for the experienced, from the researchers, engineers and scientists, to policy makers, technology developers, energy providers and consumers.

This very first volume of the book series contains six chapters. The first chapter provides an overview of energy systems, from their compositions to adverse impacts, and from the fundamental thermodynamic limitations to the problems associated with the excessive use of energy and the dominance of a single energy system; this leads naturally to the need for green energy for sustainability and energy security; and the practically feasible approach to achieving it being the concept referred to as "energy diversity", a word mimicking the more recognized term "bio-diversity", backed also by the historical and statistical perspective. The second chapter details the most useful analysis methodology for energy systems, energy and exergy analysis, rooted in the first and second laws of thermodynamics. The physical concepts and theory are developed with ample examples illustrating how they can be applied to practical energy systems, including solar and wind energy utilization.

The following four chapters are devoted to four important topical areas of wind energy, biomass energy, fuel cells and hydrogen storage. In the third chapter, a critical review is provided on the probability distribution of wind speeds, an important issue in the estimation of wind energy potential at a given site to determine the economic viability and the design of wind turbine installations at the site; the commonly used empirical distributions, Weibull and Rayleigh distribution functions, are analyzed and a theoretical distribution is developed that is formulated based on the maximum entropy principle. Chapter 4 is devoted to a critical review of the recent research and experimentation on cattle biomass (including both dairy biomass and feedlot biomass), their properties and utilization through co-combustion and gasification with coal. The polymer electrolyte fuel cell (PEFC) has become increasingly promising as a zero-emission power source for mobile and stationary applications, and it is recognized that the traditional macroscopic level modeling and simulation is not sufficient to fully understand the phenomena involved, which is essential for PEFC performance improvement and the cost reduction needed for commercialization. Chapter 5 provides a systematic account of recent efforts in pore-scale modeling to gain fundamental insight into two-phase transport along with the evaluation of the salient transport properties pertaining to the catalyst layer and the gas diffusion layer of a PEFC, and demonstrates a vertical approach encompassing a pore-scale to macro-scale modeling strategy. The practical application of PEFCs requires a compact and economic technique for hydrogen storage, which is the focus of Chapter 6, devoted to a comprehensive review of nanostructured hydrides for solid state hydrogen storage, from the fundamental hydrogen absorption/desorption characteristics to the recent progress in the nanoprocessing of solid state hydrides and high capacity hydrides.

I would like to thank all the contributors for their efforts in making their outstanding contributions to this book, and their patience in the meanwhile. There is no word of description that I can use to express my gratitude to the publisher and their editorial team whose patience, persistence and guidance made this book a reality. Finally, I would like to thank in advance all the readers for their feedback and criticism that will make future endavours of this book series more successful.

March 2009
Waterloo, Ontario, Canada

Xianguo Li

Contents

Short Biography of the Editor

Dr. Xianguo Li is a professor in the Department of Mechanical and Mechatronics Engineering at the University of Waterloo. He received his PhD in 1989 in mechanical engineering from Northwestern University in Evanston, Illinois, USA. He has a wide range of research interests, including fuel cells, liquid atomization and sprays, and energy systems. His research is both fundamental and applied, involving technologies deployable today and in the future. Dr. Li has published extensively, including over 100 journal articles. He has authored a book entitled "Principles of Fuel Cells" and edited/co-edited a number of books. His published articles have received extensive citations, including some rated as the "highly cited articles, within the top 1 % in the field", and many are listed within the top 25 hottest articles for the journals in which the articles were published.

Dr. Li has contributed actively to the progress of the profession and society. He is the founding editor in chief for the International Journal of Green Energy, and established the International Green Energy Conference series. He is the founding president of the International Association for Green Energy (IAGE). He is currently serving on the editorial board of a number of international scientific/technical journals, a book series on fuel cells and an encyclopaedia on energy engineering and technology. Dr. Li also serves as the division chair for the Advanced Energy Systems Technical Division, Canadian Society for Mechanical Engineering. He has also served on various ad hoc committees established by different levels of government. Dr. Li provides consulting services to national and international corporations and different levels of government.

List of Contributors

Kalyan Annamalai
Department of Mechanical Engineering
Texas A&M University
College Station
Texas
USA
e-mail: kannamalai@tamu.edu

Udayasarathy Arcot V.
Department of Mechanical Engineering
Texas A&M University
College Station
Texas
USA

Gerardo Gordillo Ariza
Department of Mechanical Engineering
Texas A&M University
College Station
Texas
USA

Nicholas T. Carlin
Department of Mechanical Engineering
Texas A&M University
College Station
Texas
USA

Tomasz Czujko
CanEnergy Technology Centre
Hydrogen Fuel Cells and Transportation Energy
Natural Resources Canada
1 Haanel Drive
Ottawa
Ontario K1A 1M1
Canada

Ibrahim Dincer
Faculty of Engineering
and Applied Science
University of Ontario Institute of Technology (UOIT)
2000 Simcoe Street North
Oshawa
Ontario L1H 7K4
Canada
e-mail: Ibrahim.Dincer@uoit.ca

Wyatte L. Harman
Blackland Research and Extension Center
Texas A&M University System Temple
Texas
USA

Kevin Heflin
Texas Agricultural Experiment Station
Texas A&M University System
Agricultural Research and Extension Center
Amarillo
Texas
USA

Ben Lawrence
Department of Mechanical Engineering
Texas A&M University
College Station
Texas
USA

Xianguo Li
Department of Mechanical and
Mechatronics Engineering
University of Waterloo
200 University Avenue West
Waterloo
Ontario N2L 3G1
Canada
e-mail: x6li@uwaterloo.ca

Partha P. Mukherjee
Los Alamos National Laboratory
New Mexico
USA
e-mail: partham@lanl.gov

Hyukjin Oh
Department of Mechanical Engineering
Texas A&M University
College Station
Texas
USA

Marc A. Rosen
Faculty of Engineering
and Applied Science
University of Ontario Institute
of Technology (UOIT)
2000 Simcoe Street North
Oshawa
Ontario L1H 7K4
Canada
e-mail: Marc.Rosen@uoit.ca

John M. Sweeten
Texas Agricultural Experiment Station
Texas A&M University System
Agricultural Research and
Extension Center
Amarillo
Texas
USA

Robert A. Varin
Department of Mechanical and
Mechatronics Engineering
University of Waterloo
Waterloo
Ontario N2L 3G1
Canada
e-mail: ravarin@mecheng1.uwaterloo.ca

Chao-Yang Wang
Director, Electrochemical Engine Center
Department of Mechanical Engineering
and Materials Science and Engineering
The Pennsylvania State University
338 A Reber Building
University Park
PA 16802
USA
e-mail: cxw31@psu.edu

Zbigniew S. Wronski
CanEnergy Technology Centre
Hydrogen Fuel Cells and
Transportation Energy
Natural Resources Canada
1 Haanel Drive
Ottawa
Ontario K1A 1M1
Canada

Chapter 1
Green Energy for Sustainability and Energy Security

Xianguo Li

1.1 Introduction

Worldwide energy consumption has been increasing rapidly [1], in fact almost exponentially, since the Industrial Revolution. This increasing trend of energy consumption has been accelerated by improvements in the quality of life, which almost directly relates to the amount of energy consumption as a result of the industrialization of developing nations and the population increase in the world. At present, most of the energy requirement worldwide is met by the combustion of fossil fuels (*i.e.*, coal, petroleum oils, natural gas, *etc.*) [1], which have become an essential and integral part of modern civilization, being increasingly relied upon since the Industrial Revolution. Only a very small proportion of the energy comes from nuclear and hydro power, and a much smaller portion from renewable energy sources, such as solar, wind, hydro, geothermal, tidal wave, and so on.

This almost exclusive reliance on the combustion of fossil fuels has resulted in enormous amounts of harmful pollutant emissions to our environment, has caused severe degradation of the local and global environment, and has exposed the world population (from humans to animals and from plants to all forms of life on earth) to the hazards and risks created by the extensive use of fossil fuels. For example, air pollution resulting from pollutant emissions poses a severe threat to the health of millions of people living in many of the world's urban areas. In 1998, over 113 million people in the USA were estimated to be living in areas not meeting US National Air Quality Standards [2]. Combustion of fossil fuels continues to contribute significantly to the increase in atmospheric carbon dioxide concentrations, thus intensifying the prospect of global warming and global climate variability, and

Xianguo Li
Department of Mechanical and Mechatronics Engineering, University of Waterloo,
200 University Avenue West, Waterloo, Ontario N2L 3G1, Canada
e-mail: x6li@uwaterloo.ca

threatening the very existence of civilization and mankind on planet earth. In addition to the health and environmental concerns, a steady depletion of the world's limited fossil fuel reserves also calls for alternative primary energy sources, and new energy technologies for energy conversion and power generation that are more energy efficient than the conventional combustion engine, with minimal or no pollutant emissions, and also compatible with renewable energy sources and sustainable development. Many studies have pointed to the hydrogen economy as the perfect solution to the present dilemma arising from the dominance of fossil fuels based energy systems [3, 4], and the fuel cell has been identified as the most promising potential energy technology as it meets all of the above requirements for energy security, economic growth, and environmental sustainability.

Is a hydrogen-based energy system the perfect solution to our plight or another "Pandora's Box" waiting to be opened? In this chapter, energy systems and their inevitable negative impacts on the environment, economy and society are examined, and the much-awaited hydrogen based energy system is analyzed. Lessons from history are learned and analogies with other fields made. We conclude that the best approach to the issue of energy, environment and sustainable development is to employ green energy systems, which can best be achieved through the diversification and localization of energy systems, also the best approach to the security of energy. In the following sections, we first describe energy systems and their impacts on the environment, economy and society, then the concept of green energy as a long-term target for sustainable development is described, and finally the idea of diversification and localization of energy sources and systems is developed as the only sensible and practically feasible solution to the goal of sustainability and energy security.

1.2 Energy Systems: Their Composition

Consumption of energy (or useful energy or exergy, to be thermodynamically correct) has become a daily necessity in modern civilization for the comfort and convenience of humanity, and the amount of energy consumption has served as an indicator for the standard of living and the degree of industrialization. It has long been recognized that associated with this excessive daily energy consumption is an adverse impact on the environment we live in, resulting in deterioration of the local and global environment. However, utilization of energy from different sources tends to have different kinds and different degrees of impact on the environment. For example, the severe impact of energy use from fossil fuels has been well known for decades, while the energy from renewable sources may be considered to have minimal or neutral impact on the environment, so long as the amount of their usage remains low. The adverse impact of the excessive use of bioenergy on the environment, economy and society has been felt and recognized more recently due to the worldwide hike in the price of foods over the last few years. Therefore, it is useful to look into the composition of energy systems and their adverse impacts on the environment, economy and society – the focus of this and following sections.

An energy system may be considered as composed of five interconnected components as shown in Figure 1.1 [5]. It includes the raw energy resources that are available in nature (such as coal, sunlight, wind, *etc.*), and that are harnessed and processed/refined to a form or forms convenient for distribution, storage and utilization of energy (referred to as transformer technologies in Figure 1.1). Such convenient forms of energy are often called energy carriers, or energy currencies, such as gasoline, diesel and hydrogen. The extraction and processing of energy resources and the production of energy carriers form the traditional energy sector of the industry. The technologies that deliver the services needed by individuals and society as a whole (or the service technologies) and the energy services needed for the convenience and comfort of humans make up the remaining part of the energy system. It is noticed that human needs (or energy services) dictate the energy system and its evolution – the insatiable human desire is the constant source of the driving power for the improvement and evolution of energy systems. Typically, the energy services needed and the energy resources available in nature (energy sources) remain fixed, unchanged over time, whereas what is changing is the technologies for the extraction of energy from the natural resources (transformer technologies), the carriers of energy, and the technologies that provide the energy services needed (service technologies). Numerous examples are given in Figure 1.1 for each of the five components in an energy system.

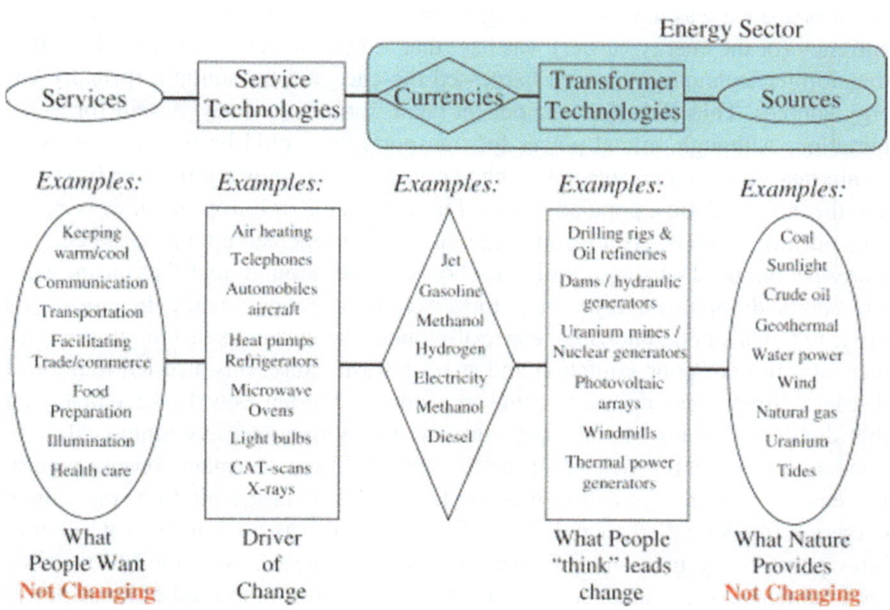

Figure 1.1 A complete energy system includes the traditional energy sector that provides the energy needed as well as the services wanted and the technologies delivering the services [5]

For a typical comparison of energy systems and their merits and disadvantages, all these five components of an energy system should be included; such an analysis is often called life cycle analysis or cradle to graveyard analysis. For example, hydrogen as a fuel (or energy carrier) is sometimes considered as clean, since its use as fuel does not emit harmful chemical pollutants and greenhouse gases at the point of use, however, the production of hydrogen from fossil fuels (often natural gas) and the transport and distribution of hydrogen involves the emission of pollutants and greenhouse gases [6]. Therefore, hydrogen as a fuel might not be completely clean.

1.3 Energy Systems: Their Adverse Impacts

An energy system is like a double-edged sword; its use would normally bring about economic growth and social advancement as a whole, and comfort and convenience for individuals. On the other hand, a persistent and large-scale use of a particular energy system will also bring about inevitable negative environmental, social and economic impact, and when this negative impact is accumulated beyond a critical threshold (the tolerance limit), permanent damage (or even catastrophe) would occur.

Let's take transportation as the energy service as an example. Traveling from one place to another was originally by using animal power, say horses or horsepower, before the arrival of combustion engines. The technology that provided the services needed was horse-drawn wagons (the service technology); and the energy currency (or the energy carrier) was hay, that fed the horses, from agriculture (the transformer technology), which harnessed the energy from sunlight (primary energy source). Thus, this five-component chain forms an energy system for transportation. Although animal power for transportation would be regarded today as an alternative and renewable approach to transportation, only a little over 100 years ago the main and most popular means of transportation in European cities were the horse-drawn wagons. As a result, large cities in Europe had a major problem: too many people used carriages pulled by horses to get around, and that meant there were horse droppings everywhere. Horse excretions not only made the city streets dirty, but also smelled horrible, especially under the burning sunshine in the summer; at a time of poor sanitation and street infrastructure, this also led to a lot of diseases. Street cleaning had become an economic burden as well. Accompanying this dominant means of transportation was the horrendous environmental, economical and societal problems of the time. At that time, the main concern regarding environmental degradation was the terrible smell in the air from the animal excretion (the so-called air pollution of the time). It was no wonder that automobiles powered by petroleum oil were hailed as a "perfect solution" to the then environmental, social and economic problems arising from the use of horse power for transportation in urban areas.

An automobile (the service technology) that provides transportation from one place to another (the energy service needed) uses gasoline or diesel as the energy carrier, and gasoline or diesel is derived from crude oil (the primary energy source) through oil exploration and oil refineries (the transformer technology) – thus a new energy system is formed for transportation based on fossil fuels. The inevitable negative impact of this energy system became obvious only a few decades after automobiles were mass introduced. It was then realized that automobiles powered by petroleum oil were not a "perfect" solution after all (such as the famous smog in Los Angeles); automobile exhaust emissions not only degrade the local environment, especially in urban areas, but also contribute to the deterioration of the global environment, and increase the health risk to humans, animals and plants. Beyond the significant economic consequences, social structure evolves as well, such as the sprawling urban areas and suburban designs – the extreme dependence of North Americans on automobiles for their daily routines. Today, the majority of transportation needs throughout the world is still being met by automobiles powered by fossil fuels. It is estimated [7] that on average, some 90 billion tons of fossil fuel pollutants, namely, CO_2, CO, SO_2, NO_x, soot and ash, are spewed out annually into the atmosphere. These pollutant emissions are the main cause of the greenhouse effect, air pollution and acid rain. The annual cost of damage caused by these pollutants to humans, to crops, to all flora and fauna and to man-made structures, in fact, to our entire environment on a worldwide basis, is around five trillion US dollars. This economic cost is equivalent to $800 US per person annually, or about 14 % of the gross world product per capita. It is to be noted that automobile emissions constitute the single largest source of emissions within the present fossil fuel based energy system, and is the dominant source of emissions and health risk for many of the world's urban areas. Closer to home, on average five tons of greenhouse gases are emitted annually per person in Canada, attributable to personal energy consumption, based on the One-Tonne Challenge program instituted by the Government of Canada in 2005 (excluding emissions related to energy consumption in industrial sectors); 50 % of these emissions are due to automobile use and another 20 % due to space heating/cooling. Based on a multi-year and multi-party study commissioned by the Government of Ontario and released in June 2005 [8], the cost of air pollution in Ontario includes 5,800 premature deaths and $ 9.6 billion in health and environmental costs annually. Thousands more Canadian suffer from respiratory illnesses such as bronchitis and asthma due to the poor air quality.

In general, air and water pollution arising from the use of fossil fuels have increased health hazards and risks, and global climate changes, which include the increase in mean temperatures on earth (better known as global warming) and fiercer fluctuations around the mean temperature (or global climate variability). Although the prospect of global warming and its effect might be argued today as being in the future, about 100 years or so away, global climate variability has already had dreadful consequences around the world as it has increased the frequency and intensity of the occurrence of natural disasters. Both the prospect of global warming and global climate variability have led to changes in health and other standards and codes, such as building codes and the design of urban areas and man-made structures, with significant social and economic implications. Diminishing fossil fuel reserves has additional adverse impacts on our economy and society; consider the effect that high crude oil price has on our economy and the associated social implications (such as higher costs for living and business, workers' demands for higher wages, businesses raising the price for their products and services, *etc.*, all leading to economic inflation), in addition to international tensions, conflicts and politics. Secured access to energy resources, energy supply and delivery to end users is topping the national interest and security agenda, and energy security has become one of the top priorities of national security measures for many nations. It is imperative to develop and deploy sustainable energy systems to ensure a sustainable future for individual nations and the world as a whole.

This transportation energy system example clearly illustrates the evolutionary nature of the environmental, social and economic impact associated with the energy system; alternatively, different energy systems have different impacts on different aspects of the environment, economy and society. Further, these negative impacts seem to become more severe and the damage caused seems to become more far-reaching with advancement of the energy technology involved.

1.4 Energy Systems: The Dilemma

As described earlier, energy is a double-edged sword; our modern civiliation cannot function properly without it, yet the future of our civiliation is being threatened by excessive use of it. So the natural question is what is the "perfect" solution to the dilemma we are facing today? Is there any chance that a balance can be achieved such that energy use will not create the observable negative impacts while harvesting its benefits to our civilization?

Maybe a more pressing question is what is the "perfect" solution to the environmental, economic and social problem caused by the present fossil fuels based energy system? Opinion abounds! It ranges from alternative fuels and renewable energy sources to the visionary hydrogen energy system as the long-term solution [3, 4]. From the Clinton-era Partnership in Next Generation Vehicles (PNGV) to the Freedom Car Program instituted by the Bush Administration, hydrogen fuelled fuel cells are being favored as the clean power source for transportation. Since hydrogen combines with oxygen in air to form the reaction product water, which is considered benign to the environment, the hydrogen energy system has been promoted by some as the "perfect" solution for sustainability and energy security. However, some studies [9] indicate that some 10–20 % of hydrogen in a hydrogen energy system would be released (or leak) to the atmosphere, along with an increase in the concentration of water vapor in the entire atmosphere. This would cause stratospheric cooling, enhancement of the heterogeneous chemistry that destroys ozone, an increase in noctilucent clouds, and changes in tropospheric chemistry and atmosphere–biosphere interactions. In short, the increased concentration of hydrogen and water vapor in the atmosphere would interfere with the atmospheric chemistry, affecting the ozone layer that protects the earth from UV radiation. Higher hydrogen concentration may even possibly affect microorganisms in soil and water because hydrogen is a microbial nutrient – with possible severe, yet currently unforeseen, consequences. Higher levels of water vapor in the atmosphere would also accelerate the corrosion of man-made structures, with a possibly uncomfortable humid environment for humans and animals. Although the above negative impact on the environment has been simulated on computers, as in [9], the negative social and economic impacts are much harder to predict exactly. However, if history is any indication, a new energy system tends to have more far-reaching and more severe impacts than the previous energy system it replaces when the impact has accumulated beyond a critical threshold (the tolerance limit), although that impact becomes less personal.

Therefore, as Henry Ford II once said "The economic and technological triumphs of the past few years have not solved as many problems as we thought they would, and, in fact, have brought us new problems we did not foresee". Such a scenario seems to be the case for the evolution of energy sources and systems. In fact, each new energy system or technology tends to have more severe and far-reaching adverse impacts on the environment, economy and society than the previous energy system it supersedes or replaces. Figure 1.2 shows the evolution of energy sources since the Industrial Revolution in the 1800s [5, 10]. It seems clear that the dominant energy sources and systems are moving in the direction of decreasing carbon content in the fuel. However, the total pollutant emissions associated with the consumption of fossil fuels (whether coals, petroleum oils, or natural gas) have been increasing due to the increase in total consumption of energy, related to the improvement in the quality of life, population growth and the industrialization of developing nations. The hazards associated with the use of nuclear energy have been recognized for some time now, especially after the Three Miles Island incident and Chernobyl accident. Further, the disposal and safe storage of

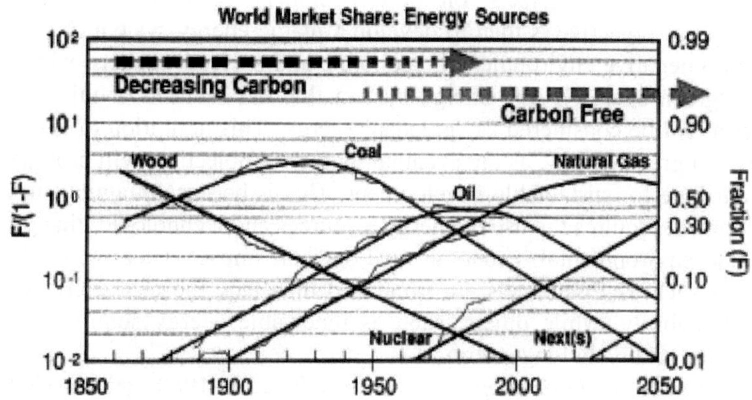

Figure 1.2 Evolution of energy sources since the Industrial Revolution in the 1800s. The energy sources are expressed in terms of the fraction of the world market share F [5, 10]

nuclear waste for hundreds and even thousands of years (most severe for long-lived radionuclides and can be of the order of one hundred thousand years) is another significant challenge, both technically and socially. As the saying goes, there is no free lunch. This is also consistent with the second law of thermodynamics, which states that degradation of useful energy always occurs in all energy systems, and that degraded energy is almost always thrown into our environment like garbage – a burden on our environment. Each energy system, as with the transportation example described earlier, weakens a particular aspect of the environment and causes an excessive burden that would eventually turn into environmental calamity if a single energy system were allowed to dominate for long. This is similar to the situation that the persistent and abusive use of antibiotic drugs in the past has led to the emergence of antibiotic-resistant bacteria. Accompanied with the environmental damage are the economic and social problems. Then once again, what is the "perfect" solution to the environmental, economic and social problems arising from the current excessive use of energy?

The second law of thermodynamics dictates that degradation of useful energy is inevitable, so is the associated adverse impact of the degraded energy dumped into the environment, as long as energy is continuously being consumed. Hence, all energy systems have negative or adverse environmental, economic and social impacts, without any exception, be it the so-called renewable energy systems, the present fossil fuel based energy systems, or the futuristic hydrogen energy systems. So the inevitable question is: are we doomed after all? What is the "perfect" solution for environmental protection, sustainable development and energy security?

1.5 Green Energy and Sustainability: The Target and Solution

Since the Industrial Revolution, increased energy use has brought about economic prosperity and an improved standard of living. It is fully expected that this trend would continue without "side effects" on the environment, economic and social growth. The target or objective is then to develop a magic energy system or systems that have no negative environmental, economic and societal impacts, which we refer to as "green energy". Any energy system that has reduced or minimal adverse impact might be considered as "greener" energy. This definition of green energy implies that green energy, as the eventual long-term objective, will provide an important attribute for sustainable development. This is because attaining sustainable development requires the use of energy resources and technologies that do not have adverse environmental, economic and societal impact. Clearly, single energy resources such as fossil fuels are finite and thus lack the characteristics needed for sustainability, while others, such as renewable energy sources, are sustainable over the relatively longer term.

The concept of *sustainable development* was introduced in 1980, gained wide publicity in the 1987 report of the World Commission on Environment and Development (the Brundtland Commission) [11], and achieved worldwide prominence at the UN Conference on Environment and Development in Rio de Janeiro in 1992. The term sustainable development was defined by the Brundtland Commission as "development that meets the needs of the present without compromising the ability of future generations to meet their own needs". The Commission noted that its definition contains two key concepts: *needs,* meaning "in particular the essential needs of the world's poor", and *limitations*, meaning "limitations imposed by the state of technology and social organization on the environment's ability to meet present and future needs" [11, 12]. Thus, sustainable development includes environmental, social and economic factors, regarded as a key to the solution of current environmental, economic and developmental problems, and has been developed into a blueprint for reconciling economic, societal and environmental necessities.

Key requirements for sustainable development include societal, economic, and environmental sustainability, all related to the sustainability of energy systems. For example, the kinds of techno-economic changes envisaged by many as necessary for long-term sustainability usually include sharp reductions in the use of fossil fuels to minimize the danger of global climate change. Alternatives to using fossil fuels include use of nuclear power, large-scale photovoltaics, intensive biomass cultivation and large-scale hydroelectric projects (in applicable regions), as well as major changes in patterns of energy consumption and conservation, although there are disputes over which of these energy alternatives is the most desirable and feasible, *etc.* [13].

The concept of sustainable development originated from the excessive use of fossil fuels and the associated environmental problems, and it has also raised the important question concerning what might be the physical or environmental limits to economic growth or the maximum carrying capacity of the environment. Sustainability-related limits on energy use include [12]:

- The rates of use of renewable resources should not exceed their rates of regeneration.
- The rates of use of non-renewable resources should not exceed the rates at which renewable substitutes are developed.
- The rates of pollutant emissions should not exceed the corresponding assimilative capacity of the environment.

Sustainability, or its opposite unsustainability, should also be considered in terms of geographic scope. Some activities may be globally unsustainable, such as those that result in global climate change or stratospheric ozone layer depletion, affecting several geographic regions, if not the entire world. Some activities may be regionally (or locally) unsustainable, such as those that lead to acid rains, destroying vegetation and resulting in famine in one region but not in other parts of the world. Some other activities might be locally unsustainable, but with global implications, *e.g.*, the deforestation of one region might cause desertification locally, but with global impact. Overall, sustainability appears to be more a global than a regional or local concern. If an environmental impact exceeds the carrying capacity of the planet, for instance, then life is threatened, but if it is beyond the carrying capacity of one area, then that area may become uninhabitable but life can probably continue elsewhere.

From the above discussion, it becomes apparent that green energy, having no negative environmental, economic and societal impact, can be regarded as sustainable energy (system). But in the light of the fundamental implication of the second law of thermodynamics, how can one develop an energy system or systems without any undesirable impact?

1.6 Diversification and Localization of Energy Systems: A Means to Sustainability and Energy Security

As pointed out earlier, the dominance of a single energy source and system, no matter how "perfect" it might be at the time, would become unsustainable in the long run. This is because the adverse impact is additive and repetitive, and once accumulated beyond a critical threshold, permanent damage ensues due to either the mere magnitude of the impact or fatigue. This becomes clear when considering the transportation energy system described earlier; a new energy system typically appears to be good when first introduced at the low level, then it becomes bad when its use becomes widespread and significant; it eventually becomes ugly when its use becomes excessive.

Before continuing our quest for a "perfect" solution for sustainable development and energy security, let us digress for a moment into other fields for enlightenment. In forestry, it is common practice that harvested areas are re-forested with a variety of species of trees interspersed – the so-called silviculture, because it is well known that bio-diversity is the key to the health of eco-systems, and that a diverse forest is best able to resist the attack and spread of pests and diseases. Other examples highlighting the success of diversity abound, and include:

- The success of democracy, which modern civilization holds dearly, could be considered as the diversification of idea development, consensus building, decision making, and governance.
- Investment companies diversify their holdings with investments ranging from high risk to conservative in order to prevent losses and guarantee healthy returns.
- The Canadian policy of multiculturalism, which recognizes and promotes the culture of immigrants, improves the social harmony of the Canadian people and contributes to the Canadian domestic economy and international trade and relations.
- Widespread computer viruses disrupt our work and often cause millions and even billions of dollars in economic losses worldwide. However, the computer viruses tend to target only one operating system, leaving computers running alternative operating systems unscathed – clearly the dominance of one single

particular computer operating system has resulted in vulnerability to the spread of computer viruses and damage in spite of painstaking efforts by the world's richest person.

Given that success in other fields arises from diversity, why should a single energy system be allowed, or even promoted preferentially, to become dominant? Diversification of energy systems should be anticipated to be healthy and beneficial for humanity and the environment as a whole, and energy diversity may be the key to sustainable development and energy security.

The dominance of a single energy source and centralized power generation are highly susceptible to disruption, failure and even sabotage, with severe consequences economically and socially, as articulated clearly and convincingly by Lovins and Lovins [14, 15] – the resulting energy insecurity and fragility is "disasters waiting to happen" [14]. In fact, even within one energy system, system resilience and energy security can be greatly increased with distributed and dispersed power generation. The reduction in energy security vulnerability can be achieved by distributing power generation through having many mini-power plants dispersed over different areas near energy consumers, instead of a few mega ones, and thereby reducing transmission losses (which amount to 10 % of electricity generated at the power plants in North America) and the vulnerability arising from having to transmit energy over wide areas.

It has been recognized that each energy system has its own drawbacks and seemingly insurmountable disadvantages. Setting aside the fossil fuels based energy systems, the renewable energy sources such as geothermal, tidal, solar, wind, hydroelectric, bio-fuels, *etc.*, all suffer from diurnal, seasonal, and yearly variations, as well as sensitivities to weather conditions and geographical locations. In addition, each energy system also has an adverse impact on a different aspect of the environment, which may only become evident when it is used at large scale and in a dominant position for a sufficiently long time. Some of the potential negative impacts are summarized in Table 1.1 for a number of energy sources and hydrogen as energy carrier. For example, harnessing hydroelectric power requires the building and maintenance of huge dams to collect and keep water for electricity generation, and the building of dams and the collection of a large body of water affect the balance of, and even damage, local eco-systems, and could even affect the local weather conditions. Some large-scale hydroelectric power generation plants may even displace local populations and damage local historical and cultural relics – thus the possible adverse impact socially and culturally, in addition to the local environment, such as the Three-Gorge project in China. The large body of water collected before the dam exerts excessive weight on the local geographical structure, potentially inciting the occurrence of earthquakes. Although a single energy system could be sustainable if the negative impact is sufficiently small, within the carrying capacity (or the tolerance limit) of the environment, economy and society, as in the pre-industrial era when a biomass-based energy system was predominant. However, in modern society the total negative impact related to the substantial amount of energy use has exceeded the tolerance limit of the environ-

Table 1.1 Energy sources and their potential negative impacts on environment

Energy source/ carrier	Potential negative impact on environment
Fossil fuel	Air pollution, acid rains, ozone depletion, global warming potentials
Hydrogen	Thermal and chemical changes in atmosphere, ozone depletion, influence on microorganisms in soils and water, accelerated corrosion of man-made structures
Wind	Landscape change, soil erosion, reduced air circulation and deterioration of local air quality
Solar	Landscape change, soil erosion, reduced solar irradiation for plants and vegetation
Hydro	Changes in local eco-systems and local weather conditions, social and cultural impact, induction of earthquake
Geothermal	Landscape change, underground water resource, accelerated cooling of earth core
Tidal/wave	Landscape change, reduced water motion/circulation and deterioration of local water quality
Biofuels	May not be CO_2 neutral, may release global warming gases like methane during the production of biofuels, landscape change, deterioration of soil productivity
Nuclear	Radiation leakage and contamination; the disposal and safe storage of nuclear waste for hundreds of years up to a hundred thousand years in geological repositories

ment; the fossil fuels based energy system is no longer sustainable for many regions of the world and for the world as a whole. As discussed in Section 1.4, nuclear energy or the proposed future hydrogen energy system will not be able to remain sustainable environmentally, economically and socially. With the ever-increasing total amount of energy needed worldwide to sustain social and economic activities, a sustainable single energy system has not been developed, and it is unlikely to emerge in the foreseeable future.

Therefore a combination of various energy systems with available energy resources must be developed to meet the energy requirement for individual nations, regions, organizations or even individuals. Since each renewable energy source has a variable output, the use of a combination of energy sources would allow for a steady, reliable source of energy supply. Stormy weather may reduce the direct collection of solar energy, but would increase the wind energy and hydropower, even the tidal and wave energy; dry, sunny weather may not be good for hydropower, but would be great for solar energy collection. Therefore, a diversity of energy sources that are locally available is essential for energy security and sustainable development – diversification and localization of energy sources and systems go hand in hand.

Even though each energy system has its own adverse impact on a particular aspect of the environment, economy and society, if that impact is small enough the environment, economy and society can tolerate or withstand (or recover from) the impact, then that particular energy system may be considered sustainable. Many different energy systems, which have an adverse impact on different aspects of the environment, economy and society (these impacts are not directly additive), are each sustainable due to their respective small impact. The aggregate of many such different energy systems may provide the large quantity of energy needed for the sustainable development of our economy and the growth of humanity with an overall undegraded environment. Taking again the transportation example, in urban areas bicycling, horse-drawn wagons, automobiles powered by petroleum oils and hydrogen fuel cells, can all be used for transportation purposes; and if each means of transportation is not highly populated, horse excretions may not be too smelly (or can be easily cleaned up), pollutant emissions from the burning of fossil fuels may be well within EPA limits for minimal or no health risks, and the total hydrogen leakage into the atmosphere from hydrogen fuel cell powered vehicles may be small enough to have minimal or no impact on the atmospheric chemical chain process.

To illustrate the importance of localization of energy systems, consider electric power generation as an example. If power generation is distributed at various levels, such as at region, city or town, district, street, or house levels, while connected through an electric transmission grid, each power generator serves fewer and nearer energy consumers. The impact of a generator failure will be greatly restricted, and the power needed can easily be shipped in through the grid from other generators. Fixing the failed generator will also require much less expense with greatly reduced down time – illustrating much less vulnerability in energy security and much less economic loss. Such a distributed power system would significantly reduce, if not totally avoid, the severe impact of blackouts on the economy and daily life, such as the one due to an ice storm in Quebec in January 1998. An analogy of the above distributed power system might be computers connected through the internet (or intranet); each computer can be operated independently, and if a particular computer develops a problem it can easily be disconnected from the internet without adversely affecting other computers connected to the same network. A real event occurred in 1965, when a power engineer in Holyoke, Massachusetts, USA, realized that the blackout that struck most of the north-east was rolling toward him, he simply disconnected the city grid from the collapsing grid, and powered the city with a local gas turbine [14, 15]. A secure and sustainable energy system should therefore consist of many relatively small and dispersed energy sources that are locally available, and the diversity of energy sources should be interconnected not at a central hub but by many short, robust links. The added economic advantage of diversified and localized energy systems is that it keeps the economic activities locally because the skills, manufacturing and maintenance capacity, *etc.* are all needed locally. That's exactly why the Lovins called it "small is profitable" in their new book promoting distributed power generation systems [16].

Clearly, the exact combination of energy sources and systems must be optimized to allow for the local and global environment, economy and society to be tolerant of the waste energy and pollutants dumped into the environment, and at the same time allow the healthy growth of humanity and economic development.

The optimal combination or mix of energy sources and systems must be considered in terms of the local situation – what local energy sources are available and what could the local environment tolerate? This localization of energy sources and systems will ensure that no nation or organization can control the world's energy supply, thus reducing, if not avoiding altogether, international tensions related to the access to, and control of, energy resources. The diversification and localization of energy sources and systems are especially meaningful for the security of energy supply and distribution to energy consumers, especially in the aftermath of the power outage from the northeastern to midwestern states of the USA and part of Canada in mid-August 2003. In summary, diversification and localization of energy sources and systems may be the "perfect" solution we humans have been searching for – a truly sustainable energy system with energy security for all nations and all people without international quarrels, tensions, and even armed conflicts over the energy resources.

Having set the goal of green energy systems for sustainability and energy security in the previous section, this section develops the notion of energy diversity with local resources as the best means to achieve the goal. Since each energy system includes the five essential components described in Section 1.2, and each component can interact with one another and with the environment, economy and society, Figure 1.3 provides an illustration of what green energy systems are about. Since green energy systems are those energy systems that have no observable (or net) negative impact on the environment, economy and society, by green energy systems we mean many essential elements that affect the impact of energy use, including the green alternative and renewable energy sources, energy carriers and related energy conversion technologies – the traditional energy sectors. In addition, it is important to have enforcible energy policies and instruments that

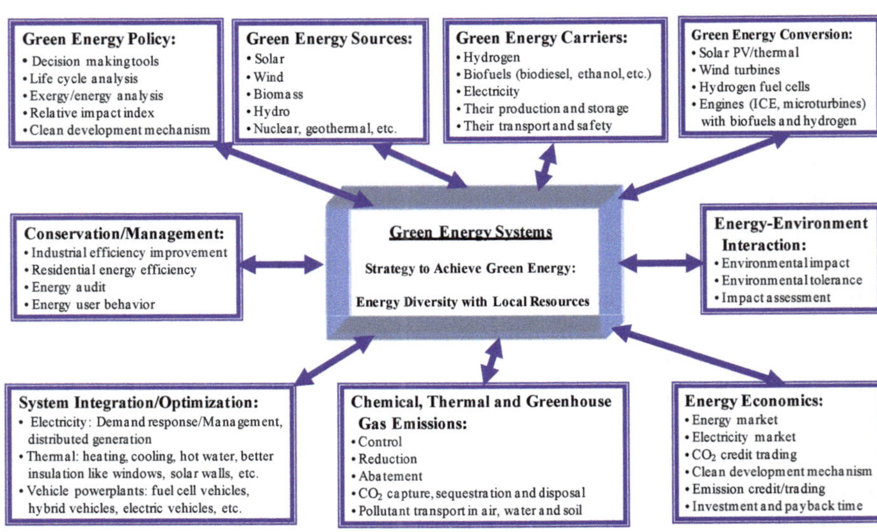

Figure 1.3 Essential elements of green energy systems

promote sustainability, feasible energy conservation and management programs for efficient and effective energy use, including changes in energy user behaviors, system integration and optimization for reduced emissions of chemical, thermal and greenhouse gas pollutants, and monitoring and assessment of environmental impact and mitigation methods. Above all, green energy technologies and systems must be economical in order to compete and win in the market place.

At this point, it should be stressed that diversified and localized energy systems can also become unsustainable if the negative impacts exceed the tolerance limits; such a situation may be a worst nightmare since permanent damage can now occur in many different areas of the environment, economy and society because of the diversity of energy systems. This issue is very much related to the maximum population of a region or nation at the local level and the maximum population the planet earth can support at the global level. Although it is not straightforward to determine the maximum sustainable population on earth due to the uncertainties associated with the estimation of the tolerance limits for different energy systems and their interactions, it is certain that diversified and localized energy systems will allow for a larger maximum sustainable population than any single energy system could support. Another issue of possible consequences for the diversification and localization of energy systems is the economics, or the potentially higher price of energy products and services due to the possibly reduced volume output due to diversification. This issue may warrant further detailed analysis, but in general terms, localization will keep the jobs and expertise developed locally. Diversity at the global level ensures that there are sufficiently large markets for the particular energy services and products developed for volume output, a key for cost reduction.

So the ultimate objective we hope to achieve through this study and many further studies in the future is to promote the idea of green (or sustainable) energy through the approach of diversification and localization of energy sources and systems, to optimize these various energy systems for nations, regions, organizations and individuals for environmental compatibility and social and economic development, to develop the technology needed for the implementation of various energy systems at reasonable costs, and to debate and establish public energy policy for the fostering of diversification and localization of energy systems so that truly sustainable development and energy security can be achieved. This is a noble goal, and cannot be achieved with the effort of an individual or a few individuals.

1.7 Summary and Outlook

The dominance of a single energy system tends to weaken a particular aspect of the environment, economy and society; and can cause permanent environmental damage or even environmental catastrophe if dominant for too long, with devastating consequences for the economy and society. Each energy system has its own adverse impact on the environment, economy and society as dictated by the second law of thermodynamics. However, if that impact is small enough and is within the tolerance range, environment, economy and society can fully withstand the adverse impact and can fully recover from it. Thus, a truly sustainable energy system (or green energy) can be achieved with the diversification and localization of energy sources and systems, which would also provide security for the energy supply and distribution. Although a single energy system can be sustainable, diversified energy systems with local resources can provide a larger amount of sustainable energy than a single energy system can, with better security of energy. Therefore, it is recommended to promote energy diversity as the sensible and practically feasible approach for sustainable development and energy security.

References

[1] IEA, Key World Energy Statistics. www.iea.org/statist/keyworld2002/keyworld2002.pdf (accessed Feb 2003).
[2] Chalk SG, Miller JF, Wagner FW. Challenges for fuel cells in transport applications. J Power Sources 2000;86:40–51.
[3] Barreto L, Makihira A, Riahi K. The hydrogen economy in the 21st century: a sustainable development scenario. Int J Hydrogen Energy 2003;28:267–284.
[4] Adamson KA. Hydrogen from renewable resources – the hundred year commitment. Energy Policy 2004;32:1231–1242.
[5] Scott DS. Interpreting the architecture of the energy system. Proceedings of World Energy Council 16th Congress, Tokyo, Japan, 1995.
[6] Zamel N, Li X. Life cycle analysis of vehicles powered by a fuel cell and internal combustion engine for Canada. J Power Sources 2006;155:297–310.
[7] Veziroglu TN. Commencement of International Journal of Green Energy 2004;1:1–2.
[8] Ontario Medical Association. The illness cost from air pollution: 2005-2026. Health and economics damage estimates; 2005. http://www.oma.org/Health/smog/report/ICAP2005_Report.pdf.
[9] Tromp TK, Shia RL, Allen M, Eiler JM, Yung YL. Potential environmental impact of a hydrogen economy on the stratosphere. Science 2003;300:1740–1742.
[10] Marchetti C, Nakicenovic N. The dynamics of energy systems and the logistic substitution model, RR-79-13. International Institute for Applied Systems Analysis, 1979.
[11] World Commission on Environment and Development (WCED). Our common future (The Brundtland Report). Oxford: Oxford University Press, 1987.
[12] OECD, Pollution Prevention and Control: Environmental Criteria for Sustainable Transport. Organisation for Economic Co-operation and Development. Report: OECD/GD(96)136, Paris, 1996.
[13] Dincer I, Rosen MA. Exergy as a driver for achieving sustainability. Int J Green Energy 2004;1:1–19.
[14] Lovins AB, Lovins LH. Brittle Power: Energy Strategy for National Security. Andover, MA, USA: Brick House Publishing Co., Inc., 1982.
[15] Lovins AB, Lovins LH. The fragility of domestic energy. The Atlantic Monthly, pp. 118–125, November 1983.
[16] Lovins AB, Lovins LH. Small Is Profitable: The Hidden Economic Benefits of Making Electrical Resources the Right Size. Rocky Mountain Institute: Snowmass, CO, 2002.

Chapter 2
Exergy Analysis of Green Energy Systems

Ibrahim Dincer and Marc A. Rosen

2.1 Introduction

Exergy analysis is a thermodynamic analysis technique based primarily on the Second Law of Thermodynamics. As an alternative to energy analysis, exergy analysis provides an illuminating means of assessing and comparing processes and systems rationally and meaningfully. Consequently, exergy analysis can assist in improving and optimizing designs. Two key features of exergy analysis are (1) it yields efficiencies which provide a true measure of how nearly actual performance approaches the ideal, and (2) it identifies more clearly than energy analysis the types, causes and locations of thermodynamic losses.

Increasing application and recognition of the usefulness of exergy methods by those in industry, government and academia has been observed in recent years in many countries. This point is backed by the fact that over the past 25 years several books on exergy analysis have been published [1–10], and many research articles have been published [11–15]. Applications of exergy have occurred in a diverse range of fields, including electricity generation and cogeneration, fuel processing, energy storage, transportation, industrial energy use, building energy systems, and others. In addition, exergy analysis has been applied to fields outside thermodynamics, such as biology and ecology, management of industrial systems and economics. The present authors, for instance, have applied exergy analysis to various

---

Ibrahim Dincer
Faculty of Engineering and Applied Science, University of Ontario Institute of Technology (UOIT), 2000 Simcoe Street North, Oshawa, Ontario L1H 7K4, Canada
e-mail: Ibrahim.Dincer@uoit.ca

Marc A. Rosen
Faculty of Engineering and Applied Science, University of Ontario Institute of Technology (UOIT), 2000 Simcoe Street North, Oshawa, Ontario L1H 7K4, Canada
e-mail: Marc.Rosen@uoit.ca

industrial systems [16–22], thermal energy storage [23, 24], countries [25, 26] and environmental impact assessments [27–30].

In this chapter, green energy systems are described and discussed. Then, theoretical and practical aspects of thermodynamics most relevant to energy and exergy analyses are described, including fundamental principles and such related issues as reference-environment selection, efficiency definition and material properties acquisition. General implications of exergy analysis results are discussed, and a step-by-step procedure for energy and exergy analyses is given. Then two case studies involving the application of exergy analysis to green energy systems are assessed: solar ponds for thermal energy storage and wind energy systems.

2.2 Green Energy and Sustainable Development

Green energy is loosely defined as the form and utilization of energy that has no or minimal negative environmental, economic and societal impact and includes such energy resources as solar, hydro, biomass, wind, geothermal and other renewables. Green energy can be utilized to reduce the negative effects of hydrocarbon energy resources and their emissions, especially greenhouse gases. Green energy provides an important option for meeting clean energy demands for both industrial and non-industrial applications, and consequently is a major factor in future sustainable development and world stability.

Green energy can contribute to energy security, sustainable development, and social, technological, industrial and governmental innovations in a country. Increasing the green energy utilization of a country often positively impacts economic growth and social development. The supply and utilization of low-priced green energy is particularly significant for global stability, since energy plays a vital role in industrial and technological development and living standards around the world.

Critical energy issues by the mid-21st century will probably include energy security for almost seven billion people, one estimate of the expected global population by that time, and global warming, mainly caused by CO_2 emissions generated from the combustion of fossil fuels [31, 32]. Fossil fuels and their extensive use in various sectors are damaging to human health and welfare. To reduce the harmful effects of fossil fuels, green energy sources and technologies need to be increasingly applied. One of the most important properties of green energy sources is their environmental compatibility. This characteristic leads many to believe that green energy sources will become the most attractive energy sources in the near future and the most promising from technological and environmental perspectives through the 21st century, particularly in the context of sustainable development.

The Brundtland Commission [33] defined sustainable development as behavior that "meets the needs of the present without compromising the ability of future generations to meet their own needs". Focusing on sustainability emphasizes that humans are an integral part of nature, and requires us to assume a custodian's accountability for resources essential to meeting our needs and stewardship for the resources required for meeting our wants [34]. Sustainable development is a concept with many levels, and embraces various intellectual and political objectives. The concept has gained a highly visible position in discussions integrating social, environmental and economic viewpoints and widespread support of its benefits exists [35].

Many factors can contribute to achieving sustainable development. One is the requirement for a supply of energy resources that is fully sustainable. A secure supply of energy resources is generally agreed to be a necessary but not sufficient requirement for development within a society. Sustainable development further requires a sustainable supply of green energy sources and effective and efficient green energy-based technologies such as solar, wind and hydraulic power plants, hydrogen production plants from non-fossil fuel sources, *etc.* In addition, sustainable development requires a supply of energy resources that is reliably available at reasonable cost and causes no or minimal negative societal impacts. Clearly, energy resources such as fossil fuels are finite and thus lack the characteristics needed for sustainability, while others such as green energy sources are sustainable over the relatively long term [36]. Low-priced green energy is particularly important for increasing sustainable technological development and industrial productivity as well as living standards in a society. Solving the energy-resource scarcities and environmental problems that we face today requires long-term actions for achieving sustainable development. Implementation of green energy sources and technologies appears to provide one of the most effective means for accelerating movement towards sustainable development. Another important means is the development and implementation of effective and sustainable green energy strategies [37, 38]. In such activities, engineering practicality, reliability, applicability, economy, scarcity of supply, and public acceptability all need to be considered.

Green energy-based sustainability is a relatively new concept in the literature. Several researchers have contributed to sustainable development aspects of various energy sources and applications [32, 34–36, 38–45]. However, fewer reports exist on the role of green energy in sustainable development and global stability.

The promotion of green energy sources and technologies for sustainability and global stability has become one of the primary goals of energy policy makers in many countries. Policy makers increasingly assign high priority to promoting green energy-based technologies because they can help mitigate climate change and pollution, increase fossil fuel source reserves, decrease dependence on imported energy, increase employment, and support remote and rural communities. Furthermore, green energy-based technologies can increase energy supply diversity, improve the national balance of trade and increase security, since most are less prone to terrorist attacks than, say, nuclear power stations or oil and gas supply infrastructure (except for large hydroelectric dams) [46]. The following criteria affect the achievement of green energy-based sustainable development in a society [43, 47, 48]: information about and public awareness of the benefits of sustainability investments; environmental education and training; appropriate green energy strategies; availability of green energy sources and technologies; reasonable supplies of financing for sustainability measures; and monitoring and evaluation tools for green energy-based sustainability. Major considerations in the development of green energy-based technologies for sustainable development and global stability are presented in Figure 2.1.

Green energy and technologies are expected to play a key role in future sustainable energy and global stability scenarios. The foremost factor that will determine the specific role of green energy and technologies will probably be energy demand. Because green energy and technologies can accelerate shifts towards sustainable development, green energy factors that directly affect sustainability need to be understood. These factors, some of which are outlined in Figure 2.2, can help in identifying and achieving green energy strategies and technologies for sustainable development. Knowledge of these parameters and their interrelations are helpful in developing optimal green energy programs and selecting the most appropriate green energy and technologies for sustainable development and global stability.

Figure 2.3 describes possible pathways for using green energy resources for green power production. Utilization of green energy sources will reduce environmental problems such as global climate change, and emissions of CO, CO_2, NOx, SOx, non-methane hydrocarbons and particulate matter. Some other potential

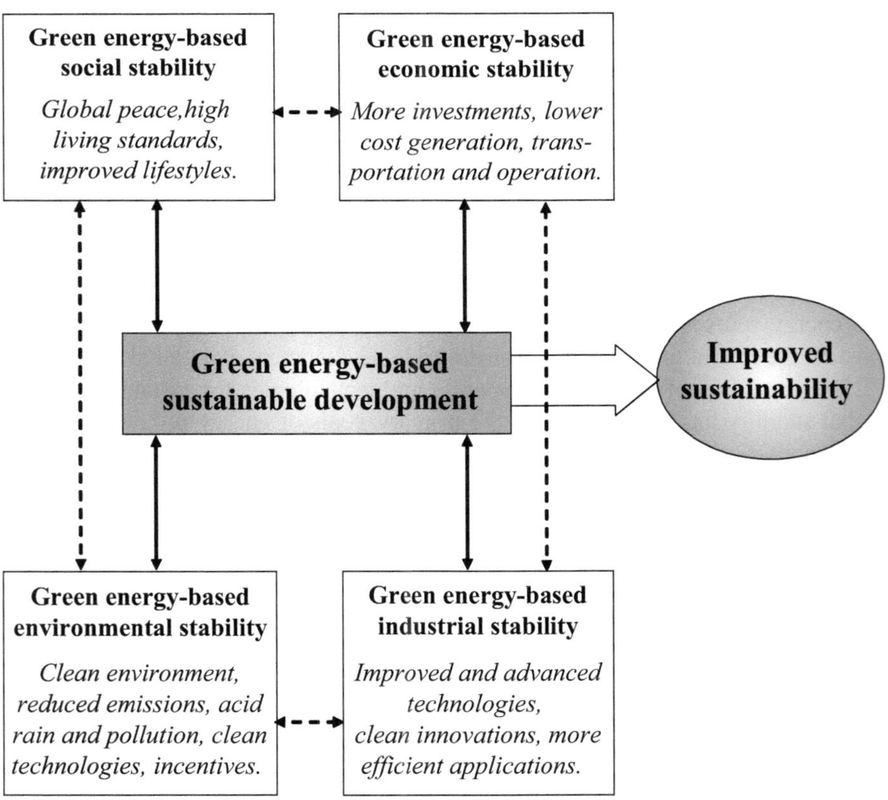

Figure 2.1 Significant considerations in the development of green energy-based technologies for sustainable development

measures to decrease the global unrest associated with harmful pollutant emissions include:

- Use of renewable energy technologies
- Energy conservation and more efficient energy utilization
- Application of cogeneration, district heating and energy storage technologies
- Use of alternative modes of transport and promotion of public transport
- Energy source switching from fossil fuels to environmentally benign energy forms
- Introduction of clean coal technologies
- Monitoring and evaluation of energy indicators
- Policy integration and the introduction of carbon or fuel taxes
- Application of techniques for reduction, re-use and recycling of resources
- Acceleration of forestation
- Substitution of greener materials for conventional materials
- Changing life styles
- Education and training for sustainable development and increasing public awareness

The importance of green energy and technologies in helping reduce world problems and achieving sustainable development should be emphasized in energy

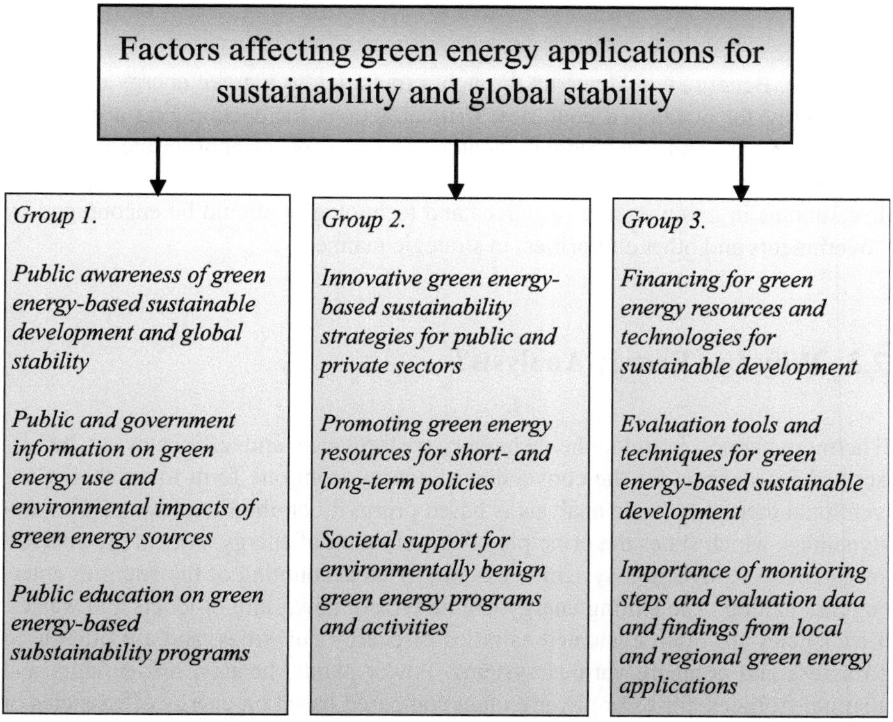

Figure 2.2 Factors affecting green energy applications for sustainability and global stability

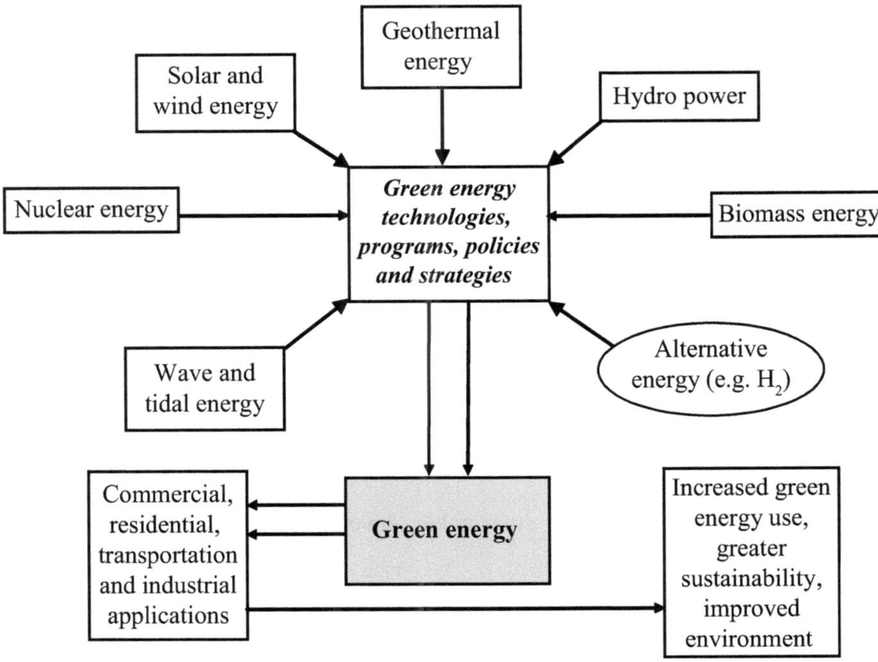

Figure 2.3 Routes to sustainable green energy-based technologies, programs and policies

strategies. Benefits would accrue through a transition to a green energy economy, particularly for developed countries. Critical actions for encouraging a transition to green energy and technologies include the provision of appropriate incentives and facilitation of interactions among countries and researchers, among others. Investments in green energy resources and technologies should be encouraged by governments and other authorities, in strategic manners.

2.3 Why Use Exergy Analysis?

Thermodynamics permits the behavior, performance and efficiency to be described for systems for the conversion of energy from one form to another. Conventional thermodynamic analysis is based primarily on the First Law of Thermodynamics, which states the principle of conservation of energy. An energy analysis of an energy-conversion system is essentially an accounting of the energies entering and exiting. The exiting energy can be broken down into products and wastes. Efficiencies are often evaluated as ratios of energy quantities, and are often used to assess and compare various systems. Power plants, heaters, refrigerators and thermal storages, for example, are often compared based on energy efficiencies or energy-based measures of merit.

However, energy efficiencies are often misleading in that they do not always provide a measure of how nearly the performance of a system approaches ideality. Further, the thermodynamic losses which occur within a system (*i.e.*, those factors that cause performance to deviate from ideality) often are not accurately identified and assessed with energy analysis. The results of energy analysis can indicate the main inefficiencies to be within the wrong sections of the system, and a state of technological efficiency different than actually exists.

Exergy analysis permits many of the shortcomings of energy analysis to be overcome. Exergy analysis, stemming from the Second Law of Thermodynamics, is useful in identifying the causes, locations and magnitudes of process inefficiencies. The exergy associated with an energy quantity is a quantitative assessment of its usefulness or quality. Exergy analysis acknowledges that although energy cannot be created or destroyed, it can be degraded in quality, eventually reaching a state in which it is in complete equilibrium with the surroundings and hence of no further use for performing tasks.

For energy storage systems, for example, exergy analysis allows one to determine the maximum potential associated with the incoming energy. This maximum is retained and recovered only if the energy undergoes processes in a reversible manner. Losses in the potential for exergy recovery occur in the real world because actual processes are always irreversible.

The exergy flow rate of a flowing commodity is the maximum rate that work may be obtained from it as it passes reversibly to the environmental state, exchanging heat and materials only with the surroundings. In essence, exergy analysis states the theoretical limitations imposed upon a system, clearly pointing out that no real system can conserve exergy and that only a portion of the input exergy can be recovered. Also, exergy analysis quantitatively specifies practical limitations by providing losses in a form in which they are a direct measure of lost exergy.

Exergy can also be applied beyond thermodynamics. Two important areas are discussed below.

Exergy has been applied to help improve environmental stewardship, ecology and sustainability [2, 6, 9, 11]. It has been suggested that the environmental impact of energy utilization and the achievement of increased efficiency are best addressed using exergy. Although exergy is a measure of usefulness, it is also a measure of potential to cause change. The latter point suggests that exergy may be, or may provide the basis for, an effective indicator of the potential of a substance or energy form to impact the environment.

Exergy has also been applied extensively in conjunction with economics, and many applications have been identified [2, 8, 10, 15]. In designing energy systems, technical methods like thermodynamics are combined with economics, with costs conventionally based on energy. Many suggest that costs are better distributed among outputs based on exergy. Exergy-based economic methods, referred to as thermoeconomics, second-law costing and exergoeconomics (1) recognize that exergy, not energy, is the commodity of value in a system, (2) assign costs and/or prices to exergy-related variables, and (3) help determine the appropriate allocation of economic resources to optimize system design and operation and system economic feasibility and profitability.

2.4 Energy and Exergy Analyses

Energy and exergy fundamentals are described in this section, and methods for applying exergy analysis are discussed.

Although a relatively standard terminology and nomenclature has evolved for conventional classical thermodynamics, there is at present no generally agreed-upon terminology and nomenclature for exergy analysis. A diversity of symbols and names exist for basic and derived quantities [3, 49]. For example, exergy is often called available energy, availability, work capability, essergy, *etc.*, and exergy consumption is often called irreversibility, lost work, dissipated work, dissipation, *etc.* The exergy analysis nomenclature used here follows that proposed by Kotas [3] as a standard exergy nomenclature.

2.4.1 Balances for Mass, Energy and Entropy

2.4.1.1 Conceptual Balances

A general balance for a quantity in a system may be written as

$$\textit{Input} + \textit{Generation} - \textit{Output} - \textit{Consumption} = \textit{Accumulation} \tag{2.1}$$

Input and output refer, respectively, to quantities entering and exiting through system boundaries. Generation and consumption refer, respectively, to quantities produced and consumed within the system. Accumulation refers to build-up (either positive or negative) of the quantity within the system.

Versions of the general balance above may be written for mass, energy, entropy and exergy. Mass and energy, being subject to conservation laws (neglecting nuclear reactions), can be neither generated nor consumed. Consequently, the general balance (Equation 2.1) for each of these quantities becomes

$$\textit{Mass input} - \textit{Mass output} = \textit{Mass accumulation} \tag{2.2}$$

$$\textit{Energy input} - \textit{Energy output} = \textit{Enegy accumulation} \tag{2.3}$$

Before giving the balance equation for exergy, it is useful to examine that for entropy:

$$\textit{Entropy input} + \textit{Entropy generation} - \textit{Entropy output} = \textit{Entropy accumulation} \tag{2.4}$$

Entropy is created during a process due to irreversibilities, but cannot be consumed.

These balances describe what is happening in a system between two instants of time. For a complete cyclic process where the initial and final states of the system are identical, the accumulation terms in all the balances are zero.

2.4.1.2 Detailed Balances

Two types of systems are normally considered: open (flow) and closed (non-flow). In general, open systems have mass, heat and work interactions, and closed systems heat and work interactions. Mass flow into, heat transfer into and work transfer out of a system are defined to be positive. Mathematical formulations of the principles of mass and energy conservation and entropy non-conservation can be written for any system, following the general physical interpretations in Equations 2.2–2.4.

Consider a non-steady flow process in a time interval t_1 to t_2. Balances of mass, energy and entropy, respectively, can be written for a control volume as

$$\sum_i m_i - \sum_e m_e = m_2 - m_1 \tag{2.5}$$

$$\sum_i (e + Pv)_i m_i - \sum_e (e + Pv)_e m_e + \sum_r (Q_r)_{1,2} - (W')_{1,2} = E_2 - E_1 \tag{2.6}$$

$$\sum_i s_i m_i - \sum_e s_e m_e + \sum_r (Q_r / T_r)_{1,2} + \Pi_{1,2} = S_2 - S_1 \tag{2.7}$$

Here, m_i and m_e denote, respectively, the amounts of mass input across port i and exiting across port e; $(Q_r)_{1,2}$ denotes the amount of heat transferred into the control volume across region r on the control surface; $(W')_{1,2}$ denotes the amount of work transferred out of the control volume; $\Pi_{1,2}$ denotes the amount of entropy created in the control volume; m_1, E_1 and S_1 denote, respectively, the amounts of mass, energy and entropy in the control volume at time t_1 and m_2, E_2 and S_2 denote, respectively, the same quantities at time t_2; and e, s, P, T and v denote specific energy, specific entropy, absolute pressure, absolute temperature and specific volume, respectively. The total work W' done by a system excludes flow work, and can be written as

$$W' = W + W_x \tag{2.8}$$

where W is the work done by a system due to change in its volume and W_x is the shaft work done by the system. The term "shaft work" includes all forms of work that can be used to raise a weight (*i.e.*, mechanical work, electrical work, *etc.*) but excludes work done by a system due to change in its volume. The specific energy e is given by

$$e = u + ke + pe \tag{2.9}$$

where u, ke and pe denote, respectively, specific internal, kinetic and potential (due to conservative force fields) energies. For irreversible processes $\Pi_{1,2} > 0$, and for reversible processes $\Pi_{1,2} = 0$.

The left sides of Equations 2.5–2.7 represent the net amounts of mass, energy and entropy transferred into (and in the case of entropy created within) the control volume, while the right sides represent the amounts of these quantities accumulated within the control volume.

For the mass flow m_j across port j,

$$m_j = \int_{t_1}^{t_2} \left[\int_j (\rho V_n dA)_j \right] dt \tag{2.10}$$

Here, ρ is the density of matter crossing an area element dA on the control surface in time interval t_1 to t_2 and V_n is the velocity component of the matter flow normal to dA. The integration is performed over port j on the control surface. One-dimensional flow (*i.e.*, flow in which the velocity and other intensive properties do not vary with position across the port) is often assumed. Then the previous equation becomes

$$m_j = \int_{t_1}^{t_2} (\rho V_n A)_j dt \tag{2.11}$$

It has been assumed that heat transfers occur at discrete regions on the control surface and the temperature across these regions is constant. If the temperature varies across a region of heat transfer,

$$(Q_r)_{1,2} = \int_{t_1}^{t_2} \left[\int_r (q dA)_r \right] dt \tag{2.12}$$

and

$$(Q_r / T_r)_{1,2} = \int_{t_1}^{t_2} \left[\int_r (q / T)_r \, dA_r \right] dt \tag{2.13}$$

where T_r is the temperature at the point on the control surface where the heat flux is q_r. The integral is performed over the surface area of region A_r.

The quantities of mass, energy and entropy in the control volume (denoted by m, E and S) on the right sides of Equations 2.5–2.7, respectively, are given more generally by

$$m = \int \rho dV \tag{2.14}$$

$$E = \int \rho e dV \tag{2.15}$$

$$S = \int \rho s dV \tag{2.16}$$

where the integrals are over the control volume.

For a closed system, $m_i = m_e = 0$ and Equations 2.5–2.7 become

$$0 = m_2 - m_1 \tag{2.17}$$

$$\sum_r (Q_r)_{1,2} - (W')_{1,2} = E_2 - E_1 \tag{2.18}$$

$$\sum_r (Q_r / T_r)_{1,2} + \Pi_{1,2} = S_2 - S_1 \tag{2.19}$$

2.4.2 *Exergy of Systems and Flows*

Several quantities related to the conceptual exergy balance are described here, following the presentations by Kotas [3] and Moran [4].

2.4.2.1 Exergy of a Closed System

The exergy $Ex_{nonflow}$ of a closed system of mass m, or the non-flow exergy, can be expressed as

$$Ex_{nonflow} = Ex_{nonflow,ph} + Ex_o + Ex_{kin} + Ex_{pot} \tag{2.20}$$

where

$$Ex_{pot} = PE \tag{2.21}$$

$$Ex_{kin} = KE \tag{2.22}$$

$$Ex_o = \sum_i (\mu_{io} - \mu_{ioo}) N_i \tag{2.23}$$

$$Ex_{nonflow,ph} = (U - U_o) + P_o (V - V_o) - T_o (S - S_o) \tag{2.24}$$

where the system has a temperature T, pressure P, chemical potential μ_i for species i, entropy S, energy E, volume V and number of moles N_i of species i. The system is within a conceptual environment in an equilibrium state with intensive properties T_o, P_o and μ_{ioo}. The quantity μ_{io} denotes the value of μ at the environmental state (*i.e.*, at T_o and P_o). The terms on the right side of Equation 2.20 represent, respectively, physical, chemical, kinetic and potential components of the non-flow exergy of the system.

Exergy is a property of the system and conceptual environment, combining the extensive properties of the system with the intensive properties of the environment.

Physical non-flow exergy is the maximum work obtainable from a system as it is brought to the environmental state (*i.e.*, to thermal and mechanical equilibrium with the environment), and chemical non-flow exergy is the maximum work obtainable from a system as it is brought from the environmental state to the dead state (*i.e.*, to complete equilibrium with the environment).

2.4.2.2 Exergy of Flows

Exergy of a matter flow. The exergy of a flowing stream of matter Ex_{flow} is the sum of non-flow exergy and the exergy associated with the flow work of the stream (with reference to P_o), *i.e.*,

$$Ex_{flow} = Ex_{nonflow} + (P - P_o)V \tag{2.25}$$

Alternatively, Ex_{flow} can be expressed following Equation 2.20 in terms of physical, chemical, kinetic and potential components:

$$Ex_{flow} = Ex_{flow,ph} + Ex_o + Ex_{kin} + Ex_{pot} \tag{2.26}$$

where

$$Ex_{pot} = PE \tag{2.27}$$

$$Ex_{kin} = KE \tag{2.28}$$

$$Ex_o = Ex_o = \sum_i (\mu_{io} - \mu_{ioo})N_i \tag{2.29}$$

$$Ex_{flow,ph} = (H - H_o) - T_o(S - S_o) \tag{2.30}$$

Exergy of thermal energy. Consider a control mass, initially at the dead state, being heated or cooled at constant volume in an interaction with some other system. The heat transfer experienced by the control mass is Q. The flow of exergy associated with the heat transfer Q is denoted by Ex_Q, and can be expressed as

$$Ex_Q = \int_i^f (1 - T_o / T)\delta Q \tag{2.31}$$

where δQ is an incremental heat transfer, and the integral is from the initial state (i) to the final state (f). This "thermal exergy" is the minimum work required by the combined system of the control mass and the environment in bringing the control mass to the final state from the dead state.

Often the dimensionless quantity in parentheses in this expression is called the "exergetic temperature factor" and denoted τ. That is,

$$\tau = 1 - T_o / T \tag{2.32}$$

The relation between the factor τ and the temperature ratio T/T_o is illustrated in Figure 2.4. In that figure, it can be seen that τ is equal to zero when $T = T_o$. For

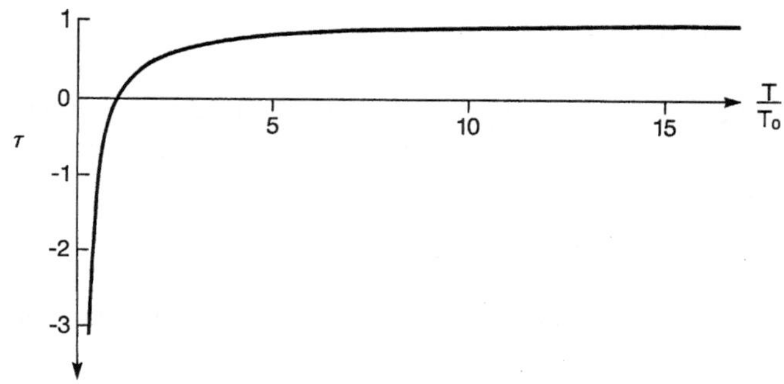

Figure 2.4 The relation between the exergetic temperature factor τ and the absolute temperature ratio T/T_o

heat transfer at above-environment temperatures (*i.e.*, $T > T_o$), $0 < \tau \leq 1$. For heat transfer at sub-environment temperatures (*i.e.*, $T < T_o$), $\tau < 0$, implying that exergy and energy flow in opposite directions in such cases. Note that the magnitude of exergy flow exceeds that of the energy flow when $\tau < -1$, which corresponds to $T < T_o/2$.

If the temperature T of the control mass is constant, the thermal exergy transfer associated with a heat transfer is

$$Ex_Q = (1 - T_o / T)Q = \tau Q \tag{2.33}$$

For heat transfer across a region r on a control surface for which the temperature may vary,

$$Ex_Q = \int_r \left[q_r \left(1 - T_o / T_r\right) \, dA_r \right] \tag{2.34}$$

where q_r is the heat flow per unit area at a region on the control surface at which the temperature is T_r.

Exergy of work. Equation 2.8 separates total work W' into two components: W_x and W. The exergy associated with shaft work Ex_W is by definition W_x.

The exergy transfer associated with work done by a system due to volume change is the net usable work due to the volume change, and is denoted by W_{NET}. Thus for a process in time interval t_1 to t_2,

$$\left(W_{NET}\right)_{1,2} = W_{1,2} - P_o \left(V_2 - V_1\right) \tag{2.35}$$

where $W_{1,2}$ is the work done by the system due to volume change $(V_2 - V_1)$. The term $P_o(V_2 - V_1)$ is the displacement work necessary to change the volume against the constant pressure P_o exerted by the environment.

Exergy of electricity. As for shaft work, the exergy associated with electricity is equal to the energy.

2.4.3 Exergy Consumption

For a process occurring in a system, the difference between the total exergy flows into and out of the system, less the exergy accumulation in the system, is the exergy consumption I, expressible as

$$I = T_o S_{gen} \tag{2.36}$$

which points out that exergy consumption is proportional to entropy creation, and is known as the Gouy–Stodola relation.

2.4.4 Exergy Balance

By combining the conservation law for energy and non-conservation law for entropy, the exergy balance can be obtained:

$$\text{Exergy input} - \text{Exergy output} - \text{Exergy consumption} = \text{Exergy accumulation} \tag{2.37}$$

Exergy is consumed due to irreversibilities. Exergy consumption is proportional to entropy creation. Equations 2.4 and 2.37 demonstrate an important difference between energy and exergy: energy is conserved while exergy, a measure of energy quality or work potential, can be consumed.

An analogous balance to those given in Equations 2.5–2.7 can be written for exergy, following the physical interpretation of Equation 2.37. For a non-steady flow process during time interval t_1 to t_2,

$$\sum_i ex_i m_i - \sum_e ex_e m_e + \sum_r \left(Ex_{Q_r}\right)_{1,2} - \left(Ex_W\right)_{1,2} - \left(W_{NET}\right)_{1,2} - I_{1,2} = Ex_2 - Ex_1 \tag{2.38}$$

where $(W_{NET})_{1,2}$ is given by Equation 2.35 and

$$\left(Ex_{Qr}\right)_{1,2} = \int_{t_1}^{t_2} \left[\int_r \left(1 - T_o / T_r\right) q_r dA_r \right] dt \tag{2.39}$$

$$I_{1,2} = T_o S_{gen,1,2} \tag{2.40}$$

$$Ex = \int \rho\, ex\, dV \tag{2.41}$$

Here, I and S_{gen}, respectively, denote exergy consumption and entropy creation, ex denotes specific exergy, and the integral for Ex is performed over the control volume. The first two terms on the left side of Equation 2.38 represent the net input of exergy associated with matter, the third term the net input of exergy associated with heat, the fourth and fifth terms the net input of exergy associated with work, and the sixth term the exergy consumption. The right side of Equation 2.38 represents the accumulation of exergy.

For a closed system, Equation 2.38 simplifies to

$$\sum_r \left(Ex_{Qr}\right)_{1,2} - \left(Ex_W\right)_{1,2} - \left(W_{NET}\right)_{1,2} - I_{1,2} = Ex_2 - Ex_1 \tag{2.42}$$

When volume is fixed, $(W_{NET})_{1,2} = 0$ in Equations 2.38 and 2.42. Also, when the initial and final states are identical, as in a complete cycle, the right sides of Equations 2.38 and 2.42 are zero.

2.4.5 Reference Environment

Exergy is evaluated with respect to a reference environment, so the intensive properties of the reference environment partly determine the exergy of a stream or system. The reference environment is in stable equilibrium, with all parts at rest relative to one another. No chemical reactions can occur between the environmental components. The reference environment acts as an infinite system, and is a sink and source for heat and materials. It experiences only internally reversible processes in which its intensive state remains unaltered (*i.e.*, its temperature T_o, pressure P_o and the chemical potentials μ_{ioo} for each of the i components present remain constant). The exergy of the reference environment is zero. The exergy of a stream or system is zero when it is in equilibrium with the reference environment.

The natural environment does not have the theoretical characteristics of a reference environment. The natural environment is not in equilibrium, and its intensive properties exhibit spatial and temporal variations. Many chemical reactions in the natural environment are blocked because the transport mechanisms necessary to reach equilibrium are too slow at ambient conditions. Thus, the exergy of the natural environment is not zero; work could be obtained if it were to come to equilibrium. Consequently, models for the reference environment are used which try to achieve a compromise between the theoretical requirements of the reference environment and the actual behavior of the natural environment.

Several classes of reference-environment models are described below.

- Natural-environment-subsystem models. These attempt to simulate realistically subsystems of the natural environment and are an important class of reference-environment models. One such model consisting of saturated moist air and liquid water in phase equilibrium was proposed by Baehr and Schmidt [50]. An extension of the above model which allowed sulphur-containing materials to be analyzed was proposed by Gaggioli and Petit [51] and Rodriguez [52]. The temperature and pressure of this reference environment (see Table 2.1) are normally taken to be 25 °C and 1 atm, respectively, and the chemical composition is taken to consist of air saturated with water vapour, and the following condensed phases at 25 °C and 1 atm: water (H_2O), gypsum ($CaSO_4{\cdot}2H_2O$) and limestone ($CaCO_3$). The stable configurations of C, O and N, respectively, are taken to be those of CO_2, O_2 and N_2 as they exist in air saturated with liquid water at T_o and P_o; that of hydrogen is taken to be in the liquid phase of water saturated with air at T_o and P_o; and that of S and Ca, respectively, are taken to be those of $CaSO_4{\cdot}2H_2O$ and $CaCO_3$ at T_o and P_o. Analyses often use the natural-environment-subsystem model described in Table 2.1, but with a temperature modified to reflect the approximate mean ambient temperature of the location of the system or process for the time period under consideration.

- Reference-substance models. Here, a "reference substance" is selected and assigned zero exergy for every chemical element. One such model in which the reference substances were selected as the most valueless substances found in abundance in the natural environment was proposed by Szargut [53]. The criteria for selecting such reference substances is consistent with the notion of simulating the natural environment, but is primarily economic in nature, and is vague and arbitrary with respect to the selection of reference substances. Part of this environment is the composition of moist air, including N_2, O_2, CO_2, H_2O and the noble gases; gypsum (for sulphur) and limestone (for calcium). Another model in this class, in which reference substances are selected arbitrarily, was proposed by Sussman [54, 55]. This model is not similar to the natural environment. Consequently absolute exergies evaluated with this model do not relate to the natural environment, and cannot be used rationally to evaluate efficiencies. Since exergy-consumption values are independent of the choice of reference substances, they can be used rationally in analyses.
- Equilibrium models. A model in which all the materials present in the atmosphere, oceans and a layer of the crust of the earth are pooled together and an equilibrium composition is calculated for a given temperature was proposed by Ahrendts [56]. The selection of the thickness of crust considered is subjective and is intended to include all materials accessible to technical processes. Ahrendts considered thicknesses varying from 1 m to 1000 m, and a tempera-

Table 2.1 A reference-environment model

Temperature:	$T_o = 298.15$ K		
Pressure:	$P_o = 1$ atm		
Composition:	(i)	Atmospheric air saturated with H_2O at T_o and P_o, having the following composition:	
		Air constituents	Mole fraction
		N_2	0.7567
		O_2	0.2035
		H_2O	0.0303
		Ar	0.0091
		CO_2	0.0003
		H_2	0.0001
	(ii)	Several condensed phases at T_o and P_o:	
		Water (H_2O)	
		Limestone ($CaCO_3$)	
		Gypsum ($CaSO_4 \cdot 2H_2O$)	

Source: Adapted from [51].

ture of 25 °C. For all thicknesses, Ahrendts [56] found that the model differed significantly from the natural environment. Exergy values obtained using these environments are strongly dependent on the thickness of crust considered, and represent the absolute maximum amount of work obtainable from a material. Since there is no technical process available that can obtain this work from materials, Ahrendts' equilibrium model does not give meaningful exergy values when applied to real processes.

- Constrained-equilibrium models. Ahrendts [56] also proposed a modified version of his equilibrium environment in which the calculation of an equilibrium composition excludes the possibility of the formation of nitric acid (HNO_3) and its compounds. That is, all chemical reactions in which these substances are formed are in constrained equilibrium, and all other reactions are in unconstrained equilibrium. When a thickness of crust of 1 m and temperature of 25 °C were used, the model was similar to the natural environment.
- Process-dependent models. A model which contains only components that participate in the process being examined in a stable equilibrium composition at the temperature and total pressure of the natural environment was proposed by Bosnjakovic [57]. This model is dependent on the process examined, and is not general. Exergies evaluated for a specific process-dependent model are relevant only to the process; they cannot rationally be compared with exergies evaluated for other process-dependent models.

Researchers have examined the characteristics of and models for reference environments [55, 56, 58], and the sensitivities of exergy values to different reference-environment models [59].

2.4.6 *Efficiencies and Other Measures of Merit*

Efficiency has always been an important consideration in decision making regarding resource utilization. Efficiency is defined as "the ability to produce a desired effect without waste of, or with minimum use of, energy, time, resources, *etc.*", and is used by people to mean the effectiveness with which something is used to produce something else, or the degree to which the ideal is approached in performing a task.

For general engineering systems, non-dimensional ratios of quantities are typically used to determine efficiencies. Ratios of energy are conventionally used to determine efficiencies of engineering systems whose primary purpose is the transformation of energy. These efficiencies are based on the First Law of Thermodynamics. A process has maximum efficiency according to the First Law if energy input equals recoverable energy output (*i.e.*, if no "energy losses" occur). However, efficiencies determined using energy are misleading because in general they are not measures of "an approach to an ideal".

To determine more meaningful efficiencies, a quantity is required for which ratios can be established which do provide a measure of an approach to an ideal. Thus, the Second Law must be involved, as this law states that maximum efficiency is attained (*i.e.*, ideality is achieved) for a reversible process. However, the Second Law must be quantified before efficiencies can be defined.

The "increase of entropy principle", which states that entropy is created due to irreversibilities, quantifies the Second Law. From the viewpoint of entropy, maximum efficiency is attained for a process in which entropy is conserved. Entropy is created for non-ideal processes. The magnitude of entropy creation is a measure of the non-ideality or irreversibility of a process. In general, however, ratios of entropy do not provide a measure of an approach to an ideal.

A quantity which has been discussed in the context of meaningful measures of efficiency is negentropy [60]. Negentropy is defined such that the negentropy consumption due to irreversibilities is equal to the entropy creation due to irreversibilities. As a consequence of the increase of entropy principle, maximum efficiency is attained from the viewpoint of negentropy for a process in which negentropy is conserved. Negentropy is consumed for non-ideal processes. Negentropy is a measure of order. Consumptions of negentropy are therefore equivalent to degradations of order. Since the abstract property of order is what is valued and useful, it is logical to attempt to use negentropy in developing efficiencies. However, general efficiencies cannot be determined based on negentropy because its absolute magnitude is not defined.

Negentropy can be further quantified through the ability to perform work. Then, maximum efficiency is attainable only if, at the completion of a process, the sum of all energy involved has an ability to do work equal to the sum before the process occurred. Exergy is a measure of the ability to perform work, and from the viewpoint of exergy, maximum efficiency is attained for a process in which exergy is conserved. Efficiencies determined using ratios of exergy do provide a measure of an approach to an ideal. Exergy efficiencies are often more intuitively rational than energy efficiencies because efficiencies between 0 % and 100 % are always obtained. Measures which can be greater than 100 % when energy is considered, such as coefficient of performance, normally are between 0 % and 100 % when exergy is considered. In fact, some researchers [61] call exergy efficiencies "real" or "true" efficiencies, while calling energy efficiencies "approximations to real" efficiencies.

Energy (η) and exergy (ψ) efficiencies are often written for steady-state processes occurring in systems as

$$\eta = \frac{Energy\ in\ product\ outputs}{Energy\ in\ inputs} = 1 - \frac{Energy\ loss}{Energy\ in\ inputs} \tag{2.43}$$

$$\psi = \frac{Exergy\ in\ product\ outputs}{Exergy\ in\ inputs} = \frac{Exergy\ loss\ plus\ consumption}{Exergy\ in\ inputs} \tag{2.44}$$

Two other common exergy-based efficiencies for steady-state devices are as follows:

$$\textit{Rational efficiency} = \frac{\textit{Total exergy output}}{\textit{Total exergy input}} = 1 - \frac{\textit{Exergy consumption}}{\textit{Total exergy input}} \quad (2.45)$$

$$\textit{Task efficiency} = \frac{\textit{Theoretical minimum exergy input required}}{\textit{Actual exergy input}} \quad (2.46)$$

Exergy efficiencies often give more illuminating insights into process performance than energy efficiencies because (i) they weigh energy flows according to their exergy contents, and (ii) they separate inefficiencies into those associated with effluent losses and those due to irreversibilities. In general, exergy efficiencies provide a measure of potential for improvement.

2.4.7 Energy and Exergy Properties

Material properties are needed for energy and exergy analyses of processes. Conventional property data are widely available for many substances (*e.g.*, steam, air and combustion gases and chemical substances).

Energy values of heat and work flows are absolute, while the energy values of material flows are relative. Enthalpies are evaluated relative to a reference level. Since energy analyses are typically concerned only with energy differences, the reference level used for enthalpy calculations can be arbitrary. For the determination of some energy efficiencies, however, the enthalpies must be evaluated relative to specific reference levels, *e.g.*, for energy-conversion processes, the reference level is often selected so that the enthalpy of a material equals its higher heating value (*HHV*).

If, however, the results from energy and exergy analyses are to be compared, it is necessary to specify reference levels for enthalpy calculations such that the enthalpy of a compound is evaluated relative to the stable components of the reference environment. Thus, a compound that exists as a stable component of the reference environment is defined to have an enthalpy of zero at T_o and P_o. Enthalpies calculated with respect to such conditions are referred to as "base enthalpies" [52]. The base enthalpy is similar to the enthalpy of formation. While the latter is the enthalpy of a compound (at T_o and P_o) relative to the elements (at T_o and P_o) from which it would be formed, the former is the enthalpy of a component (at T_o and P_o) relative to the stable components of the environment (at T_o and P_o). For many environment models, the base enthalpies of material fuels are equal to their *HHV*s.

Base enthalpies for many substances, corresponding to the reference-environment model in Table 2.1, are listed in Table 2.2 [52].

Table 2.2 Base enthalpy and chemical exergy values of selected species

Species	Specific base enthalpy (kJ/g-mol)	Specific chemical exergy* (kJ/g-mol)
Ammonia (NH_3)	382.585	$2.478907 \ln y + 337.861$
Carbon (graphite) (C)	393.505	410.535
Carbon dioxide (CO_2)	0.000	$2.478907 \ln y + 20.108$
Carbon monoxide (CO)	282.964	$2.478907 \ln y + 275.224$
Ethane (C_2H_6)	1,564.080	$2.478907 \ln y + 1{,}484.952$
Hydrogen (H_2)	285.851	$2.478907 \ln y + 235.153$
Methane (CH_4)	890.359	$2.478907 \ln y + 830.212$
Nitrogen (N_2)	0.000	$2.478907 \ln y + 0.693$
Oxygen (O_2)	0.000	$2.478907 \ln y + 3.948$
Sulphur (rhombic) (S)	636.052	608.967
Sulphur dioxide (SO_2)	339.155	$2.478907 \ln y + 295.736$
Water (H_2O)	44.001	$2.478907 \ln y + 8.595$

* y represents the molal fraction for each of the respective species.
Source: Compiled from data in [51, 52].

It is necessary for chemical exergy values to be determined for exergy analysis. Many researchers have developed methods for evaluating chemical exergies, and tabulated values [52, 54]. Included are methods for evaluating the chemical exergies of solids, liquids and gases. For complex materials (*e.g.*, coal, tar, ash), approximation methods have been developed. By considering environmental air and gaseous process streams as ideal gas mixtures, chemical exergy can be calculated for gaseous streams using component chemical exergy values (*i.e.*, values of $(\mu_{io}-\mu_{ioo})$ listed in Table 2.2).

2.4.8 *Implications of Results of Exergy Analyses*

The results of exergy analyses of processes and systems have direct implications on application decisions and on research and development directions. Further, exergy analyses more than energy analyses provide insights into the "best" directions for R&D effort. Here, best is loosely taken to mean "most promising for significant efficiency gains". There are two main reasons for this statement:

- Exergy losses represent true losses of the potential that exists to generate the desired product from the given driving input. This is not true in general for energy losses. Thus, if the objective is to increase efficiency, focusing on exergy losses permits R&D to focus on reducing losses that will affect the objective.
- Exergy efficiencies always provide a measure of how nearly the operation of a system approaches the ideal, or theoretical upper limit. This is not in general true for energy efficiencies. By focusing R&D effort on those plant sections or processes with the lowest exergy efficiencies, the effort is being directed to those areas that inherently have the largest margins for efficiency improvement.

By focusing on energy efficiencies, on the other hand, one can expend R&D effort on topics for which little margins for improvement, even theoretically, exist.

Exergy analysis results typically suggest that R&D efforts should concentrate more on internal rather than external exergy losses, based on thermodynamic considerations, with a higher priority for the processes having larger exergy losses. Although this statement suggests focusing on those areas for which margins for improvement are greatest, it does not indicate that R&D should not be devoted to those processes having low exergy losses, as simple and cost-effective ways to increase efficiency by reducing small exergy losses should certainly be considered when identified.

More generally, it is noted that application and R&D allocation decisions should not be based exclusively on the results of energy and exergy analyses, even though these results provide useful information to assist in such decision making. Other factors must be considered also, such as economics, environmental impact, safety, and social and political implications.

2.4.9 *Procedure for Energy and Exergy Analyses*

A simple procedure for performing energy and exergy analyses involves the following steps:

- Subdivide the process under consideration into as many sections as desired, depending on the depth of detail and understanding desired from the analysis.
- Perform conventional mass and energy balances on the process, and determine all basic quantities (*e.g.*, work, heat) and properties (*e.g.*, temperature, pressure).
- Based on the nature of the process, the acceptable degree of analysis complexity and accuracy, and the questions for which answers are sought, select a reference-environment model.
- Evaluate energy and exergy values, relative to the selected reference-environment model.
- Perform exergy balances, including the determination of exergy consumptions.
- Select efficiency definitions, depending on the measures of merit desired, and evaluate values for the efficiencies.
- Interpret the results, and draw appropriate conclusions and recommendations, relating to such issues as design modifications.

2.5 Case Study 1: Exergy Analysis of Solar Ponds

Solar ponds are devices for capturing solar energy and storing it as thermal energy. Many energy-based investigations of solar ponds have been carried out [62–67]. Here an exergetic performance analysis of a solar pond (with a surface area 4 m^2 and a depth of 1.5 m) built at Cukurova University in Adana, Turkey is described. Exergy analysis appears to be a potential tool for design, analysis, evaluation, and performance improvement of solar pond systems. Further details are presented by Karakilcik and Dincer [68]. Information on the experimental system and measurement details, as well as some thermophysical properties of materials and fluids, are available elsewhere [69, 70].

2.5.1 Solar Pond Model

Generally solar ponds are bodies of water with three zones:

- The first zone or upper convective zone (UCZ) is the fresh water layer at the top of the pond. This zone is fed with fresh water of a density as close as possible to the density of fresh water in the upper part. This zone protects the cleanliness of the pond, and makes up for lost water due to evaporation.
- The second (middle) zone or non-convective zone (NCZ) is composed of salty water layers whose brine density gradually increases towards the lower convective zone (LCZ). The NCZ is the key to the working of a solar pond. It allows an extensive amount of solar radiation incident on the solar pond to penetrate into the storage zone. The NCZ provides a cost-effective method of collecting and storing energy and prohibiting the propagation of long-wave radiation solar energy on a large scale because water is opaque to infrared radiation.
- The third zone or heat storage zone (HSZ) or LCZ is composed of salty water with the highest density. A considerable part of the solar energy is absorbed and stored by this bottom region. The LCZ has the highest temperature and, hence, the strongest thermal interaction occurs between this zone and the insulated bottom wall (IBW) and insulated side walls (ISW) surrounding it.

Figure 2.5 shows each of the zones and the respective energy and exergy flows.

2.5.2 Energy Analysis

The energy efficiencies are presented for each zone of the pond shown in Figure 2.5. These are subsequently compared with the corresponding exergy efficiencies.

2.5.2.1 Energy Efficiency for UCZ

The energy efficiency for the UCZ can be expressed as

$$\eta_{ucz} = 1 - \frac{\left\{A_{01,UCZ} R_{ps} \left(T_{UCZ} - T_{sw,UCZ}\right) + U_{wa} A\left(T_{UCZ} - T_a\right)\right\}}{\left\{\beta E A_{UCZ}\left[1-(1-F)h(X_1-\delta)\right] + \frac{kA}{X_1}\left(T_{NCZ} - T_{UCZ}\right)\right\}} \tag{2.47}$$

where T_a is the average ambient air temperature, X_1 is the thickness of the UCZ; $A_{01,UCZ}$ is the surface area of the painted metal sheet on the side walls (and taken as $8 \times 0.10 = 0.8\,\text{m}^2$); δ is the thickness of the layer in the UCZ absorbing long-wave solar incident radiation; E is the total solar radiation incident on the pond surface, A is the upper surface area of the pond; ρ is the density of the layers in the UCZ; C is the specific heat of the layers in the UCZ; k is the thermal conductivity of the layers in the UCZ; R_{ps} is the thermal resistance of the painted metal sheet surrounding the first layer, respectively, and can be written as $R_{ps} = \dfrac{k_p k_s}{S_p k_s + S_s k_p}$. Here, k_p and k_s are thermal conductivities of the paint and iron sheet, and S_p and S_s are the corresponding thicknesses.

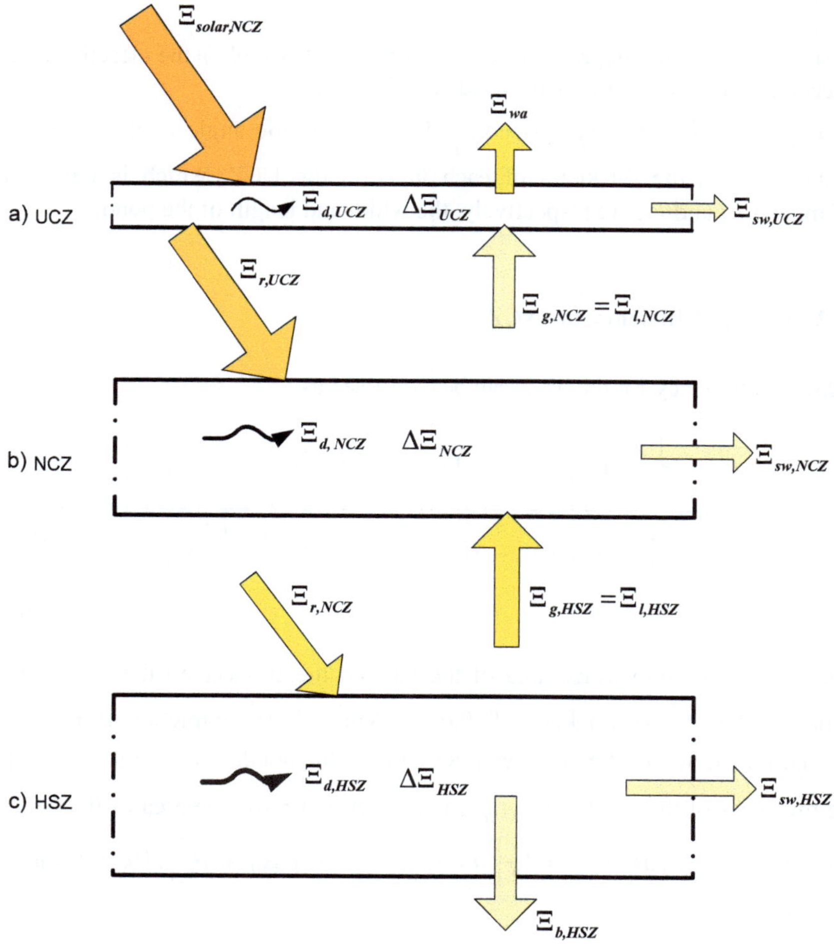

Figure 2.5 Energy and exergy flows in the inner zones of the solar pond

Also, β is the fraction of the incident solar radiation that enters the pond, and is written using an expression by Hawlader [71] as

$$\beta = 1 - 0.6\left[\frac{\sin\left(\theta_i - \theta_r\right)}{\sin\left(\theta_i + \theta_r\right)}\right]^2 - 0.4\left[\frac{\tan\left(\theta_i - \theta_r\right)}{\tan\left(\theta_i + \theta_r\right)}\right]^2$$

Here θ_i and θ_r are the angles of incident and reflected solar radiation.

The ratio of the solar energy reaching the bottom of layer I to the total solar radiation incident on the surface of the pond (h) is given by Bryant and Colbeck [72] as

$$h_I = 0.727 - 0.056 \ln\left[\frac{\left(X_1 - \delta\right)}{\cos\theta_r}\right]$$

Also, A_{UCZ} is the net upper surface area of the UCZ, which is the effective area that receives incident solar radiation and is defined as $A_{UCZ} = L_W\left[\,L_L - \left(\delta + \left(I-1\right)\Delta x\right)\tan\theta_r\right]$. Here, θ_r is the angle of the reflected incidence, Δx is the thickness of each layer in the UCZ, which is taken as 0.005 m, and L_w and L_L, are respectively, the width and length of the pond.

2.5.2.2 Energy Efficiency for NCZ

The energy efficiency for the NCZ can be expressed as

$$\eta_{NCZ} = 1 - \frac{\left\{\frac{kA}{\Delta X}\left(T_{NCZ} - T_{UCZ}\right) + A_{01,NCZ} R_{ps}\left(T_{NCZ} - T_{sw,UCZ}\right)\right\}}{\left\{\beta E A_{NCZ}\left[(1-F)[h(X_1 - \delta) - h(X_1 - \delta + \Delta x)]\right] + \frac{kA}{\Delta X}\left(T_{HSZ} - T_{NCZ}\right)\right\}} \tag{2.48}$$

where $A_{01,NCZ}$ is the surface area of the painted metal sheet on the side walls surrounding the NCZ (and taken as $8\times0.60 = 4.8\,m^2$); F is the fraction of the incident solar radiation absorbed by the upper layer of the pond; $\Delta X_{NCZ} = \left(X_2 - X_1\right)$ is the thickness of the NCZ; and A_{NCZ} is the net upper surface area of the NCZ, which can be written as $A_{NCZ} = L_w\left[\,L_L - \left(X_1 + \left(I-1\right)\Delta x\right)\tan\theta_r\right]$. Here I varies from 2 to 14.

2.5.2.3 Energy Efficiency for HSZ

The energy efficiency for the HSZ can be expressed as

$$\eta_{NCZ} = 1 - \frac{\left\{ AR_{ps}\left(T_{HSZ} - T_b\right) + \dfrac{kA}{\Delta X_{HSZ}}\left(T_{HSZ} - T_{NCZ}\right) + A_{01,HSZ} R_{ps}\left(T_{HSZ} - T_{sw,UCZ}\right)\right\}}{\left\{\beta EA_{HSZ}\left[(1-F)h(X_3 - \delta)\right]\right\}} \tag{2.49}$$

where $A_{01,HSZ}$ is the surface area of the painted metal sheet on the side walls surrounding the HSZ (and taken as $8 \times 0.80 = 6.4\ \text{m}^2$) and $\Delta X_{HSZ} = \left(X_3 - X_2\right)$ is the thickness of the HSZ of the pond. Note that $A_{HSZ,I} = A_{NCZ,I}$ with I varying from 15 to 30.

2.5.3 *Exergy Analysis*

An exergy analysis is presented for each zone. The exergy flows are outlined in Figure 2.5.

2.5.3.1 Exergy Analysis for UCZ

With the UCZ exergy flows illustrated in Figure 2.5a, an exergy balance can be written for that zone as

$$\Xi_{solar} + \Xi_{g,NCZ} = \Xi_{r,UCZ} + \Xi_{d,UCZ} + \Xi_a + \Xi_{sw,UCZ} \tag{2.50}$$

where Ξ_{solar} is the exergy of solar radiation reaching the UCZ surface, $\Xi_{g,NCZ}$ is the exergy gained from the NCZ, $\Xi_{r,UCZ}$ is the recovered exergy from the UCZ for the NCZ, $\Xi_{d,UCZ}$ is the exergy destruction in the UCZ, $\Xi_{a,UCZ}$ is the exergy loss from the UCZ to the ambient air and $\Xi_{sw,UCZ}$ is the exergy loss through the side walls.

With Equation 2.50, $\Xi_{r,UCZ}$ can be written as

$$\Xi_{r,UCZ} = \Xi_{ti} - \Xi_{tl} = \left(\Xi_{solar} + \Xi_{g,NCZ}\right) - \left(\Xi_{d,UCZ} + \Xi_a + \Xi_{sw,UCZ}\right) \tag{2.51}$$

where Ξ_{tl} is the total exergy losses, including exergy destruction and Ξ_{ti} is the total exergy input to the UCZ.

The exergy of solar radiation can be written, using a modified form of an expression given by Petela [73], as

$$\Xi_{solar} = E_{net}\left[1 - \frac{4T_0}{3T} + \frac{1}{3}\left(\frac{T_0}{T}\right)^4\right] A_{UCZ} \tag{2.52}$$

The exergy gained from the NCZ can be expressed as

$$\Xi_{g,NCZ} = m_{NCZ} C_{p,NCZ}\left[\left(T_{m,NCZ} - T_{UCZ}\right) - T_0\left(\ln\frac{T_{m,NCZ}}{T_{UCZ}}\right)\right] \tag{2.53}$$

where E_{net} is the net incident solar radiation reaching the UCZ surface; A_{UCZ} is the net surface area of the UCZ; T is the surface temperature of the sun, which is taken to be 6000 K [73]; $m_{NCZ} = \rho_{NCZ} V_{NCZ}$ is the mass of salty water in the NCZ; ρ_{NCZ} is the average density; and V_{NCZ} is the volume of salty water in the NCZ, which for the present system is V_{NCZ} = 2.4 m^3.

The exergy destruction in the UCZ can be written as

$$\Xi_{d,UCZ} = T_0 (\Delta S_{net}) \tag{2.54}$$

where ΔS_{net} is the net entropy change of the UCZ, which is $\Delta S_{net} = \Delta S_{sys} + \Delta s_{surr}$. After substituting expressions for each of the entropy change terms, Equation 2.54 becomes

$$\Xi_{d,UCZ} = T_0\left[m_{UCZ} C_{p,UCZ} \ln\frac{T_{UCZ}}{T_0} - \left(\frac{Q_{wa}}{T_{UCZ}} + \frac{Q_{sw,UCZ}}{T_0}\right) + \left(\frac{Q_{g,NCZ}}{T_{NCZ}} + \frac{Q_{sw,UCZ}}{T_0}\right)\right] \tag{2.55}$$

In addition, we can write the exergy losses to the ambient air and through the side walls in the following manner:

$$\Xi_{a,UCZ} = m_{UCZ} C_{p,UCZ}\left[\left(T_{UCZ} - T_a\right) - T_0\left(\ln\frac{T_{UCZ}}{T_a}\right)\right] \tag{2.56}$$

and

$$\Xi_{sw,UCZ} = m_{UCZ} C_{p,sw}\left[\left(T_{UCZ} - T_{sw,UCZ}\right) - T_0\left(\ln\frac{T_{UCZ}}{T_{sw,UCZ}}\right)\right] \tag{2.57}$$

where $m_{UCZ} = \rho_{UCZ} V_{UCZ}$ is the mass of salty water in the UCZ; ρ_{UCZ} is the average density; and V_{UCZ} is the volume of salty water in the UCZ, which is 0.4 m^3. Also, $C_{p,UCZ}$ and $C_{p,sw}$ are the respective specific heats of the UCZ and insulating material; T_a and T_0 are the ambient temperature and the reference environment temperature, respectively; and T_{UCZ}, $T_{sw,UCZ}$ and $T_{m,NCZ}$ denote the average temperatures of the UCZ, the side wall and the NCZ, respectively.

The exergy efficiency of the UCZ can be written as the ratio of the exergy recovered from the UCZ to the total exergy input to the UCZ:

$$\psi_{UCZ} = \frac{\Xi_{r,UCZ}}{\Xi_{ti}} = 1 - \frac{\Xi_{d,UCZ} + \Xi_{a} + \Xi_{sw,UCZ}}{\Xi_{solar} + \Xi_{g,NCZ}} \tag{2.58}$$

2.5.3.2 Exergy Analysis for NCZ

With the NCZ exergy flows illustrated in Figure 2.5b, an exergy balance can be written for that zone as

$$\Xi_{r,UCZ} + \Xi_{g,HSZ} = \Xi_{r,NCZ} + \Xi_{d,NCZ} + \Xi_{l,NCZ} + \Xi_{sw,NCZ} \tag{2.59}$$

where $\Xi_{r,UCZ}$ is the exergy recovered from the UCZ; $\Xi_{g,HSZ}$ is the exergy gained from the HSZ, $\Xi_{r,NCZ}$ is the recovered exergy from the NCZ for the HSZ, $\Xi_{d,NCZ}$ is the exergy destruction in the NCZ, $\Xi_{l,NCZ}$ is the exergy loss from the NCZ to the UCZ and is equivalent to $\Xi_{g,NCZ}$, and $\Xi_{sw,NCZ}$ is the exergy loss through the side walls.

The term $\Xi_{r,NCZ}$ in Equation 2.59 can be written as

$$\Xi_{r,NCZ} = \Xi_{ti,NCZ} - \Xi_{tl,NCZ} = \left(\Xi_{r,UCZ} + \Xi_{g,HSZ}\right) - \left(\Xi_{d,NCZ} + \Xi_{l,NCZ} + \Xi_{sw,NCZ}\right) \tag{2.60}$$

where

$$\Xi_{g,HSZ} = m_{HSZ} C_{p,HSZ}\left[\left(T_{HSZ} - T_{NCZ}\right) - T_0\left(\ln\frac{T_{HSZ}}{T_{NCZ}}\right)\right] \tag{2.61}$$

and where $m_{HSZ} = \rho_{HSZ} V_{HSZ}$ is the mass of salty water in the HSZ; ρ_{HSZ} is the average zone density; and V_{HSZ} is the volume of salty water in the HSZ, which is $3.2\,m^3$.

The exergy destruction in the NCZ can be written as

$$\Xi_{d,NCZ} = T_0(\Delta S_{net,NCZ}) \tag{2.62}$$

where $\Delta S_{net,NCZ}$ is the net entropy change of the NCZ, which is $\Delta S_{net,NCZ} = \Delta S_{sys} + \Delta S_{surr}$.

The exergy losses, including the exergy destruction within the NCZ, can be expressed as follows:

$$\Xi_{d,NCZ} = T_0\left[m_{NCZ} C_{p,NCZ} \ln\frac{T_{m,NCZ}}{T_0} - \left(\frac{Q_{g,NCZ}}{T_{m,NCZ}} + \frac{Q_{sw,NCZ}}{T_0}\right) + \left(\frac{Q_{g,HSZ}}{T_{m,NCZ}} + \frac{Q_{sw,NCZ}}{T_0}\right)\right] \tag{2.63}$$

$$\Xi_{l,NCZ} = m_{NCZ} C_{p,NCZ} \left[\left(T_{m,NCZ} - T_{UCZ} \right) - T_0 \left(\ln \frac{T_{m,NCZ}}{T_{UCZ}} \right) \right] \quad (2.64)$$

$$\Xi_{sw,NCZ} = m_{NCZ} C_{p,sw} \left[\left(T_{m,NCZ} - T_{sw,NCZ} \right) - T_0 \left(\ln \frac{T_{m,NCZ}}{T_{sw,NCZ}} \right) \right] \quad (2.65)$$

where $C_{p,NCZ}$ is the specific heat at constant pressure of the NCZ and T_{HSZ} is the temperature of the HSZ.

The exergy efficiency of the NCZ can be expressed as the ratio of the exergy recovered from the NCZ to the total exergy input to the NCZ:

$$\psi_{NCZ} = \frac{\Xi_{r,NCZ}}{\Xi_{ti}} = 1 - \frac{\Xi_{d,NCZ} + \Xi_{l,NCZ} + \Xi_{sw,NCZ}}{\Xi_{r,UCZ} + \Xi_{g,HSZ}} \quad (2.66)$$

2.5.3.3 Exergy Analysis for HSZ

An exergy balance can be written for the HSZ using the exergy flows illustrated in Figure 2.5c as

$$\Xi_{r,NCZ} - \left(\Xi_{d,HSZ} + \Xi_{l,HSZ} + \Xi_{sw,HSZ} + \Xi_{b,HSZ} \right) = \Delta\Xi_{st} \quad (2.67)$$

where $\Xi_{r,NCZ}$ is the recovered exergy from the NCZ for the HSZ, $\Xi_{d,HSZ}$ is the exergy destruction in the HSZ, $\Xi_{l,HSZ}$ is the exergy loss from the HSZ to the NCZ, $\Xi_{sw,HSZ}$ is the exergy loss through the side walls, $\Xi_{b,HSZ}$ is the exergy loss through the bottom wall, and $\Delta\Xi_{st}$ is the exergy stored in the HSZ.

In Equation 2.67, $\Xi_{d,HSZ}$ is the exergy destruction in the HSZ, which can be written as

$$\Xi_{d,HSZ} = T_0 (\Delta S_{net,HSZ}) \quad (2.68)$$

Here, $\Delta S_{net,HSZ}$ is the net entropy change of the HSZ, expressible as $\Delta S_{net,HSZ} = \Delta S_{sys} + \Delta S_{surr}$.

The exergy losses, including exergy destruction within the NCZ, can be written as follows:

$$\Xi_{d,HSZ} = T_0 \left[m_{HSZ} C_{p,HSZ} \ln \frac{T_{HSZ}}{T_0} - \left(\frac{Q_{g,HSZ}}{T_{HSZ}} + \frac{Q_{sw,HSZ}}{T_0} \right) + \left(\frac{Q_b}{T_0} \right) \right] \quad (2.69)$$

$$\Xi_{l,HSZ} = m_{HSZ} C_{p,HSZ} \left[\left(T_{HSZ} - T_{m,NCZ} \right) - T_0 \left(\ln \frac{T_{HSZ}}{T_{m,NCZ}} \right) \right] \quad (2.70)$$

$$\Xi_{sw,HSZ} = m_{HSZ} C_{p,sw} \left[\left(T_{HSZ} - T_{sw,HSZ} \right) - T_0 \left(\ln \frac{T_{HSZ}}{T_{sw,HSZ}} \right) \right] \tag{2.71}$$

where $C_{p,HSZ}$ is the specific heat at constant pressure of the salty water in the HSZ. Note that $\Xi_{b,HSZ} = \Xi_{sw,HSZ}$ due to the fact that both side wall and bottom layer have the same insulating materials and are surrounded by ambient air.

The exergy efficiency for the HSZ can be expressed as the ratio of the exergy stored in the HSZ to the total exergy input to the HSZ, which is essentially the exergy recovered from the NCZ:

$$\psi_{HSZ} = \frac{\Delta\Xi_{st}}{\Xi_{r,NCZ}} = 1 - \frac{\left\{ \Xi_{d,HSZ} + \Xi_{l,HSZ} + \Xi_{sw,HSZ} + \Xi_{b,HSZ} \right\}}{\Xi_{r,NCZ}} \tag{2.72}$$

2.5.4 Numerical Efficiency Values

We now present and compare the efficiencies obtained with the model for the zones in the solar pond using experimental data, highlighting how the results demonstrate that exergy is crucial for determining true magnitudes of the losses and efficiencies.

Zone temperatures in the pond were measured throughout each month and averaged to determine the average monthly temperature values. The UCZ temperature reaches a maximum of 35.0 °C in August, a minimum of 10.4 °C in January, and 27.9 °C in May. Similarly, the NCZ temperature attains a maximum of 44.8 °C in August, a minimum of 13.9 °C in January, and 37.9 °C in May, while the HSZ temperature reaches a maximum of 55.2 °C in August, a minimum of 16.9 °C in January, and 41.1 °C in May. The zone temperatures clearly vary with month, depending on the environment temperature and incoming solar incidence. The temperatures of the zones generally increase with increasing incident solar energy per unit area of surface. Heat is lost from each zone and these losses are largest in the heat storage zone, decreasing storage performance significantly.

Note that the stability of the salt density distribution in a solar pond is of great significance to its performance [68]. Reducing the salt gradient region decreases the ability of the pond to store heat and increases molecular diffusion flow. The primary reason for monthly differences in efficiency is probably the seasonal temperature distribution, which affects the thermophysical properties of salty water, heat losses from the pond to the air, and the absorption and reflection of incident solar radiation on the surface. Fluctuations are observed in the saline density in the UCZ and NCZ, with increases in saline density caused by evaporation of water at the upper region. These increases can be reduced by continuously adding fresh water to the top of the pond. Note that significant changes occur in the non-convective and upper convective regions during one month due to not using one of the salt gradient protection systems for cleaning purposes.

Figure 2.6 illustrates the variations in average energy and exergy contents of the three zones of the pond with month of the year, based on the experimental data. The figure shows the solar radiation reaching the surface for the UCZ, the energy or exergy recovered from the UCZ for the NCZ, and the energy or exergy recovered from the NCZ for the HSZ. The exergy contents are seen to be lower than the corresponding energy contents due to the fact that, although energy is conserved, exergy is not. So, in addition to the exergy losses to the surrounding air, some exergy is destroyed in the each zone (Figure 2.7). The lowest exergy contents are observed in January and the highest in July. The temperature of the surroundings is significant since energy and exergy losses are rejected to the ambient air. Note that the shapes of the energy and exergy content monthly distributions in Figure 2.6 mirror the solar irradiation monthly profile.

Figure 2.8 shows the variations of energy and exergy efficiencies of the zones of the pond with month. The maximum and minimum energy and exergy efficiencies are observed to be 4.2 % and 3.0 % in August, and 0.9 % and 0.7 % in January for the UCZ; 13.8 % and 12.6 % in August and 3.2 % and 3.0 % in January for the NCZ; and 28.1 % and 27.5 % in August and 9.7 % and 10.0 % for the HSZ, respectively. As expected, the highest efficiencies are observed for the HSZ due to its heat storage capability, and the lowest for the UCZ due to the large heat losses to the surroundings from the UCZ surface (including the reflected portion of the solar irradiation). Variations with month of exergy destruction and loss for the solar pond zones are shown in Figure 2.7, and relate directly to the efficiency variations shown in Figure 2.8.

Useful insights have been provided by this case study of a solar pond, by contrasting the exergy efficiencies for each zone with the corresponding energy efficiencies. As expected, the exergy efficiencies are lower than the corresponding

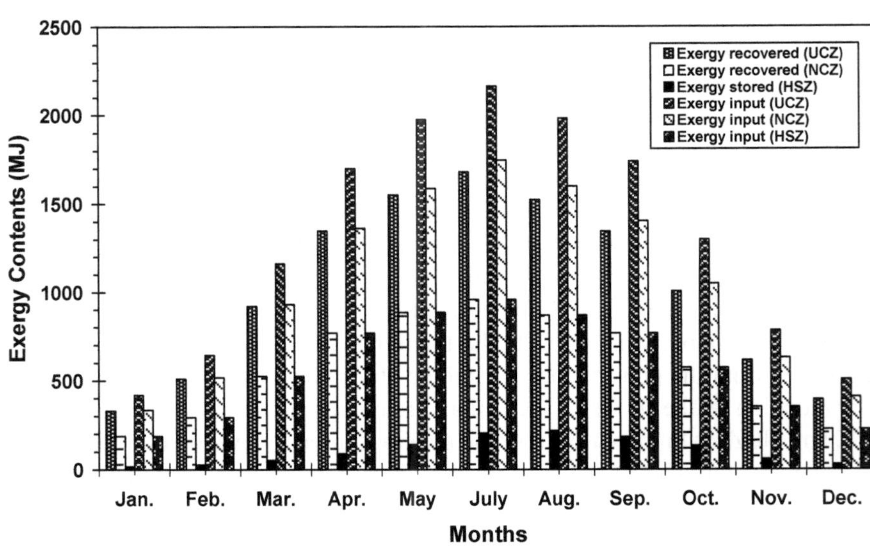

Figure 2.6 Variations of the exergy input, recovered and stored for the solar pond zones

energy efficiencies due to exergy destructions in the zones and losses from them to the surroundings. It is important to understand these destructions and losses in order to improve efficiency.

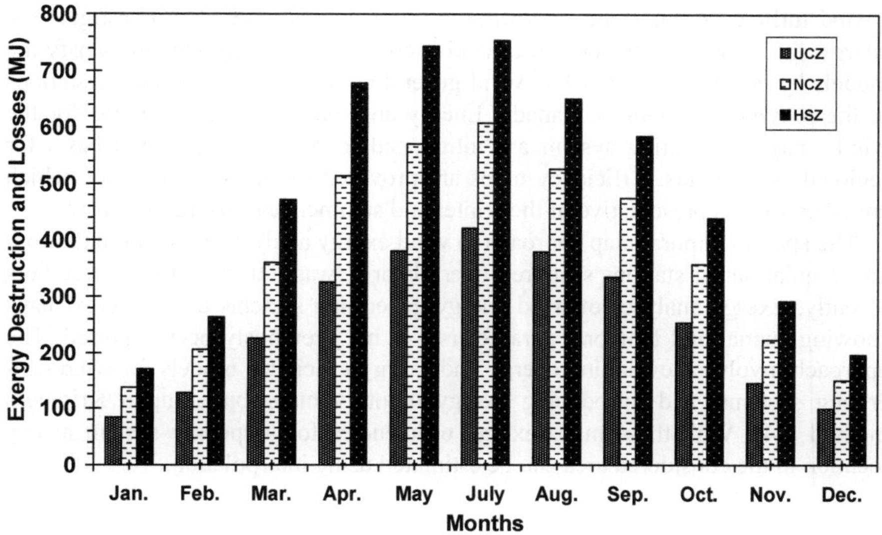

Figure 2.7 Variations of exergy destruction and loss for the solar pond zones

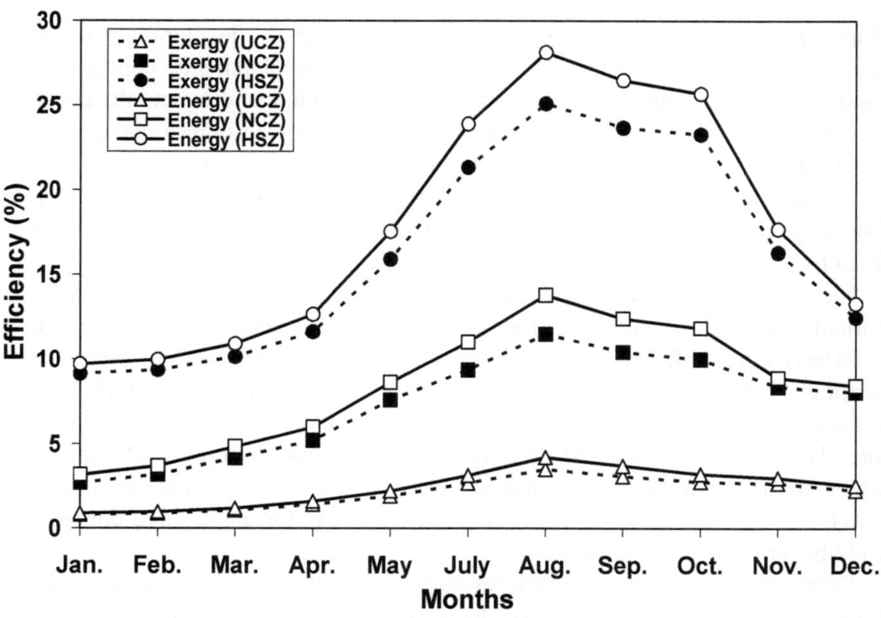

Figure 2.8 Variations of energy and exergy efficiencies of the zones of the solar pond

2.6 Case Study 2: Exergy Analysis of Wind Energy Systems

In this case study, energy and exergy characteristics of wind energy are described, and the effects of wind speed and air temperature and pressure at the inlet of a wind turbine on wind chill temperature are examined. We also investigate the energy and exergy efficiencies of a wind energy generating system and verify the models by considering a 100 kW wind generating system for 21 climatic stations in the province of Ontario, Canada. Energy and exergy efficiency maps for the wind energy generating system are introduced to provide a common basis for regional assessments. Efficiency maps are provided for January and July, which are taken to be representative of the winter and summer seasons, respectively.

The spatio-temporal map approach to wind exergy analysis, based on data from an irregular set of stations scattered over Ontario, was introduced by the authors recently. Exergy analyses of wind energy generating systems using exergy maps showing spatial and temporal parameters had not previously been reported. The approach involves developing energy and exergy efficiency models for wind generating systems, and introducing exergy monthly maps based upon Kriging's method [74]. With these maps exergy efficiencies for a specific system at any location in the considered area can be estimated using interpolation.

2.6.1 Background

Wind energy is among the most significant and rapidly developing renewable energy sources in the world. Recent technological developments, concerns over fossil fuel demands and the corresponding environmental effects, and the continuous increase in the consumption of conventional energy resources, have reduced wind energy costs to economically acceptable levels in many locations. Wind energy farms, which have been installed and operated in some instances for more than 25 years, are being considered as an alternative energy source in many jurisdictions.

Meteorological variables such as temperature, pressure and moisture play important roles in the occurrence of wind. Pressure forces lead to kinetic energy that is observed as wind. Petersen *et al.* [75], consider wind meteorology while investigating wind power. The dynamic behavior of the atmosphere generates spatio-temporal variations in such parameters as pressure, temperature, density and moisture. These parameters can be described by expressions based on continuity principles, the first law of thermodynamics, Newton's law and the state law of gases. Mass, energy and momentum conservation equations for air in three dimensions yield balances for the atmosphere.

During the last decade, wind energy applications have grown in Germany, Denmark and Spain and such successes have encouraged other countries to consider wind energy as a component of their electricity generation systems. Today, total operational wind power capacity worldwide has reached approximately 46,000 MW [76]. Wind turbine rotor efficiency increased from 35 % in the early 1980s to 48 % by the mid-1990s, and the technical availability of such systems has increased to 98 % [77].

Koroneos *et al.* [78] apply exergy analysis to renewable energy sources including wind power. But Koroneos *et al.* [78] treat the exergy efficiency of wind turbines for wind speeds above 9 m/s as zero and only consider the exergy of the wind turbine depending on electricity generation with no entropy generation analysis. Jia *et al.* [79] perform an exergy analysis of wind energy and consider wind power for air compression systems operating over specified pressure differences. Jia *et al.* estimate exergy components and show pressure differences, by considering two systems (a wind turbine and an air compressor) as a united system. Dincer and Rosen [43] investigate thermodynamic aspects of renewable energy and describe relations between exergy and sustainable development. Goff *et al.* [80], investigate wind speed thermodynamic characteristics by using the cooling capacity of wind as a renewable energy source (*i.e.*, using the wind chill effect for a heat pump system).

There is a need to assess accurately the behavior of wind scientifically, as some of the thermodynamic characteristics of wind energy are not yet clearly understood. The efficiency of a wind turbine often is taken as the ratio of the electricity generated to the wind potential within the area swept by the wind turbine. In this definition only the kinetic energy component of wind is considered, but other wind properties, such as temperature differences and pressure effects, are neglected. Usually the thermodynamic properties of wind are considered only from an energy perspective and in a limited fashion. By considering exergy, a more comprehensive understanding is obtained of the thermodynamic characteristics of wind and wind turbines. Maps of spatial and temporal energy and exergy efficiencies facilitate simple assessments for all parts of a geographic area.

2.6.2 Analysis

2.6.2.1 Wind Chill and Pressure

People sense whether air is warm or cool based on not only air temperature, but also wind speed and humidity. During cold weather, higher wind speed makes the air feel colder because it removes heat from our bodies faster. Wind chill is a measure of this effect, and is the hypothetical air temperature in calm conditions (air speed $V = 0$) that causes the same heat flux from the skin as occurs for the actual air speed and temperature. The heat transfer for air flow over a surface is slightly modified in some versions of the wind-chill expression [81]. The present wind-chill expression is based on the approaches of Osczevski [82] and Zecher [83], and was presented at a meeting of the Joint Action Group for Temperature Indices

(JAG/TI) [84]. The results of trials have been used to verify and improve the accuracy of the JAG/TI expression, which is given as

$$T_{windch} = 35.74 + 0.6215T_{air} - 35.75(V^{0.16}) + 0.4274T_{air}(V^{0.16}) \tag{2.73}$$

where the wind chill temperature T_{windch} is in °F and wind speed V is in mph.

Another wind speed factor is wind pressure, which affects a wind turbine and its blades. When wind approaches an obstacle, the air flows around it. However, one of the streamlines that hits the obstacle decelerates from the upstream velocity V_1 to a final velocity of zero (or to some lower velocity) (Figure 2.9). The pressure (dynamic pressure) at this stagnation point is higher than the free stream pressure (static pressure) well away from the obstacle. The dynamic pressure can be calculated from Bernoulli's equation. For flow at constant altitude, the only two variable terms in Bernoulli's equation are kinetic energy and pressure.

2.6.2.2 Energy and Analyses

To evaluate entropy generation, system inlet and outlet temperature and pressure differences are needed. Here we use the wind chill effect to determine the changes in energy capacities of wind.

Wind energy E is the kinetic energy of a flow of air of mass m at a speed V. The mass m is difficult to measure and can be expressed in terms of its volume **V** and density ρ via the relation $\rho = m/\mathbf{V}$. The volume can be expressed as $\mathbf{V} = AL$ where A is the cross-sectional area perpendicular to the flow and L is the horizontal distance. Physically, $L = Vt$ and wind energy can be expressed as

$$E = \frac{1}{2}\rho AtV^3 \tag{2.74}$$

Here, Betz [85] applied to a windmill the simple momentum theory established by Froude [86] for a ship propeller. The retardation of wind passing through a windmill occurs in two stages, before and after its passage through the windmill rotor. Provided that a mass m of air passes through the rotor per unit time, the rate of momentum change is $m(V_1 - V_2)$, which equals the resulting thrust. Here, V_1 and V_2 represent upwind and downwind speeds [87, 88] at a considerable distance from the rotor, as shown in Figure 2.9.

The total electricity generated is related to the decrease in wind potential. Subtracting the generated power from the total potential gives the wind turbine back-side wind potential. With the input and output variables shown in Figure 2.9 and the continuity relation, we can determine the back-side wind speed from the remaining potential as

$$V_2 = \sqrt[3]{\frac{2(E_{potential} - E_{generated})}{\rho At}} \tag{2.75}$$

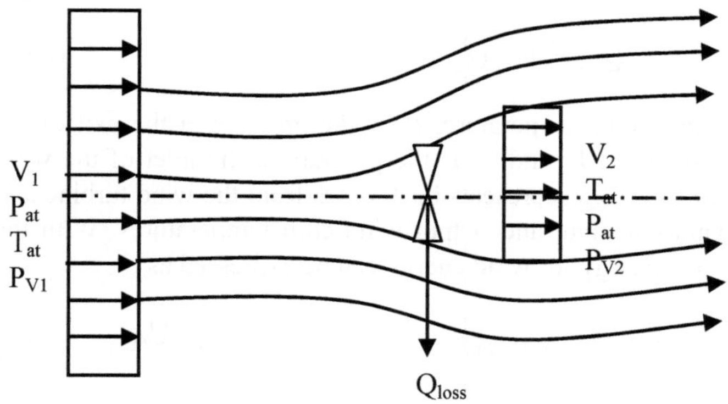

Figure 2.9 Wind energy input and output parameters for a wind turbine system

The total kinetic energy difference gives the generated electricity, which can be written as

$$\Delta K_e = E_{generated} \tag{2.76}$$

The air mass flow rate depends on density and wind speed, and can be expressed as

$$\dot{m} = \rho A V \tag{2.77}$$

The exergy of a matter flow is defined as the maximum work that can be acquired when the air flows from state (T_2, P_2) to the ambient state (T_1, P_1). The expressions of specific enthalpy, entropy and exergy at state 1 and state 2 are given below:

$$\Delta H = \dot{m} C_p (T_2 - T_1) \tag{2.78}$$

where ΔH is the change in enthalpy, h is the specific enthalpy, $\dot{m}$ is mass flow rate of air, and T_1 and T_2 are the wind chill temperatures input to and output from the wind turbine, respectively. The entropy difference ΔS for the system can be written as

$$\Delta S = \Delta S_{system} + \Delta S_{surr} \tag{2.79}$$

$$\Delta S = \dot{m} T_{at} \left(C_p \ln\left(\frac{T_2}{T_1}\right) - R \ln\left(\frac{P_2}{P_1}\right) - \frac{Q_{loss}}{T_{at}}\right) \tag{2.80}$$

where

$$P_i = P_{at} \pm \frac{\rho}{2} V^2 \tag{2.81}$$

and

$$Q_{loss} = \dot{m}C_p(T_{at} - T_{average}) \tag{2.82}$$

Here, T_{at} is the atmospheric temperature, P_2 is the pressure at the exit of the wind turbine for a wind speed V_2 and P_1 is the pressure at the inlet of the wind turbine for a wind speed V_1, Q_{loss} represents heat losses from the wind turbine and $T_{average}$ is the mean value of input and output wind chill temperatures. With the above equations, the total exergy for wind energy can be expressed as

$$\dot{E}x = \dot{E}_{generated} + \dot{m}C_p(T_2 - T_1) + \dot{m}T_{at}(C_p \ln(\frac{T_2}{T_1}) - R\ln(\frac{P_2}{P_1}) - \frac{\dot{Q}_{loss}}{T_{at}}) \tag{2.83}$$

where on the right side the first term represents the generated electricity and the second and third terms the enthalpy and entropy contributions, respectively.

2.6.3 Energy and Exergy Efficiencies

The expressions for energy (η) and exergy (ψ) efficiencies for the principal types of processes considered in the case study are based on expressions presented in Section 2.4.6. Note that the exergy efficiency frequently gives a finer understanding than the energy efficiency, since the exergy efficiency weights energy flows in terms of usefulness. The exergy efficiency stresses that both external losses and internal irreversibilities need to be addressed to improve efficiency. In many cases, the internal irreversibilities are more significant and more difficult to deal with than external losses.

Efficiency expressions for three types of processes are relevant to the case study. For electric work production, the efficiencies for producing shaft work W from electricity with a wind energy system are

$$\eta_{e,m} = \dot{W}_e / \dot{W} \tag{2.84}$$

$$\psi_{e,m} = \dot{E}^{W_e} / \dot{E}^{W} = \dot{W}_e / \dot{W} = \eta_{e,m} \tag{2.85}$$

Similarly, the efficiencies for electricity generation with a wind energy system can be written as

$$\eta_{e,f} = \dot{W}_e / \dot{W}_p \tag{2.86}$$

$$\psi_{e,f} = \dot{E}^{W_e} / \dot{E}^{W_p} \cong \eta_{e,f} \tag{2.87}$$

The exergy efficiencies for electricity generation are equivalent to the corresponding energy efficiencies. It is also instructive to consider processes for kinetic energy production. The efficiencies for kinetic energy generation from fossil fuels and wind, which produce a change in kinetic energy Δke in a stream of matter m_s, are as follows:

$$\eta_{ke,f} = m_s \Delta ke_s / m_f H_f \tag{2.88}$$

$$\psi_{ke,f} = m_s \Delta ke_s / m_f \gamma_f H_f \cong \eta_{ke,f} \tag{2.90}$$

Note that the exergy efficiency can be written as a function of the corresponding energy efficiency. Note also that kinetic energy is dominant in a wind turbine and there is no potential energy change or chemical component. Removing these components yields the general specific exergy equation as

$$ex = \left[ke + (h - h_0) - T_0(s - s_0)\right] \tag{2.91}$$

2.6.4 Application to Ontario

Energy and exergy efficiencies are estimated using measured generated power data. Here, these data are obtained from a Denmark group [89]. Petersen *et al.* [89] recommend wind turbine power curve measurements be used to determine the wind turbine required in relation to technical requirements and for approval and certification of wind turbines.

In this application, output electrical power data are used for a 100 kW wind turbine with a rotor diameter at 18 m and hub height 30 m. Such a turbine is more convenient for rural settings, Cut-in and cut-out wind speed values are 3.5 m/s and 21 m/s, respectively. The capacity factor of this wind turbine system is high (approximately 45 % for wind speeds of 8–11 m/s). Energy and exergy efficiencies of this system are compared by Sahin *et al.* [90], who show that values of capacity factor and energy efficiencies are generally higher than exergy efficiencies.

Maps of estimated efficiencies are presented for 21 stations in Ontario (Table 2.3). Exergy efficiencies maps can help inform companies and investors of the characteristics of wind systems before making decisions about investments. For these stations, 30-year average wind speed, temperature and pressure data from Ontario Weather Data [91] are used. Wind speeds are interpolated from 10 m to 30 m. The 100 kW wind turbine is selected to minimize wind speed interpolation errors. This region includes a lake so interactions between water and land surfaces are significant. Low temperatures with high wind speeds, leading to high wind chill temperatures, are prevalent in the region.

Geostatistical spatio-temporal maps are presented and discussed for January (representing winter) and July (representing summer). These maps show the differences between energy and exergy efficiencies of a specific wind turbine system at the same conditions. The maps indicate how efficiently wind energy is used and the magnitudes and locations of losses. By using these maps, wind industry engi-

Table 2.3 Topographical characteristics of selected meteorological stations in Ontario

Station	Latitude (degrees)	Longitude (degrees)	Altitude (m)
Atikokan	48.45	91.37	395
Big Trout Lake	53.50	89.52	220
Dryden Airport	49.50	92.45	413
Kapuskasing	49.25	82.28	227
Kenora	49.47	94.22	407
Kingston	44.13	76.36	93
London	43.02	81.09	278
Moosonee	51.16	80.39	10
North Bay	46.21	79.26	358
Ottawa	45.19	75.40	116
Red Lake	51.04	93.48	375
Simcoe	46.29	84.30	187
Sault Ste Marie	42.51	80.16	241
Sioux Lookout	50.07	91.54	398
Sudbury	46.37	80.48	348
Thunder Bay	48.22	89.19	199
Timmins	48.34	81.22	295
Toronto Pearson	43.40	79.38	173
Trenton	44.07	77.32	85
Wiarton	44.45	81.06	222
Windsor	42.16	82.58	190

Source: [91].

neers and decision makers can try to improve systems by reducing losses and irreversibilities. The 21 stations considered are scattered throughout each map in Figures 2.10 and 2.11, and the scale of each map is given at the right side. The bottom right of each map shows Lake Ontario, where climatological data are not measured, so this area is not considered.

2.6.4.1 Winter

A wind speed map for Ontario in January is shown in Figure 2.10a. In that map, and subsequent maps, the axes sloping up to the right and left, respectively, show longitude and latitude, while the vertical axis shows the variable being considered, By allowing interpolation to be used to estimate parameter values in regions for which there are no measured data, the maps are useful for practical engineering applications.

In January, the monthly maximum average wind speed observed in southwestern Ontario is 9–10 m/s. Low wind speeds are observed in the east and north parts of Ontario. The monthly minimum average value observed in Atikokan in this month is below the typical wind-turbine cut-in wind speed and as a result there is no electricity generation.

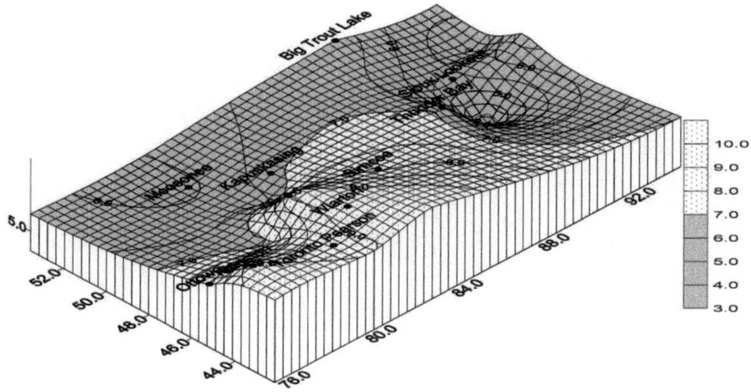

(a)

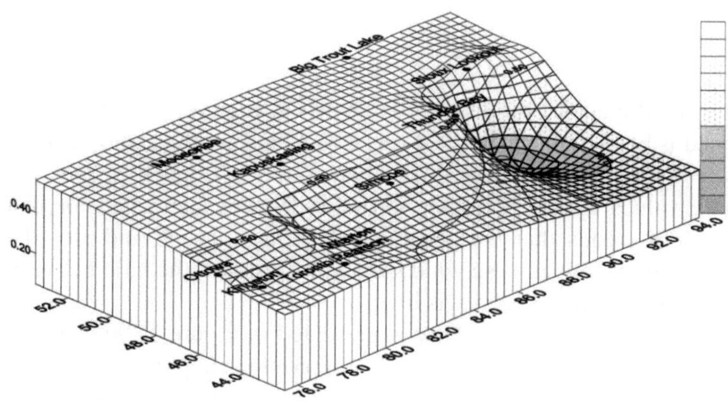

(b)

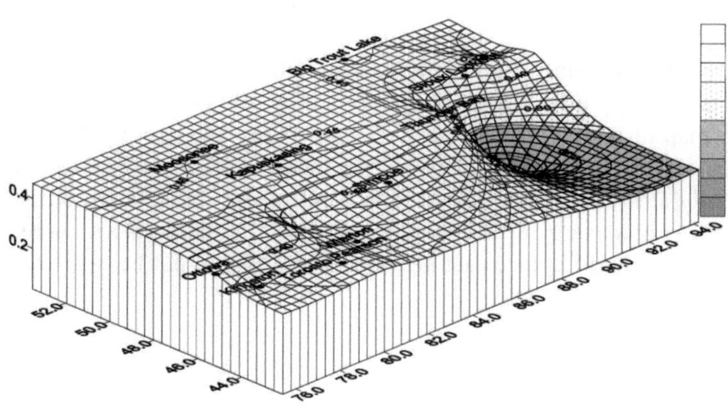

(c)

Figure 2.10 **a** Map of wind speed (m/s) at a height of 30 m for Ontario for January. **b** Energy-efficiency map for Ontario for January. **c** Exergy-efficiency map for Ontario for January. **d** Map of relative differences (in %) between energy and exergy efficiencies for Ontario for January

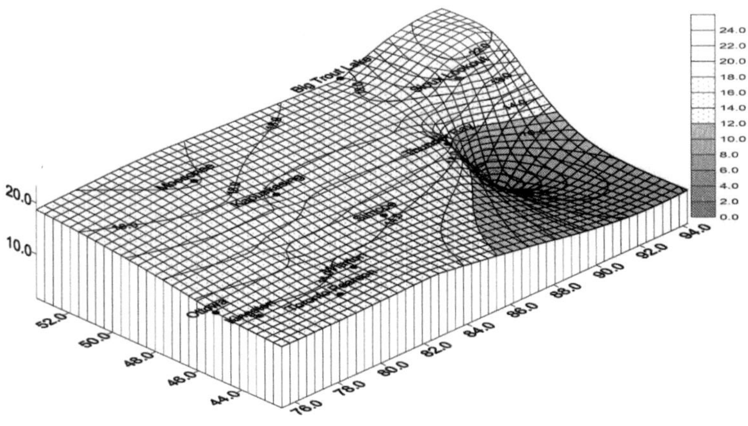

(d)

Figure 2.10 (continued)

A map of energy efficiencies for January is shown in Figure 2.10b. Since the average wind speed is below the cut-in wind speed in Atikokan, the station energy efficiency is zero. At low wind speeds, efficiencies are high, but this does not mean that at these values the wind turbine is more efficient than rated for that wind speed. Rather, it means that the generated electricity is low and also the potential of wind energy is low at these wind speeds. As a result, the ratio between generated electricity and potential energy is high.

A map of exergy efficiencies for January is presented in Figure 2.10c. The average exergy efficiency value is 40 %. The same observations apply for exergy as for energy, although the contours for exergy efficiency are lower than those for energy efficiency for all regions.

For meaningful comparisons of energy and exergy efficiencies, the wind speed maps should be considered together. The relative error method is a helpful engineering approach for comparing energy and exergy data sets. Here, differences between energy and exergy efficiencies are multiplied by 100 and divided by the highest value. A map of relative differences between energy and exergy efficiencies is shown in Figure 2.10d. Large relative differences in efficiency values are observed, especially at low wind speeds. At high wind speeds, the relative differences between energy and exergy efficiencies are smaller. But these values exceed 10 % at all stations. These differences are significant and should not be neglected in energy planning and management.

2.6.4.2 Summer

Maps corresponding to those in Figure 2.10a–2.10d are presented in Figure 2.11a–2.11d, respectively, for summer. Wind speeds for July (Figure 2.11a) exhibit different clusters as a result of topographical effects in summer. The strong heating during this month creates unstable surface conditions. The average wind speed at one station is lower than the cut-in value and as a result the corresponding energy and exergy efficiencies are zero. The highest wind speed for this month is the lowest of the maximums of the other months.

The energy efficiencies in July (Figure 2.11b) are approximately 40–50 %, while the exergy efficiencies (Figure 2.11c) are about 40 %, except for the eastern regions of Ontario. The spatial distribution for energy efficiencies exhibit three clusters. There is an area of high energy efficiency in northwest Ontario but the exergy efficiencies are lowest in this area. The relative differences between energy and exergy efficiencies in July (Figure 2.11d) are relatively low.

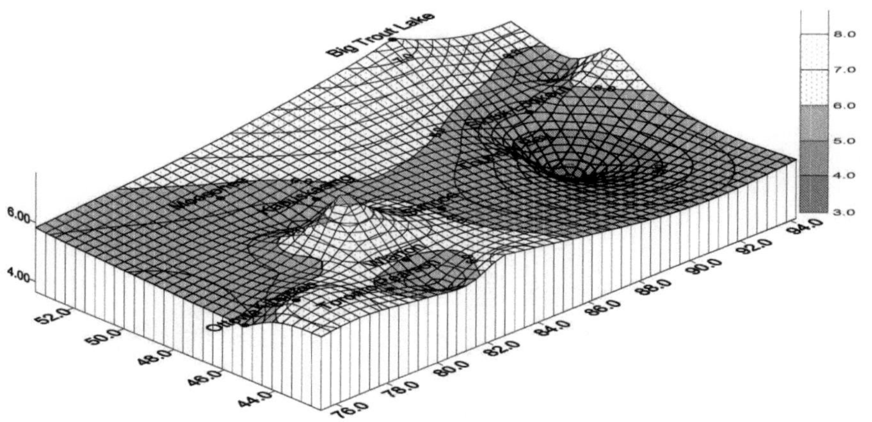

(a)

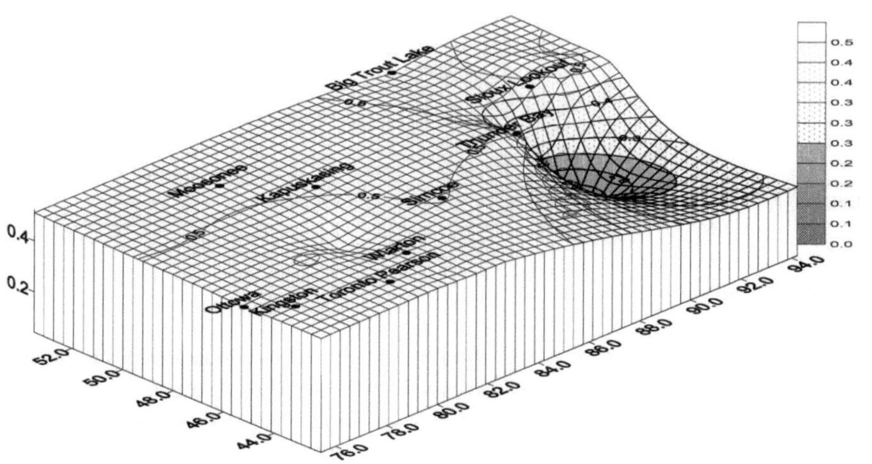

(b)

Figure 2.11 **a** Map of wind speed (m/s) at a height of 30 m for Ontario for July. **b** Energy-efficiency map for Ontario for July. **c** Exergy-efficiency map for Ontario for July. **d** Map of relative differences (in %) between energy and exergy efficiencies for Ontario for July

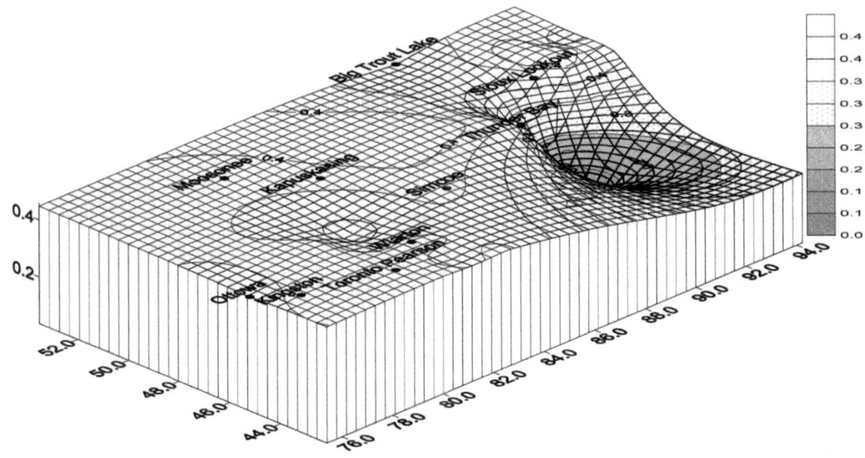

(c)

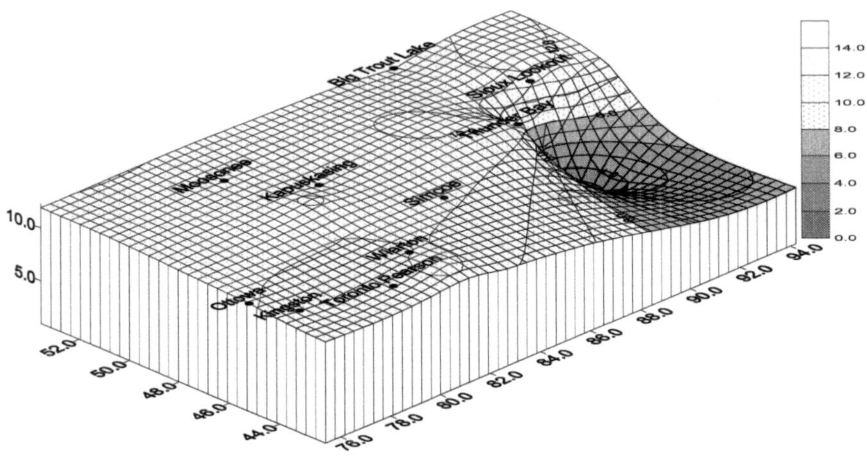

(d)

Figure 2.11 (continued)

2.6.4.3 Discussion

Energetic and exergetic aspects of wind energy have been described and spatio-temporal energy and exergy maps presented. The use of such maps enhances understanding of the thermodynamic behavior of wind energy systems and reduces the complexity of analyses, thus helping to facilitate practical applications. The exergy maps provide more meaningful and useful information than energy analysis regarding the efficiency, losses and performance for wind turbines.

The exergy and energy efficiencies presented in the form of geostatistical maps for two months in Ontario allow some important observations to be made. The relative differences between the energy and exergy efficiencies are seen to be highest in winter and lowest in summer. These differences are approximately 20–24 % at low wind speeds and 10–15 % at high wind speeds. Also, exergy efficiencies are lower than energy efficiencies for each station for every month considered. This sample application demonstrates that the tools presented here for approximating the efficiencies of wind energy systems could have widespread applications. In particular, exergy-related information could assist designers and decision makers, and thereby increase the application of wind systems, help optimize the design of such systems and their components, and identify appropriate applications and optimal arrangements of these systems.

2.7 Closing Remarks

In this chapter, energy and exergy analyses are described, with a focus on green energy systems. Implications of exergy analysis results are discussed, and procedures for energy and exergy analyses are described. The advantages of applying exergy analysis in place of or in concert with energy analysis are explained and illustrated. Two case studies are assessed involving the application of exergy analysis to green energy systems: solar ponds for thermal energy storage and wind energy systems.

Symbols

A	area
C	specific heat
C_p	specific heat at constant pressure
e	specific energy
E	energy; total solar energy reaching solar pond; wind energy
$E_{potential}$	wind potential energy
$E_{generated}$	electricity generated by wind turbine
ex	specific exergy
Ex	exergy
Ex^Q	exergy associated with heat Q
Ex^W	exergy associated with work W
F	absorbed energy percentage at δ-thickness region
h	specific enthalpy; ratio
H	enthalpy
I	number of the layers; irreversibility
k	thermal conductivity
ke	specific kinetic energy
KE	kinetic energy

L	horizontal distance; length
m	mass
M	mass
N	number of moles
P	pressure
P_1	pressure at inlet to wind turbine
P_2	pressure at exit from wind turbine
pe	pecific potential energy
PE	potential energy
q	heat transfer per unit area
Q	heat
Q_r	heat transfer into system across region r on system boundary
R	thermal resistance
s	specific entropy
S	entropy; thickness
S_{gen}	entropy generation
ΔS	entropy change
t	time
T	temperature
T_{air}	air temperature
T_{windch}	wind chill temperature
u	specific internal energy
U	internal energy; heat loss from pond surface to air
v	specific volume
$\mathbf{V}$	volume
V	wind speed
V_1	upwind speed far from wind turbine rotor
V_2	downwind speed far from wind turbine rotor
W	work; shaft work
x	mass fraction
Δx	thickness of horizontal layers
X	zone thickness
y	mole fraction

Greek Letters

β	incident beam rate entering water
δ	thickness where long-wave solar energy is absorbed
η	energy efficiency
θ	angle
μ	chemical potential
Ξ	exergy
$\Delta\Xi$	stored exergy
Π	entropy creation

ρ	density
τ	exergetic temperature factor
ψ	exergy efficiency

Subscripts

a	ambient air
at	atmosphere
b	bottom
e	exit
f	final
g	gained
HSZ	heat storage zone
i	incident; inlet; initial
kin	kinetic
L	length
LCZ	lower convective zone
net	net irradiation
NCZ	non-convective zone
o	reference-environment state
oo	dead state
p	paint
ph	physical
pot	potential
ps	painted sheet
r	reflectance
s	sheet-iron
st	stored
sw	side wall
ti	total input
tl	total lost
UCZ	upper convective zone
w	width
wa	water to air
01	painted inner surface area of side wall

References

[1] Brodyanski VM, Sorin MV, Le Goff P. The Efficiency of Industrial Processes: Exergy Analysis and Optimization. London: Elsevier, 1994.
[2] Edgerton RH. Available Energy and Environmental Economics. Toronto: DC Heath, 1082.
[3] Kotas TJ. The Exergy Method of Thermal Plant Analysis, Reprint edn. Malabar, Florida: Krieger, 1995.

[4] Moran MJ. Availability Analysis: A Guide to Efficient Energy Use, Revised edn. New York: American Society of Mechanical Engineers, 1989.
[5] Sato N. Chemical Energy and Exergy: An Introduction to Chemical Thermodynamics for Engineers. Oxford: Elsevier, 2005.
[6] Szargut J. Exergy Method: Technical and Ecological Applications. Southampton, WIT Press, 2005.
[7] Szargut J, Morris DR, Steward FR. Exergy Analysis of Thermal, Chemical and Metallurgical Processes. New York: Hemisphere, 1988.
[8] Yantovskii EI. Energy and Exergy Currents: An Introduction to Exergonomics. New York: Nova Science Publishers, 1994.
[9] Dincer I, Rosen MA. Exergy: Energy, Environment and Sustainable Development, Oxford: Elsevier, 2007.
[10] El-Sayed Y. The Thermoeconomics of Energy Conversion. Amsterdam: Elsevier, 2004.
[11] Hammond GP. Engineering sustainability: thermodynamics, energy systems, and the environment. Int J Energy Res 2004; 28(7):613–639.
[12] Wall G. Exergy tools. Proceedings of the Institution of Mechanical Engineers, Part A. J Power Energy 2003; 217(2):125–136.
[13] Moran MJ, Sciubba E. Exergy analysis: principles and practice. J Eng Gas Turbines Power 1994; 116:285–290.
[14] Rosen MA. Second law analysis: approaches and implications. Int J Energy Res 1999; 23:415–429.
[15] Sciubba E. Exergo-economics: thermodynamic foundation for a more rational resource use. Int J Energy Res 2005; 29:613–636.
[16] Rosen MA, Le MN, Dincer I. Efficiency analysis of a cogeneration and district energy system. Appl Thermal Eng 2005; 25(1):147–159.
[17] Rosen MA, Dincer I. Exergoeconomic analysis of power plants operating on various fuels. Appl Thermal Eng 2003; 23(6):643–658.
[18] Rosen MA, Dincer I. Thermoeconomic analysis of power plants: an application to a coal-fired electrical generating station. Energy Convers Manage 2003; 44(17):2743–2761.
[19] Rosen MA, Dincer I. A study of industrial steam process heating through exergy analysis. Int J Energy Res 2004; 28(10):917–930.
[20] Rosen MA, Etele J. 2004. Aerospace systems and exergy analysis: applications and methodology development needs. Int J Exergy 2004; 1(4):411–425.
[21] Rosen MA, Scott DS. 1998. Comparative efficiency assessments for a range of hydrogen production processes. Int J Hydrogen Energy 1998; 23:653–659.
[22] Rosen MA, Horazak DA. Energy and exergy analyses of PFBC power plants. In Alvarez Cuenca M, Anthony EJ (eds), Pressurized Fluidized Bed Combustion. London: Chapman and Hall, Chapter 11, pp. 419–448, 1995.
[23] Dincer I, Rosen MA. Thermal Energy Storage: Systems and Applications. Chichester: Wiley, 2002.
[24] Rosen MA, Tang R, Dincer I. Effect of stratification on energy and exergy capacities in thermal storage systems. Int J Energy Research 2004; 28:177–193.
[25] Rosen MA, Dincer I. Sectoral energy and exergy modeling of Turkey. ASME J. Energy Resources Technol 1997; 119:200–204.
[26] Rosen MA. Evaluation of energy utilization efficiency in Canada using energy and exergy analyses. Energy – Int J 1992; 17:339–350.
[27] Crane P, Scott DS, Rosen MA. Comparison of exergy of emissions from two energy conversion technologies, considering potential for environmental impact. Int J Hydrogen Energy 1992; 17:345–350.
[28] Rosen MA, Dincer I. On exergy and environmental impact. Int J Energy Res 1997; 21:643–654.
[29] Rosen MA, Dincer I. Exergy analysis of waste emissions. Int J Energy Res 1999; 23(13):1153–1163.

[30] Gunnewiek LH, Rosen MA. Relation between the exergy of waste emissions and measures of environmental impact. Int J Environ Pollut 1998; 10(2):261–272.

[31] Midilli A, Ay M, Dincer I, Rosen MA. On hydrogen and hydrogen energy strategies–I: current status and needs. Renewable Sustainable Energy Rev 2005; 9(3):255–271.

[32] Midilli A, Dincer I, Ay M. Green energy strategies for sustainable development. Energy Policy 2006; 34(18):3623–3633.

[33] Bruntland Commission. Our Common Future, World Commission on Environment and Development. Oxford: Oxford University Press, 1987.

[34] Ruth M. A quest for the economics of sustainability and the sustainability of economics. Ecol Econ 2006; 56(3):332–342.

[35] Jefferson M. Sustainable energy development: performance and prospects. Renewable Energy 2006; 31(5):571–582.

[36] Dincer I, Rosen MA. Exergy as a driver for achieving sustainability. Int J Green Energy 2004; 1(1):1–19.

[37] Midilli A, Ay M, Dincer I, Rosen MA. On hydrogen energy strategy: the key role in this century and beyond. Proceedings of the First Cappadocia International Mechanical Engineering Symposium (CMES–04), 2004; 1:211–217, Cappadocia, Turkey.

[38] Midilli A, Ay M, Dincer I, Rosen MA. On hydrogen and hydrogen energy strategies–I: current status and needs. Renewable Sustainable Energy Rev 2005; 9(3):255–271.

[39] Suganthi L, Samuel AA. Exergy based supply side energy management for sustainable energy development. Renewable Energy 2000; 19(1–2):285–290.

[40] Momirlan M, Veziroglu TN. Current status of hydrogen energy. Renewable Sustainable Energy Rev 2002; 6(1–2):141–179.

[41] Momirlan M, Veziroglu TN. The properties of hydrogen as fuel tomorrow in sustainable energy system for a cleaner planet. Int J Hydrogen Energy 2005; 30(7):795–802.

[42] Afgan NH, Carvalho MG. Sustainability assessment of hydrogen energy systems. Int J Hydrogen Energy 2004; 29(13):1327–1342.

[43] Dincer I, Rosen MA. Thermodynamic aspects of renewables and sustainable development. Renewable Sustainable Energy Rev 2005; 9:169–189.

[44] Midilli A, Ay M, Dincer I, Rosen MA. On hydrogen and hydrogen energy strategies–II: future projections affecting global stability and unrest. Renewable Sustainable Energy Rev 2005; 9(3):273–287.

[45] Newman L. Change, uncertainty, and futures of sustainable development. Futures 2006; 38(5):633–637.

[46] Madlener R, Stagl TS. Sustainability–guided promotion of renewable electricity generation. Ecol Econ 2005; 53:147–167.

[47] Dincer I, Renewable energy and sustainable development: a crucial review. Renewable Sustainable Energy Rev 2000; 4(2):157–75.

[48] Dincer I. The role of exergy in energy policy making. Energy Policy 2002; 30:137–149.

[49] Lucca G. The exergy analysis: role and didactic importance of a standard use of basic concepts, terms and symbols. A Future for Energy: Proceedings of Florence World Energy Research Symposium, pp. 295–308, 1990.

[50] Baehr HD, Schmidt EF. Definition und Berechnung von Brennstoffexergien (Definition and calculation of fuel exergy). Brennst–Waerme–Kraft 1963; 15:375–381.

[51] Gaggioli RA, Petit PJ. Use the second law first. Chemtech 1977; 7:496–506.

[52] Rodriguez LSJ. Calculation of available–energy quantities. In Thermodynamics: Second Law Analysis. ACS Symposium Series 1980; 122:39–60.

[53] Szargut J. Grenzen fuer die Anwendungsmoeglichkeiten des Exergiebegriffs (limits of the applicability of the exergy concept). Brennst.–Waerme–Kraft 1967; 19:309–313.

[54] Sussman MV. Steady–flow availability and the standard chemical availability. Energy – Int J 1980; 5:793–804.

[55] Sussman MV. Second law efficiencies and reference states for exergy analysis. Proceedings of. 2nd World Congress Chemical Engineering, Canadian Society of Chemical Engineers, Montreal, pp. 420–421, 1981.

[56] Ahrendts J. Reference states. Energy – Int J 1980; 5:667–678.
[57] Bosnjakovic F. Bezugszustand der Exergie eines reagiernden Systems (Reference states of the exergy in a reacting system). Forsch. Ingenieurw 1963; 20:151–152.
[58] Wepfer WJ, Gaggioli RA. Reference datums for available energy. In Thermodynamics: Second Law Analysis. ACS Symposium Series 122, American Chemical Society, Washington, pp. 77–92, 1980.
[59] Rosen MA, Dincer I. Effect of varying dead–state properties on energy and exergy analyses of thermal systems. Int J Thermal Sci 2004; 43(2):121–133.
[60] Hafele W. Energy in a Finite World: A Global Systems Analysis. Toronto: Ballinger, 1981.
[61] Gaggioli RA. Second law analysis to improve process and energy engineering. In Efficiency and Costing: Second Law Analysis of Processes. ACS Symposium Series 235, pp. 3–50, Washington, DC: American Chemical Society, 1983.
[62] Jaefarzadeh MR. Thermal behavior of a small salinity–gradient solar pond with wall shading effect. Solar Energy 2004; 77:281–290.
[63] Kurt H, Halici F, Korhan Binark A. Solar pond conception–experimental and theoretical studies. Energy Convers Manage 2000; 41, 9:939–951.
[64] Ouni M, Guizani A, Lu,H, Belghith A. Simulation of the control of a salt gradient solar pond in the south of Tunisia. Solar Energy 2003; 75(2):95–101.
[65] Ramadan MRI, El-Sebaii AA, Aboul-Enein S, Khallaf AM. Experimental testing of a shallow solar pond with continuous heat extraction. Energy Buildings 2004; 36:955–964.
[66] Sozen A. Effect of irreversibilities on performance of an absorption heat transformer used to increase solar pond's temperature. Renewable Energy 2003; 29:501–515.
[67] Yi Li X, Kanayama K, Baba H. Spectral calculation of the thermal performance of a solar pond and comparison of the results with experiments. Renewable Energy 2000; 20:371–373.
[68] Karakilcik M, Dincer I. Exergetic performance analysis of a solar pond, International Journal of Thermal Sciences 2008; 47(1):93–102.
[69] Karakilcik M, Dincer I, Rosen MA. Performance investigation of a solar pond. Appl Thermal Eng 2006; 26:727–735.
[70] Karakilcik M, Kıymaç K, Dincer I. Experimental and theoretical distributions in a solar pond. Int J Heat Mass Transfer 2006; 49:825–835.
[71] Hawlader MNA. The influence of the extinction coefficient on the effectiveness of solar ponds. Solar Energy 1980; 25:461–464.
[72] Bryant HC, Colbeck,I. A solar pond for London. Solar Energy 1977; 19:321.
[73] Petala R. Exergy of undiluted thermal radiations. Solar Energy 2003; 74:469–488.
[74] Krige DG. A statistical approach to some basic mine evaluation problems on the Witwateround. J Chim Min Soc South-Africa 1951; 52:119–139.
[75] Petersen EL, Mortensen NG, Landberg L, Hojstrup J, Frank HP. Wind power meteorology – part I: climate and turbulence. Wind Energy, pp. 25–45, 1998.
[76] Wind power monthly. http://www.windpower–monthly.com/WPM: WINDICATOR
[77] Şahin AD. Progress and recent trends in wind energy. Prog Energy Combust Sci 2004; 30:501–543.
[78] Koroneos C, Spachos N, Moussiopoulos N. Exergy analysis of renewable energy sources. Renewable Energy 2003; 28:295–310.
[79] Jia GZ, Wang XY, Wu GM. Investigation on wind energy–compressed air power system. J Zhejiang University Sci 2004; 5(3):290–295.
[80] Goff LH, Hasert UF, Goff PL. A "new" source of renewable energy: the coldness of the wind. Revue Generale de Thermique 1999; 38(10):916–924.
[81] Stull R. Meteorology for Scientists and Engineers, 2nd edn. Pacific Grove: Brooks/Cole Thomson Learning, 2000.
[82] Osczevski RJ. Windward cooling: an overlooked factor in the calculation of wind chill. Bull Am Met Soc 2000; 81(12):2975–2978.
[83] Zecher JBM. A new approach to an accurate wind chill factor. Bull Am Met Soc 1999; 80(9):1893–1899.

[84] JAG/TI. New wind chill equation. August, Toronto, Canada, 2001.
[85] Betz A. Windenergie und ihre Ausnutzung durch Windmühlen. Göttingen: Vandenhoek and Ruprecht, Göttingen, 1946.
[86] Froude RE. On the part played in propulsion by differences of fluid pressure. Trans. Inst. Naval Architects 1889; 30:390.
[87] Golding EW. The Generation of Electricity by Wind Power. London: E.&F. N. Spon, 1955.
[88] Spera DA. Wind Turbine Technology. New York: ASME, 1998.
[89] Petersen TF, Petersen SM, Paulsen US, Fabian O, Pedersen BM, Velk P, Brink M, Gjerding J, Frandsen S, Olesen J, Budtz L, Nielsen MA, Stiesdal H, Petersen KØ, Danwin PL, Danwin LJ, Friis P. Recommendation for wind turbine power curve measurements to be used for type approval of wind turbines in relation to technical requirements for type approval and certification of wind turbines in Denmark, Danish Energy Agency, 1992.
[90] Şahin AD, Dincer I, Rosen MA. Thermodynamic analysis of wind energy. Int J Energy Res 2006; 30:553–566.
[91] Ontario Weather Data. http://www.theweather network.com/weather/stats/north_america.htm, 2004.

Chapter 3
Wind Speed Distribution – A Theoretical Approach to Probability Density Function

Xianguo Li

3.1 Introduction

Provided overwhelmingly by conventional fossil fuels, global primary energy consumption has been increasing as a result of improved living standards, industrialization of developing countries, and population increase. However, fossil fuels are limited, and unsustainable; their widespread and excessive use degrades the local and global environment as a result of rising pollution levels, and increases global climate changes due to the emissions of "greenhouse gases" such as carbon dioxide from fossil fuel combustion. The uneven distribution of known fossil fuel reserves has intensified geopolitics and international tensions and conflicts. Energy security has risen to the top of the national security agenda for most countries around the world. High fossil fuel prices and extreme fossil fuel price fluctuations have imposed significant strains on economy and society.

Alternatives to conventional fossil fuels, especially renewable, sustainable, and environmentally friendly energy, are becoming increasingly attractive all over the world due to their almost inexhaustible and non-polluting characteristics. The renewable energy resources include wind, wave, hydro, tidal, solar, geothermal, and bio-energy. All these renewable energy resources are abundant, available locally and distributed more evenly, and the technologies, if well established, can provide complete security of energy supply. It is now evident that renewable energy technologies play a strategic role in achieving the goals of sustainable economic development and environmental protection. Attempts have been made to explore new energy resources to meet the ever-increasing demand by modern

Xianguo Li
Department of Mechanical and Mechatronics Engineering, University of Waterloo,
200 University Avenue West, Waterloo, Ontario N2L 3G1, Canada
e-mail: x6li@uwaterloo.ca

civilization, to conserve fossil fuels for their better end use, and to make a substantial contribution to the improvement of the environment.

Among renewable energy resources, wind energy seems to be the most suitable and cost effective power source for electricity production. Wind energy, first used more than 3500 years ago in Egypt, has some key advantages such as cleanliness, abundant in most parts of the world, low cost, sustainability, safety, popularity, *etc.* Wind turbine technology has been the subject of intensive development over the last two decades [1, 2]. There are no significant unsolvable technical barriers to the widespread implementation of wind energy, and the safety of wind energy is of much less concern. The cost of wind-generated electricity has dropped substantially due to reduction in the installed cost of turbines. Another advantage of a wind energy system lies in its flexibility. It can be constructed into grid-connected or stand-alone wind power generation systems (distributed generation or mini-grid). This eliminates the need for high voltage transmission lines running through rural and urban landscapes. Due to the above mentioned and other merits, ample attention has been directed toward the use of renewable wind energy, especially after the major energy crisis in the 1970s, and wind energy continues to be the fastest growing power generating technology in the world, in terms of percentage yearly growth of installed capacity per technology source [3]. The wind energy industry celebrated a near-record breaking year in 2003, adding more than 8000 megawatts (MW) of wind energy capacity in more than two dozen nations. The total world capacity of 39,294 MW provides enough to power the equivalent of 9 million average American homes [4].

According to the commercially international system of classification by Elliott and Schwartz [5] from the Pacific Northwest Laboratory (PNL) shown in Table 3.1, it is popularly accepted that wind power Class 4 and above are suitable for large-scale electricity generation with modern wind turbine technology. It has already been proven that large-scale wind turbines with good enough wind sources are cost competitive and can become one of the least-cost power sources. With the development of future generation wind turbine technology, Class 2 areas and under may be viable for large-scale applications in the near future. In his study, Celik [6] states that wind electricity by medium-scale wind turbines is preferable in remote locations as it is socially valuable and economically competitive. As for small-scale turbines, when sized properly and used at optimal working conditions, these also could be reliable energy sources producing socio-economically valuable energy. Other than electricity generation, wind with speeds ranging from 2.6 m/s to about 4 m/s at 10 m above the ground level is utilizable for wind pumping and other mechanical conversion systems. In all cases, wind resources have specific uses; lower-to-moderate wind speed for small-scale utilization, and high wind speeds for community electricity supplies. To supply reliable electricity at a reasonable cost in a given location, detailed examination of wind characteristics and accurate assessment of wind energy have to be carried out.

Before wind turbines are installed a suitable site needs to be selected to harness the wind energy. Site selection is normally based on the wind energy potential assessment at the site. Wind energy potential assessment is typically relied on the

Table 3.1 Commercially international system of classification for wind [5] (classes of mean wind speed and wind power density at 10, 30, and 50 m above the ground)

Wind power class	10 m above ground		30 m above ground		50 m above ground	
	Wind power density (W/m^2)	Mean wind speed (m/s)	Wind power density (W/m^2)	Mean wind speed (m/s)	Wind power density (W/m^2)	Mean wind speed (m/s)
1	≤ 100	≤ 4.4	≤ 160	≤ 5.1	≤ 200	≤ 5.6
2	≤ 150	≤ 5.1	≤ 240	≤ 5.9	≤ 300	≤ 6.4
3	≤ 200	≤ 5.6	≤ 320	≤ 6.5	≤ 400	≤ 7.0
4	≤ 250	≤ 6.0	≤ 400	≤ 7.0	≤ 500	≤ 7.5
5	≤ 300	≤ 6.4	≤ 480	≤ 7.4	≤ 600	≤ 8.0
6	≤ 400	≤ 7.0	≤ 640	≤ 8.2	≤ 800	≤ 8.8
7	≤ 1000	≤ 9.4	≤ 1600	≤ 11.0	≤ 2000	≤ 11.9

measurements of (absolute) wind speed at a fixed interval (such as every 15 minutes, or half an hour, or even an hour) at the particular site over a period of several years. To avoid the time and expense associated with processing multiple year data records of wind speed data measured at a regular interval, it is very important to describe the variation of wind speeds in terms of probability density functions or statistical functions for optimizing the design of the systems. The wind speed distribution, one of the wind characteristics, is of great importance for the assessment of the wind energy potential and predominantly determines the performance of wind energy systems in a given location.

Empirical wind speed distributions, such as the Weibull and the Rayleigh functions, have long been applied to fit the wind speed variations to create wind speed frequency distribution over a period of time. The two-parameter Weibull function is accepted as the best one [7] and the application of this function for different sites can be found in many publications [8–11]. However, one main limitation of the Weibull distribution is that it does not accurately represent the probabilities of observing zero and very low wind speeds. Recently, Li and Li [12, 13] developed a theoretical approach to the analytical determination of the wind speed distributions based on the Maximum Entropy Principle (MEP) and this model was shown capable of describing not only the actual data more accurately than the Weibull distribution, but also a much wider range of data types.

The main purpose of this chapter is to develop a theoretical model to predict wind speed distributions. The model is constructed by application of the MEP, along with the conservation of mass, momentum, and energy in the flowing wind stream as the constraint equations. Validation of the model is made by comparing the theoretical prediction with measured data obtained from different sources, as well as the corresponding Weibull distribution. It is found that the present theoretical predictions not only agree well with the measured data, but are also better than the correspondingly fitted empirical Weibull distribution.

3.2 Analysis of the Wind Speed Data

Normally absolute wind speed is measured at a given site as a mean value over a measurement interval (such as 15 minutes, or half an hour, or even an hour); such measurements span a period of several years with enormous data records. The measured data is processed statistically so that it is represented by a frequency distribution of wind speed at the give site. The wind speed distribution function is used for many purpose, such as the calculation of wind-induced forces on man-made structures (*e.g.*, buildings and bridges), in addition to the wind energy potential assessment that is the interest of the present chapter.

3.2.1 Frequency Distribution of Wind Speed

To represent the measured wind speed data in a simple and straightforward manner for easy use of the information, the measured wind speed data is first grouped into velocity bins (V_i, V_{i+1}) with $i = 1, 2, \ldots, I$ representing the ith speed bin, and I the total number of velocity bins (or groups), the number of wind speeds N_i falling in the ith speed interval $V_i < V < V_{i+1}$ is tallied. Then the frequency of the occurrence of wind speed in the velocity interval $V_i < V < V_{i+1}$ can be calculated as follows:

$$f_i = \frac{N_i}{N} \tag{3.1}$$

where N is the total number of wind speed measurements for a given site. The above equation represents the frequency distribution of wind speed measured, and

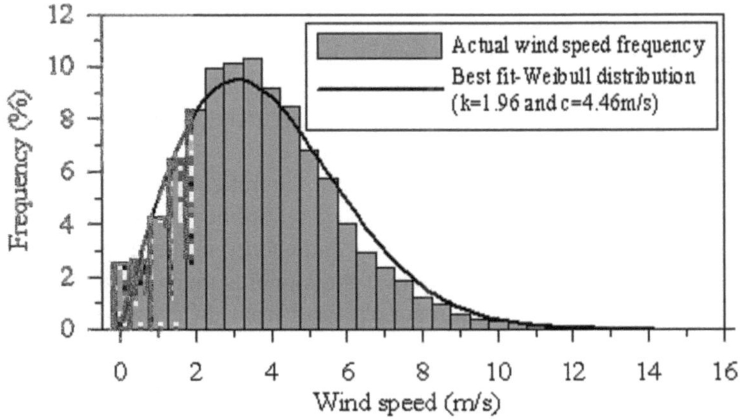

Figure 3.1 Bar diagram representing the actual wind speed frequency distribution and the corresponding best fit by the Weibull function [14]

the processed data can easily be shown in a bar diagram for a clear representation, as illustrated in Figure 3.1. Clearly, the total of the frequency distribution so defined in Equation 3.1 should be equal to unity, or

$$\sum_{i=1}^{I} f_i = 1 \tag{3.2}$$

3.2.2 Mean Wind Speeds

The monthly, seasonal, yearly or the overall mean wind speed for the entire measurement duration can easily be calculated as follows

$$V_m = \frac{1}{N}\sum_{j=1}^{N} V_j \tag{3.3}$$

where $j = 1, 2, \ldots, N$ represents the actual series of the data measured, N is the total number of measurements, and V_j is the actual wind speed measured over the time.

Another parameter that measures the characteristics of the wind speeds measured, or the spread, is the standard deviation, defined as

$$\sigma = \left[\frac{1}{N-1}\sum_{j=1}^{N}\left(V_j - V_m\right)^2\right]^{1/2} \tag{3.4}$$

Now referring to the wind speed grouping and the frequency distribution given in Equation 3.1, the mean wind speed can be determined as

$$V_m = \frac{1}{N}\sum_{i=1}^{I} N_i V_i = \sum_{i=1}^{I} f_i V_i \tag{3.5}$$

That is, the mean wind speed is simply the frequency distribution weighted average.

In this spirit, various other mean wind speeds can be defined as well:

$$V_{rq}^{r-q} = \frac{\sum_{i=1}^{I} f_i V_i^r}{\sum_{i=1}^{I} f_i V_i^q} \tag{3.6}$$

Therefore, when $r = 1$ and $q = 0$, the above equation defines the mean wind speed, hence, $V_{10} = V_m$. The practically important mean wind speeds, V_{20} and V_{30}, are then expressed as

$$V_{20}^2 = \sum_{i=1}^{I} f_i V_i^2 \tag{3.7}$$

$$V_{30}^3 = \sum_{i=1}^{I} f_i V_i^3 \tag{3.8}$$

The mean wind speeds, V_{10}, V_{20}, and V_{30}, are important parameters for the determination of the mean rate of wind mass, momentum, and kinetic energy flow, as illustrated in the next section.

3.2.3 Wind Energy Potential

The rate of wind (air) mass, momentum, and kinetic energy flowing at the speed V through a blade sweep area A of a wind turbine can be estimated as

$$\dot{m} = \rho A V \tag{3.9}$$

$$\dot{M} = \dot{m}V = \rho A V^2 \tag{3.10}$$

$$\dot{E} = \frac{1}{2}\dot{m}V^2 = \frac{1}{2}\rho A V^3 \tag{3.11}$$

Because the wind speed changes over time, it is desirable to determine the mean rate of wind mass, momentum, and kinetic energy flow. Following the descriptions given in the previous sections, we have

$$\dot{m}_m = \sum_{i=1}^{I} \rho A\left(f_i V_i\right) = \rho A V_{10} \tag{3.12}$$

$$\dot{M}_m = \dot{m}V = \sum_{i=1}^{I} \rho A\left(f_i V_i^2\right) = \rho A V_{20}^2 \tag{3.13}$$

$$\dot{E}_m = \sum_{i=1}^{I} \frac{1}{2}\rho A\left(f_i V_i^3\right) = \frac{1}{2}\rho A V_{30}^3 \tag{3.14}$$

It is clear that the mean wind speeds, V_{10}, V_{20}, and V_{30}, are the mean values for the determination of the mean rate of wind mass, momentum, and energy flowing through the turbine blade seep area. As a result, their determination and specific values have significant practical importance, as the rate of momentum flow is related to the force wind exerted on the turbine structure and the rate of kinetic energy flowing through the turbine represents the wind energy potential.

In practice, the rate of kinetic energy flow is often referred to as the wind power; and the wind power per unit of turbine blade sweep area A is often called the wind power density. The mean wind power density is hence defined as

$$P = \frac{\dot{E}_m}{A} = \sum_{i=1}^{I} \frac{1}{2}\rho\left(f_i V_i^3\right) = \frac{1}{2}\rho V_{30}^3 \tag{3.15}$$

The above equation represents the wind energy potential on average. The actual electricity that a wind turbine can generate can then be estimated by considering the turbine's energy conversion efficiency as well as the lower and upper cutoff wind speeds for the particular turbine design.

3.3 Empirical and Continuous Wind Speed Distribution Functions

Although the measured wind speed data is discrete, as described in the previous sections, it is common practice that the discrete and statistically processed wind speed distribution data is fitted with a continuous distribution function, or probability density function. This is because in reality wind speed variation is continuous, and a continuous distribution function tends to represent the wind speed better.

3.3.1 Mean Wind Speeds and Wind Power Density

Assuming $f(V)$ is a continuous wind speed distribution function, determined by fitting the measured wind distribution data by the methods to be described in the next section, then the mean wind speeds given in Equation 3.6 in the discrete form can be determined in terms of $f(V)$ by replacing the summation with the integration, noticing that the (absolute) wind speed can vary from zero to a very large value:

$$V_{rq}^{r-q} = \frac{\int_0^\infty f(V)V^r dV}{\int_0^\infty f(V)V^q dV} \tag{3.16}$$

Therefore, we have

$$V_{10} = \int_0^\infty f(V)VdV \tag{3.17}$$

$$V_{20}^2 = \int_0^\infty f(V)V^2 dV \tag{3.18}$$

$$V_{30}^3 = \int_0^\infty f(V)V^3 dV \tag{3.19}$$

Similarly, the mean wind power density is determined as

$$P = \int_0^\infty \frac{1}{2}\rho V^3 f(V)dV \tag{3.20}$$

Clearly, once the wind speed distribution function $f(V)$ is known, the mean wind speeds and mean wind power density can be determined relatively easily. The wind speed is the most critical data needed to appraise the power potential of a candidate site. However, wind power has a cubic relation with wind speed, so that accurate determination of the wind speed distribution function is essential to provide an accurate assessment of wind power potential.

3.3.2 *Empirical Distribution Functions*

The current method of modeling wind speed distribution is almost entirely empirical. Since wind speed at a fixed location changes over day and night, over seasons and years, actual measurements are usually taken over many years, even decades, in order to obtain a statistically meaningful, *i.e.*, stationary, distribution. Therefore, empirical determination of the wind speed distribution, $f(V)$, is extremely time-consuming and laborious as well.

Once the empirical data are available, an empirical curve fitting approach is almost always used to determine the constant parameters in the empirical correlation functions for the wind speed distribution. There exist a number of empirical functions for wind speed distributions. Two of the commonly used functions for fitting a measured wind speed data in a given location over a period of time are the Weibull and Rayleigh distributions.

3.3.2.1 Weibull Distribution Function

Statistical analysis of measured wind speed data indicates that among all the empirical distributions ever considered for description of the wind speed distribution, the Weibull distribution function offers the best agreement with a variety of experimental data analyzed [7, 15–17]. It has been used extensively to assess the wind potential for different regions in different countries [9, 18–23]. The general Weibull distribution can be written as follows [24, 25]:

$$f(V) = \left(\frac{k}{c}\right)\left(\frac{V-\mu}{c}\right)^{k-1} \exp\left[-\left(\frac{V-\mu}{c}\right)^{k}\right] \quad V \geq \mu;\ (k,c) > 0\ ; -\infty < \mu < \infty \tag{3.21}$$

where k is the shape parameter (dimensionless), μ is the location parameter and c is the scale parameter having the dimension of speed – all these three parameters are determined from the measured wind speed at a particular site of interest. Because the (absolute) wind speed V is always higher or equal to zero ($V \geq 0$), $\mu = 0$ is taken; then the general Weibull distribution commonly used in wind speed analysis becomes a two-parameter function [7, 26–28]:

$$f(V) = \left(\frac{k}{c}\right)\left(\frac{V}{c}\right)^{k-1} \exp\left[-\left(\frac{V}{c}\right)^{k}\right] \quad V \geq 0;\ (k,c) > 0 \tag{3.22}$$

The distributions will take different shapes with different values of k, the shape parameter. Typical values of k are larger than one ($k > 1$) and is mostly around the value 2. A typical Weibull distribution fitting the measured wind speed data is shown in Figure 3.1.

The corresponding cumulative distribution $F(V)$, representing the probability of finding wind speed in the range zero to V, can easily be determined as

$$F(V) = \int_0^V f(V)dV = 1 - \exp\left[-\left(\frac{V}{c}\right)^k\right] \tag{3.23}$$

For the Weibull distribution shown in Equation 3.22, it is easily shown that it possesses the following characteristics:

(1) The mean wind speed: Equation 3.17 yields

$$V_{10} = c\Gamma\left(1 + \frac{1}{k}\right) \tag{3.24}$$

where Γ is the Gamma function defined as

$$\Gamma(n) = \int_0^\infty e^{-x} x^{n-1} dx \tag{3.25}$$

(2) The higher order mean wind speeds: Equation 3.16 with $q = 0$ results in

$$V_{r0}^r = c^r\Gamma\left(1 + \frac{r}{k}\right) \tag{3.26}$$

Thus, Equation 3.24 is recovered when r in Equation 3.26 is set to 1.

(3) The standard deviation:

$$\sigma = \left[\int_0^\infty (V - V_{10})^2 f(V)dV\right]^{1/2} = c\sqrt{\Gamma\left(1 + \frac{2}{k}\right) - \Gamma^2\left(1 + \frac{1}{k}\right)} \tag{3.27}$$

(4) The most probable wind speed: this represents the most frequent wind speed, *i.e.*, $f(V) = \max$ when $V = V_{mp}$. By setting $df/dV = 0$, we obtain

$$V_{MaxP} = c\left(\frac{k-1}{k}\right)^{1/k} \tag{3.28}$$

(5) The maximum energy carrying wind speed: this represents the wind speed at which the wind carries the maximum amount of wind kinetic energy. Similar to (4) above, we can derive

$$V_{MaxE} = c\left(\frac{k+2}{k}\right)^{1/k} \tag{3.29}$$

(6) The mean wind power density: substituting Equation 3.26 into Equation 3.20 with $r = 3$ leads to

$$P = \frac{1}{2}\rho\, c^3 \left(1 + \frac{3}{k}\right) \tag{3.30}$$

Finally, it is important to point out that Weibull distribution as given in Equation 3.22 or 3.23 with the shape parameter $k > 1$ usually yields a zero probability at the wind speed $V = 0$. This suggests that the Weibull distribution cannot properly take care of the situation where calm periods exist for a particular site (*i.e.*, $V = 0$). Attempts to deal with this restriction of no calm periods in the distribution have been made, and one of the approaches is to consider the following hybrid distribution function [30] (the other is deferred to Section 3.3.2.3 with a better approach given in Section 3.4):

$$f_H(V) = F_0\delta(V) + (1 - F_0) f_W(V), \qquad V \geq 0 \tag{3.31}$$

where F_0 is the probability of observing zero wind speed at a given site, $\delta(V)$ is the Dirac delta function and f_W is the usual Weibull distribution function given earlier. The corresponding cumulative distribution function is

$$F_H(V) = F_O + (1 - F_O) F_W(V), \quad V \geq 0 \tag{3.32}$$

where $F_W(V)$ is the usual Weibull cumulative distribution function (for $V > 0$). This approach is plausible when the cumulative distribution function, Equation 3.32, is plotted; however, Equation 3.31 contains a Delta function, which would show a spike in the distribution function when plotted. Thus, it would not be able to take the calm spells into account properly. In practice, it has been identified that in order for the Weibull distribution to fit the measured wind speed data well, the measured data set must satisfy certain characteristics, as pointed out by Tuller and Brett [30].

3.3.2.2 Weibull Distribution Function: Methods for Parameter Determination

There are several methods for estimating Weibull parameters: the shape and scale parameters k and c based on the measured wind speed data of different characteristics at different sites. The following three methods provide accurate, effective, and easy methods for Weibull parameter determination.

Method 1: Least-squares regression method [31]
The cumulative Weibull distribution, Equation 3.23, can be rearranged into the following form:

$$\ln\left[-\ln(1 - F(V))\right] = k \ln V - k \ln c \tag{3.33}$$

Now considering a linear function defined as

$$y = ax + b; \; y = \ln[-\ln(1 - F(V))], \; x = \ln V, \; a = k, \; b = -k \ln c \tag{3.34}$$

where y and x are determined by the measured wind speed data. Then the standard least square regression method can be used to determine the slope a and the intercept b. Finally, the Weibull parameters can be obtained as

$$k = a \text{ and } c = e^{-b/k} \tag{3.35}$$

The cumulative distribution function $F(V)$ can be calculated based on the measured data directly, or estimated easily using an estimator, the median rank, according to Benard's approximation [32]:

$$F(V) = \frac{i - 0.3}{n + 0.4} \tag{3.36}$$

where i is the number of wind speed measurements and n is the total number of observations.

Method 2: Mean wind speed and standard deviation method [31]
From the mean wind speed V_{10} and the standard deviation σ given in Equations 3.23 and 3.26, we have the following relation:

$$\left(\frac{\sigma}{V_{10}}\right)^2 = \frac{\Gamma\left(1 + \frac{2}{k}\right)}{\Gamma^2\left(1 + \frac{1}{k}\right)} - 1 \tag{3.37}$$

The mean wind speed and the standard deviation are calculated from the measured wind speed data, the Weibull parameters c and k are then determined from Equations 3.24 and 3.37, respectively, as follows

$$c = \frac{V_{10}}{\Gamma\left(1 + \frac{1}{k}\right)} \tag{3.38}$$

Since Equation 3.37 is not linear in k, the following approximate relation can be used to estimate the shape parameter

$$k = \left(\frac{\sigma}{V_{10}}\right)^{-1.086} \tag{3.39}$$

Method 3: Maximum likelihood estimation [32, 33]
The maximum likelihood method has been used to estimate the shape parameter k by solving the following equation iteratively [33]:

$$k = \left[\frac{\sum_{j=1}^{N} V_j^k \ln V_j}{\sum_{j=1}^{N} V_j^k} - \frac{1}{N} \sum_{j=1}^{N} \ln V_j \right]^{-1} \tag{3.40}$$

where N is the total number of wind speed measurements and V_j is the measured wind speed value for the jth measurement. It is noticed that the iterative calculation of k by Equation 3.40 is not trivial, in fact, it is rather time-consuming and computationally intensive when compared with the previous two methods, which are much simpler to use.

Once the shape parameter k is obtained, the scale parameter c is then found by using the shape parameter k as follows

$$c = \left(\frac{1}{N} \sum_{j=1}^{N} \ln V_j^k \right)^{1/k} \tag{3.41}$$

Method 4: Alternative (non-iterative) to maximum likelihood method [32]
The maximum likelihood method given earlier is iterative in the determination of the shape parameter k, it is computationally demanding as convergence is achieved quite slowly. As a result, a simpler method has been developed [34] to estimate the shape parameter k as follows, with the scale parameter c still determined by Equation 3.41

$$k = \frac{\pi}{\sqrt{6}} \left[\frac{N(N-1)}{N\left(\sum_{j=1}^{N} \ln^2 V_j\right) - \left(\sum_{j=1}^{N} \ln V_j\right)^2} \right]^{0.5} \tag{3.42}$$

In practice, Methods 1, 2, and 4 are frequently used, although many other methods have been developed and tried. Each method tends to produce values of Weibull parameters that are different for the same data set, and that fit some data sets better than others [35]. This is one of the problems for the empirical Weibull distribution function.

3.3.2.3 Rayleigh Distribution Function

As pointed out earlier, the two-parameter Weibull distribution function may not be able to fit all the different types of wind speed data measured at different locations around the world. The Rayleigh distribution function is another popular empirical distribution function that has been in use.

The Rayleigh distribution is a special case of the Weibull distribution when the shape parameter k in the Weibull distribution is assumed to be equal to 2. Therefore, the Rayleigh distribution is really a one-parameter Weibull distribution. From Equation 3.24 we have $c^2 = 4V_{10}/\pi$. As a result, the Rayleigh distribution can be written for the probability density and the cumulative distribution as follows:

$$f(V) = \frac{\pi}{2}\frac{V}{V_{10}^2}\exp\left[-\left(\frac{\pi}{4}\right)\left(\frac{V}{V_{10}}\right)^2\right] \tag{3.43}$$

$$F(V) = 1 - \exp\left[-\left(\frac{\pi}{4}\right)\left(\frac{V}{V_{10}}\right)^2\right] \tag{3.44}$$

Similarly, the mean wind power density for the Rayleigh distribution can be obtained as

$$P = \frac{3}{\pi}\rho\, V_{10}^3 \tag{3.45}$$

From the above equations, it is clear that the advantage the Rayleigh distribution has is that it can easily be determined from the mean value of the wind speed data. The Rayleigh distribution has also been widely used in practice to fit the measured probability density distribution, and its validity has been shown for various locations [15, 36]. In fact, it sometimes even provides a better fitting than the Weibull distribution having two parameters [15]. In general, the Weibull distribution is more versatile with two parameters while the Rayleigh distribution is simpler to use because of only one parameter involved. The Rayleigh distribution has been found to have a bias towards low wind speeds.

3.4 Maximum Entropy Principle – A Theoretical Approach

For the last many decades, the MEP has been successfully applied to many problems arising in a wide variety of disciplines, such as statistical mechanics, thermodynamics, data mining, systems analysis, finance, to name but a few. Kapur and Kesavan [37] described the basic concept and derivation of MEP and summarized the applications of MEP in a wide range of fields, from engineering and beyond.

The concept of entropy was originally associated with thermodynamics and the inquiry field is macroscopic phenomena. In 1948, Shannon [38] discovered a measure of uncertainty for any probability distribution and proposed the concept of information entropy as defined by the following expression:

$$S = -k\sum P_i \ln P_i \tag{3.46}$$

where P_i is the probability of occurrence of the state i and k is the Boltzmann constant.

Thermodynamic entropy is defined in terms of temperature in macroscopic view, while its microscopic definition has the same form as the information entropy. The latter is contended to predict microscopic behaviors on the basis of insufficient data pertaining to a few well-chosen macroscopic parameters. It was Jaynes [39] who extended this concept into a now well-known method of maximum entropy formalism in the realm of statistical mechanics, which can be applied to problems that involve probability. As a statistical tool, the MEP allows one to determine the least biased probability distribution function when the information available is limited to some macroscopic constraints.

To utilize the MEP, it is recognized that many physical systems can be described by averages which may be known for the particular system. These observations can be expressed mathematically by the following constraints:

$$\sum_{i=1}^{n} P_i g_{r,i} = \langle g_r \rangle \qquad r = 1, 2, 3, \ldots, m \tag{3.47}$$

where m is the number of physical constraints for the particular system, $g_{r,i}$ is some function evaluated at state i, $\langle g_r \rangle$ is the expectation of average value of the function g over the entire system.

The additional constraint is the definition of probability. That is,

$$\sum_{i=1}^{n} P_i = 1 \tag{3.48}$$

In most cases, $m+1 < n$, the number of equations is less than the number of unknowns. These equations are not sufficient to determine the probabilities uniquely. Some missing information results in uncertainty. Jaynes suggested that the most likely probability distribution should be the one which maximizes Shannon's entropy, subject to the given information. Based on Lagrange's method, the most likely distribution maximizing the entropy function under the constraints is

$$P_i = \exp\left(-\alpha_0 - \sum_{r=1}^{m} \alpha_r g_{r,i} \right) \tag{3.49}$$

where the multipliers α_0 and α_r must satisfy the constraint Equations 3.47 and 3.48. The multiplier α_0 can be obtained by substituting Equation 3.49 into Equation 3.48, that is,

$$\alpha_0 = \ln \sum_{i=1}^{n} \exp\left(-\sum_{r=1}^{m} \alpha_r g_{r,i} \right) \tag{3.50}$$

Once the α_r are determined, the α_0 can be determined from Equation 3.50. In order to obtain α_r, Equation 3.49 is substituted back into the constraint equation, Equation 3.47.

$$\sum_{i=1}^{n} g_{r,i} \exp\left(-\alpha_0 - \sum_{r=1}^{m} \alpha_r g_{r,i}\right) = \langle g_r \rangle \tag{3.51}$$

Equations 3.50 and 3.51 yield a set of $m+1$ equations with $m+1$ unknown Lagrange multipliers, α_0 and α_r, $r = 1, 2, 3, \ldots, m$. The solution to the set of equations results in the most probable probability distribution that maximizes entropy under the constraints given by Equations 3.47 and 3.48. As indicated by Kapur and Kesavan, the MEP method does not necessarily yield the right distribution under some improper constraints. Therefore, determining the suitable set of constraint equations for a particular problem is not trivial when using the MEP method.

3.5 MEP-based Wind Speed Distribution

In this section, the MEP described above will be used to develop a mathematical model for the wind speed distribution. The MEP-based distribution will be compared with a variety of typical wind speed data measured around the world. Comparison will also be made with the Weibull distribution fitting the same data sets, so that the superior performance of the MEP-based distribution will be highlighted.

3.5.1 Mathematical Formulation

The MEP method can be used to predict the most likely probability distribution, or probability density distribution (pdf), for a particular physical problem under a set of constraints expressing the available information related to the distribution sought. When the MEP is applied to the wind energy field to determine the wind speed distribution, the constraint equations imposed must be based on the physical principles, or the conservation of mass, momentum, and energy for the air stream flowing with the wind, and can be expressed as follows:

Mass conservation: $$\dot{m} = \sum_i \rho P_i V_i A = \rho A V_{10} \tag{3.52}$$

Momentum conservation: $$\dot{M} = \sum_i (\rho P_i V_i A) \cdot V_i = \rho A V_{20}^2 \tag{3.53}$$

Energy conservation: $$\dot{E} = \sum_i (\rho P_i V_i A) \cdot \left(\frac{1}{2} V_i^2\right) = \frac{1}{2} \rho A V_{30}^3 \tag{3.54}$$

where V_{10}, V_{20}, and V_{30} are the mean wind speeds, defined mathematically through

$$V_{rq}^{r-q} = \frac{\sum_i P_i V_i^r}{\sum_i P_i V_i^q} \tag{3.55}$$

where r equals 1, 2, and 3, respectively, corresponding to V_{10}, V_{20}, and V_{30} when q is set to 0.

Therefore, V_{10} is the same as the mean speed commonly used; V_{20} and V_{30} represent the wind speed at which the wind flowing through the rotor produces the same force or energy, respectively, as the wind flowing at variable speeds. These mean velocities can be obtained from measured data. They also may be determined if the measurement of the wind mass flow rate, momentum flow rate (force), and the energy flow rate (or power) is available. It is evident that it is much easier to measure these three mean wind velocities than to measure the instantaneous wind speed over a long period of time followed by a tedious data analysis.

If the air density is assumed constant and the intercepting area unchanged, Equations 3.52–3.54 can be simplified to

$$\sum_i P_i V_i = V_{10} \tag{3.56}$$

$$\sum_i P_i V_i^2 = V_{20}^2 \tag{3.57}$$

$$\sum_i P_i V_i^3 = V_{30}^3 \tag{3.58}$$

Another constraint arises from Equation 3.48 because of the definition of probability. By maximizing Shannon's entropy, the probability is derived as, similar to Equation 3.49

$$P_i = \exp\left(-\alpha_0 - \alpha_1 V_i - \alpha_2 V_i^2 - \alpha_3 V_i^3\right) \tag{3.59}$$

For continuous variables, such as wind speed, the subscripts can be dropped and the summation can be replaced by integrals with the corresponding limits from minimum to maximum. The continuous probability density function can be obtained as [37]

$$f(V) = \exp\left(-\alpha_0 - \alpha_1 V - \alpha_2 V^2 - \alpha_3 V^3\right) \tag{3.60}$$

Substituting the above probability density function into all the constraint equations, Equations 3.48 and 3.56–3.58, yield the final set of equations for determination of the Lagrange multipliers, α_i, needed in Equation 3.60:

$$\int_{V_{min}}^{V_{max}} \exp\left\{-\alpha_0 - \alpha_1 V - \alpha_2 V^2 - \alpha_3 V^3\right\} dV = 1 \tag{3.61}$$

$$\int_{V_{\min}}^{V_{\max}} V \exp\{-\alpha_0 - \alpha_1 V - \alpha_2 V^2 - \alpha_3 V^3\} dV = V_{10} \tag{3.62}$$

$$\int_{V_{\min}}^{V_{\max}} V^2 \exp\{-\alpha_0 - \alpha_1 V - \alpha_2 V^2 - \alpha_3 V^3\} dV = V_{20}^2 \tag{3.63}$$

$$\int_{V_{\min}}^{V_{\max}} V^3 \exp\{-\alpha_0 - \alpha_1 V - \alpha_2 V^2 - \alpha_3 V^3\} dV = V_{30}^3 \tag{3.64}$$

where the minimum wind speed can be set to zero, while the maximum wind speed can either correspond to the maximum wind speed measured for a given site, or conventionally it is taken as infinitely large.

The above set of equations is highly non-linear with the unknown α_i appearing in the exponent, and the integrals involving exponential function adding additional difficulty to the numerical solution process. In the present study, a modified Newton–Raphson method is used and details on the numerical methods can be found in [40]. Once the unknown α_i are obtained, the wind speed distribution can be described with Equation 3.60. Therefore, Equations 3.60–3.64 form the complete mathematical model for the probabilistic distribution of wind speed.

3.5.2 Fitting Criteria for Comparison

The coefficient of determination (COD) is used here to evaluate the performance of the MEP-based distributions and the Weibull distributions. This coefficient expressed as a percentage indicates how much of the total variation in the dependent variable can be accounted for by the theoretical or empirical distribution. A higher COD represents a better fit using the theoretical or empirical function. A complete fit has the COD. value of 100 %. The COD is defined as

$$\text{C.O.D} = R^2 = 1 - \frac{\sigma_{y,x}^2}{\sigma_y^2} \tag{3.65}$$

where R is the correlation coefficient [41] and σ_y is the standard deviation of the measured data y from its own mean value y_m, and is conventionally defined as

$$\sigma_y = \left[\frac{1}{N-1} \sum_{j=1}^{N} (y_j - y_m)^2 \right]^{1/2}$$

and similarly

$$\sigma_{y,x} = \left[\frac{1}{N-2} \sum_{j=1}^{N} (y_j - y_{jc})^2 \right]^{1/2}$$

where y_j are the actual values of y as measured, and y_{jc} are the values computed from the correlation equation for the same independent value of x. In the present context, y represents the probability density, $f(V)$, while the wind speed acts as the independent parameter x.

Two other goodness-of-fit parameters in statistics analysis, chi-square and root mean square error (RMSE), are also considered as additional evaluation criteria for the fitting performance of the MEP and Weibull distributions. They are defined, respectively, as follows:

$$\chi^2 = \sum_{j=1}^{N} \frac{\left(y_j - y_{jc}\right)^2}{y_j} \tag{3.66}$$

$$\text{RMSE} = \left[\frac{1}{N}\sum_{j=1}^{N}\left(y_j - y_{jc}\right)^2\right]^{1/2} \tag{3.67}$$

The smaller the values of these two parameters are, the better the proposed distribution function approximates the measured data (or the better the curve fits). In the ideal case, the values should be zero for these two parameters.

3.5.3 Comparison with the Measured Data and Weibull Distribution

To demonstrate the value of, and to validate, the present theoretical model for the probabilistic distribution of wind speed based on the MEP method, comparisons have been made between the model predictions and some measured data [10, 28, 42]. The three mean velocities, V_{10}, V_{20}, and V_{30}, needed for the MEP distribution, are calculated from the measured tabulated data. The lower integration limit in Equations 3.61–3.64 is set as $V_{\min} = 0$ according to the measurement, and the upper limit, $V_{\max}$, is set to the maximum wind speed measured for the corresponding data set. The Weibull distribution was used to curve fit the same measured data sets. Therefore, comparison of the predictions from the MEP distribution will also be made with these curve-fit Weibull distributions as well.

The actual wind data for Dodge City and Kansas City in the USA for the year 1970 are tabulated in the forms of probability density function and cumulative distribution in [28]. At that time, these speed data were recorded in knots (kn). The shape parameters of the Weibull distribution for Dodge City and Kansas City are obtained as $k = 2.110$ and $k = 1.776$, respectively, and the scale parameters are $c = 11.96$ kn and $c = 7.65$ kn, respectively. To be consistent with the actual data, all the mean wind speeds are calculated based on the actual data in knots and input to the numerical program. Therefore, the unit of wind speed in MEP distribution function is also in knots. The computed Lagrangian multipliers for the two cities are listed in Table 3.2. Notice that α_1, α_2, and α_3 are dimensional as well and given to nine significant digits after the decimal point in Table 3.2. This may be necessary to guarantee the accuracy of numerical computation because these parameters appear in the exponent of the exponential function, and small variations may be magnified by the exponential function.

Figure 3.2 shows a comparison between the present MEP distribution and the actual data for Dodge City. The curve for the Weibull distribution is also shown in the figure. It is clear that the two curves almost overlap when wind speed is greater than 15 kn. The MEP distribution is narrower and has a higher peak than the Weibull distribution. The location of the peak shifts a little bit to the right for the MEP distribution, which is more consistent with the actual data points. Another salient feature of the MEP distribution is that the probability density function starts at a non-zero value when the wind speed is zero. As Johnson [28] pointed out, one of the important features of the speed–frequency curves is that the intercept on the vertical axis is always greater than zero in practice due to the existence of calm spells at any site. In this respect, the present MEP distribution compares with the practical distribution feature much better than the empirical Weibull distribution. This is because the Weibull distribution always gives a zero value for the probability density function whenever the wind speed is zero in any case, or it is unable to predict the existence of the calm spells at all.

Table 3.2 The computed Lagrangian multipliers for the MEP distribution for Dodge City and Kansas City, USA [28]

	a_0	a_1	a_2	a_3
Dodge City	2.760022423	–6.554945748	4.694611511	–0.7584568700
Kansas City	1.030199731	–0.9806489367	0.2034485615	0.2332288684

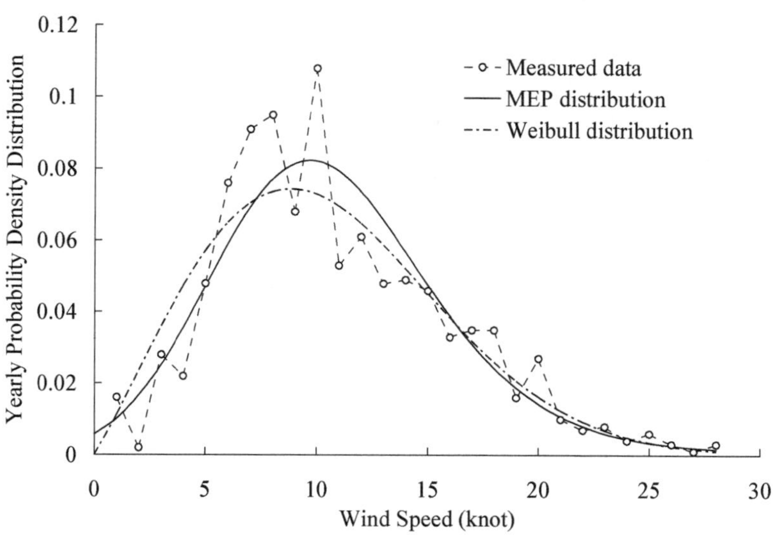

Figure 3.2 The comparison of the present MEP distribution with the measured data [28] and the Weibull distribution for Dodge City, USA

The parameters for the statistical analysis: the coefficient of determination (COD), RMSE, and chi-square error, are given in Table 3.3 to indicate the degree of fitting for both distributions. The larger value of COD and the smaller values of the RMSE and chi-square error indicate that the present MEP distribution is better than the Weibull distribution in describing this set of wind speed data.

The comparison between the present MEP and the Weibull distributions for the measured data for Kansas City is shown in Figure 3.3. The data points for Kansas City have a very large fluctuation for successive wind speeds and exhibit an irregular pattern. Due to the measurement technique used in the 1960s when the data were measured, the data were actually recorded by human observation of a wind speed indicator, which was continuously changing. Johnson [28] pointed out that there is a human tendency to favor even integers and multiples of 5 when reading such an indicator. Therefore, the data points are rather scattered. The actual data have a peak at the small wind speed region and zigzag at the medium wind speed region. The second peak appears at a speed of 10 kn or so and then the next point goes directly to a very low value followed by another high value point. From the comparison made for the Weibull distribution with the actual data, Johnson concluded that the Weibull density function does a reasonable job in fitting the scattered data points. From Figure 3.3, it is obvious that the present MEP distribu-

Table 3.3 Fitting criteria for the comparison of MEP and Weibull distributions with the wind speed data for Dodge City, USA [28]

Fitting criteria	MEP distribution	Weibull distribution
COD	0.8406	0.8281
RMSE	0.0119	0.0123
χ^2	0.1779	0.3146

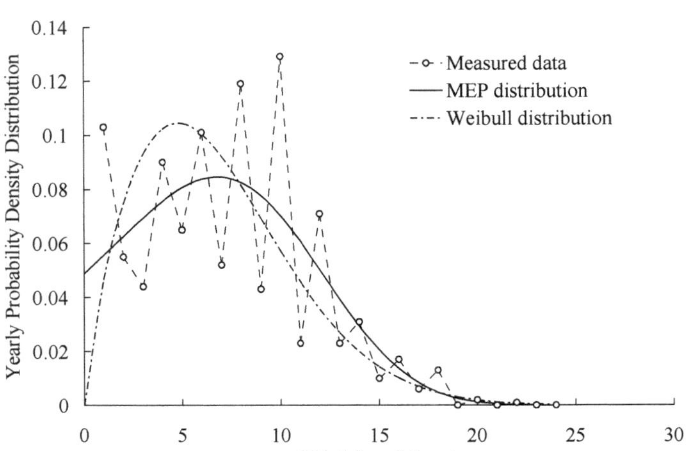

Figure 3.3 The comparison of the present MEP distribution with the measured data [28] and the Weibull distribution for Kansas City, USA

tion is much better than the Weibull distribution when compared with the actual data points. Further, the present MEP distribution predicts the existence of physically realistic calm spells, and the distribution peak location also agrees much better than the Weibull distribution.

The statistical analysis shown in Table 3.4 clearly substantiates the above qualitative conclusion. It is seen that the COD for the present MEP distribution is significantly larger (29 %) than that for the Weibull distribution, whereas the RMSE and chi-square error are considerably smaller (16 % and 15 %, respectively) for the present MEP distribution. This indicates quantitatively that the present MEP distribution gives significant improvement over the Weibull distribution in describing the quite scattered measured data. In summary, the present MEP distribution is more physical and accurate than the Weibull distribution.

Some measured data for Oman are available in [42]. These data are summarized based on the monthly weather data from 1986 to1998 for four stations in Marmul, Masirah, Sur, and Thumrait, respectively. However, the available data are tabulated in the form of a cumulative distribution, instead of the original probability density distribution. To calculate the mean speeds needed for the MEP inputs, the backward differencing scheme is used to transform the cumulative distribution into pdf distribution. The computed Lagrangian multipliers for the four cases are listed in Table 3.5. Three methods were compared by Dorvlo [42] in an effort to estimate the Weibull distribution parameters k and c, and the chi-square method was found to be the best in providing the numerical estimate of k and c for the measured data. Therefore, the Weibull distributions calculated with this method are cited in this study. Figure 3.4 shows the comparison of the present MEP distribution with the measured data and the corresponding Weibull distribution for the measurement taken at the station in Marmul. The probability density function distributions are presented in Figure 3.4a and the cumulative distributions in Figure 3.4b. It is seen that the present MEP distribution fits the lower and higher speed regions better than the Weibull distribution. Both of the distributions almost have the same peak location, while the Weibull distribution has a higher peak value, which is consistent with the measured data. It is further noticed from Figure 3.4 that the pdf value is predicted to be zero at the zero wind speed by the present MEP distribution, consistent with the trend of the measured data. This might imply that the measurement location is quite windy with few calm spells. The statistical analysis for both the pdf and the cumulative distributions is given in Table 3.6. Comparison of the COD and RSME indicates that the present MEP distribution fits better for the pdf data, while the Weibull distribution fits better for

Table 3.4 Fitting criteria for the comparison of MEP and Weibull distributions with the wind speed data for Kansas City, USA [28]

Fitting criteria	MEP distribution	Weibull distribution
COD	0.6548	0.5090
RMSE	0.0233	0.0278
χ^2	0.2316	0.2719

the cumulative data, although the difference here is very small. However, the chi-square test shows that the present MEP distribution fits better for both cases. Therefore, it might be concluded that both the MEP and Weibull distributions fit the data equally well for this case.

Comparisons for the three other stations are shown in Figures 3.5–3.7. Similar behaviors to the Marmul case were observed when comparing the proposed MEP distribution with the Weibull distribution and the wind speed data measured. However, compared with the Marmul station, the differences in peak values between the MEP distributions and the Weibull distributions are smaller. The statistical analyses for both the pdf and the cumulative distributions for the three sta-

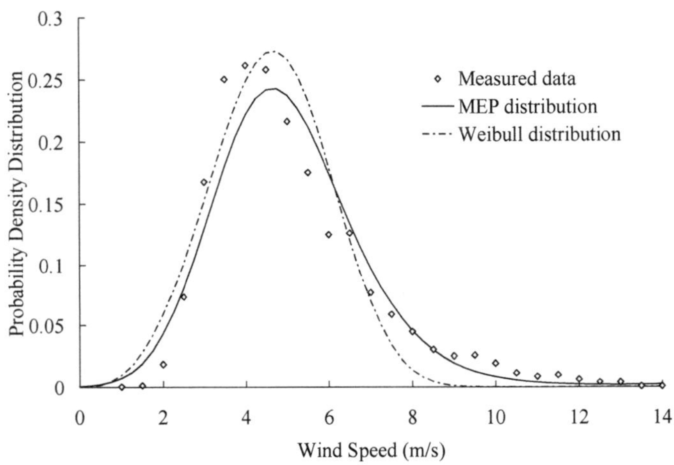

(a)

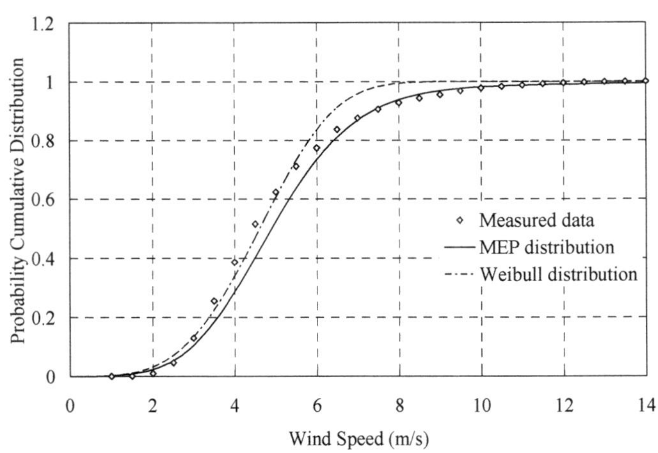

(b)

Figure 3.4 Comparison of the present MEP distribution with the measured data [42] and the Weibull distribution for Marmul station, Oman: **a** probability density distribution; **b** cumulative distribution

tions are given in Table 3.7. It seems that the Weibull distribution fits better for both the pdf data and the cumulative data from the COD and the RMSE. However, the differences are quite small, and chi-square tests always show that the MEP distributions fit better than the Weibull distributions. Therefore, it can be concluded that the MEP distributions and the Weibull distributions have equivalent goodness-of-fit for these cases.

It might be emphasized that the common features of the wind speed data measured at the four stations in Oman are that they have large long-term average wind speeds and almost no calm spells. For this type of data, the Weibull distribution function fits the data equally as well as the MEP distribution.

Table 3.5 Computed Lagrangian multipliers for the MEP distribution for the four locations in Oman [42]

	α_0	α_1	α_2	α_3
Marmul	5.815457532	–15.02578190	11.18700688	–2.151536979
Masirah	3.990322417	–9.921765328	7.162570418	–1.252442450
Sur	4.077375576	–10.41696824	7.765834632	–1.442037016
Thumrait	3.760308488	–9.629987525	7.245598066	–1.351451852

Table 3.6 Fitting criteria for the comparison of MEP and Weibull distributions with the wind speed data measured at Marmul station, Oman [42]

	Density distribution		Cumulative distribution	
Fitting criteria	MEP	Weibull	MEP	Weibull
COD	0.9277	0.9043	0.9865	0.9882
RMSE	0.0230	0.0265	0.0419	0.0392
χ^2	0.5827	0.9932	0.2190	0.2657

Table 3.7 Fitting criteria for the comparison of MEP and Weibull distributions with the wind speed data measured at Masirah, Sur and Thumrait stations, Oman [42]

	Density distribution		Cumulative distribution	
Fitting criteria	MEP	Weibull	MEP	Weibull
Masirah				
C.O.D	0.8264	0.8362	0.9846	0.9899
RMSE	0.0286	0.0278	0.0436	0.0352
χ^2	0.3684	0.5915	0.2746	0.3346
Sur				
C.O.D	0.9415	0.9436	0.9913	0.9948
RMSE	0.0156	0.0153	0.0323	0.0249
χ^2	0.1644	0.3365	0.1676	0.1887
Thumrait				
C.O.D	0.7112	0.7154	0.9806	0.9810
RMSE	0.0344	0.0342	0.0484	0.0480
χ^2	1.4296	2.5848	1.4849	1.8807

However, for sites where calm spells are significant, the MEP distribution has a clear advantage over the Weibull distribution. This kind of data, for example, can be found in Izmir, Turkey [10]. Based on the tabulated data over a 5-year period from 1995 to 1999, the wind speed frequency distributions are calculated and listed in Table 3.8. It can be seen that data for the low speed range accounted for a relatively large percentage of the overall data. The annual average, 3 m/s or so, is much lower than the Oman cases. Based on these data, the Lagrangian multipliers for these six cases are calculated and listed in Table 3.9. Figure 3.8a–f shows the predictions from the MEP distribution and the Weibull distribution with the corresponding measured data for the six cases. Figure 3.8a–e are for each year 1995 to 1999, respectively, and Figure 3.8f is for the 5-year data. It can be seen from the

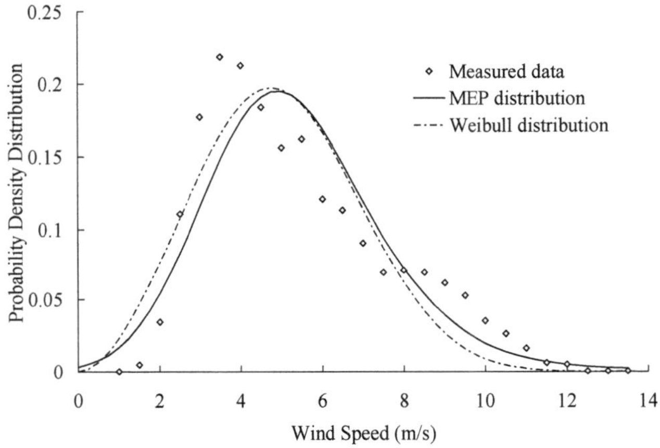

(a)

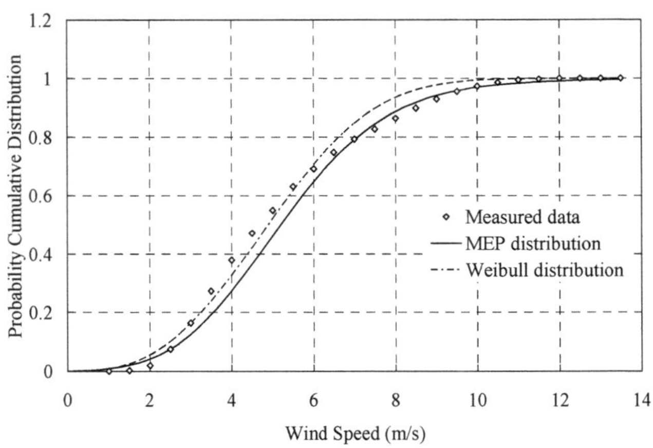

(b)

Figure 3.5 Comparison of the MEP distribution with the measured data [42] and the Weibull distribution for Masirah station, Oman: **a** probability density distribution; **b** cumulative distribution

figures that the MEP distributions closely match the measured data points, much better than the corresponding Weibull distributions for all the speed range. As pointed out earlier, the finite probability of zero wind speed is predicted by the MEP distribution for the significant amount of calm spells measured, whereas the Weibull distribution cannot fit the existence of calm spells at all.

In summary, when the frequency of calm spells is not zero and low wind speeds account for a relatively large proportion of the yearly data, the Weibull distribution cannot provide a suitable fit to the data, especially in the low wind speed region. The proposed MEP distribution can accommodate a variety of different characteristic data, and exhibits more adaptability than the Weibull distribution, even when the measured data have considerable scatter arising from the measurement technique. For the wind speed data measured at Izmir, Turkey, Fig-

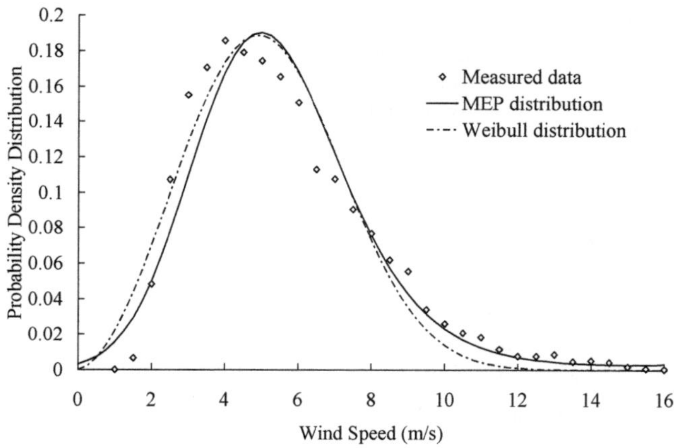

(a)

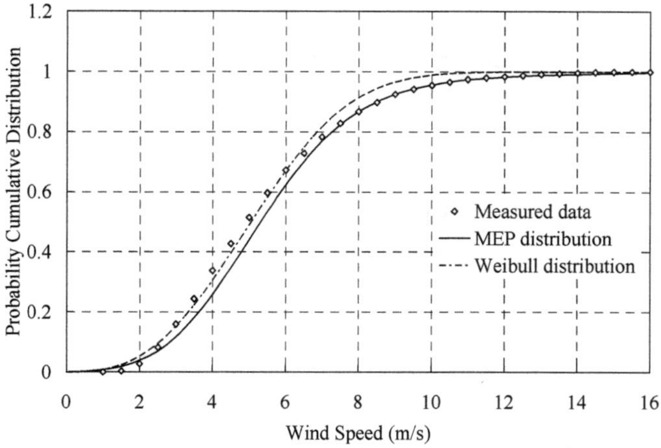

(b)

Figure 3.6 Comparison of the MEP distribution with the measured data [42] and the Weibull distribution for Sur station, Oman: **a** probability density distribution; **b** cumulative distribution

ure 3.8 shows the excellent agreement of the MEP distribution with the data, much better than the corresponding empirical Weibull distribution.

Table 3.10 presents a quantitative measure for the extent of agreement. The CODs are 0.9864, 0.9878, 0.9922, 0.9737, 0.9880, and 0.9879 for the 1995 to 1999 data and the 5-year data, respectively, for the MEP distribution; this is in sharp contrast with only 0.8955, 0.9266, 0.9328, 0.8777, 0.9161, and 0.9151, respectively, for the Weibull distribution. The RMSE for the MEP distribution is significantly reduced, about 64 %, 59 %, 66 %, 53 %, 62 %, and 62 %, respectively for the 1995 to 1999 data and the 5-year data, much smaller than the corresponding Weibull distribution. The reduction in the chi-square error is even more striking, about 90 %, 95 %, 90 %, 91 %, 86 %, and 89 % for all the sets of data shown.

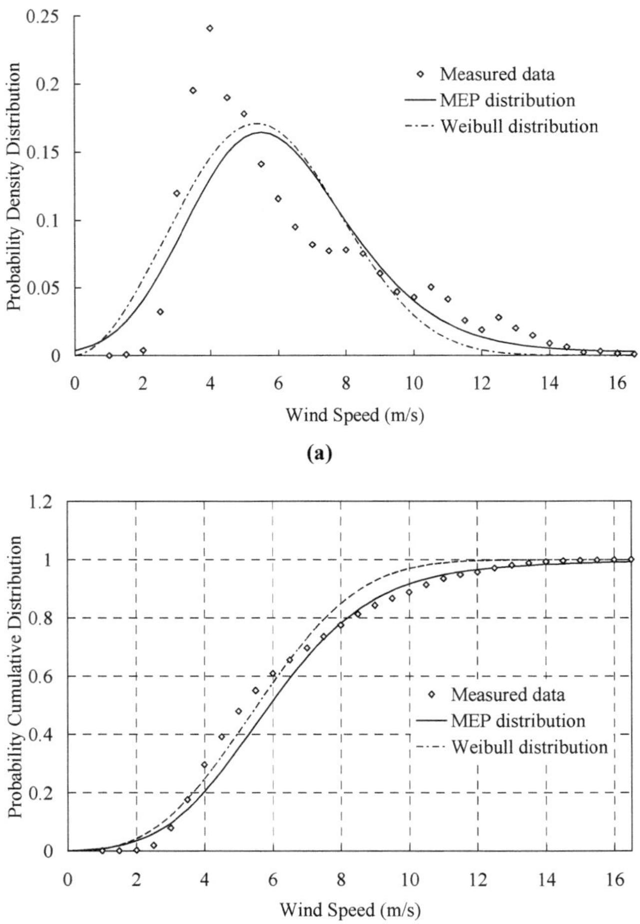

Figure 3.7 Comparison of the MEP distribution with the measured data [42] and the Weibull distribution for Thumrait station, Oman: **a** probability density distribution; **b** cumulative distribution

It is thus evident from the above comparisons and analyses that the capability and accuracy of the MEP distribution in describing a variety of measured wind speed data have been amply demonstrated, and its advantage over the empirical Weibull distribution has been clearly illustrated.

Table 3.8 Yearly wind speed frequency distribution in Izmir, Turkey [10]

Wind speed	1995	1996	1997	1998	1999	5-year
0–1	0.1659	0.1806	0.1742	0.2248	0.1821	0.1855
1–2	0.1759	0.1839	0.1893	0.1660	0.1860	0.1802
2–3	0.1612	0.1748	0.1911	0.1869	0.1579	0.1744
3–4	0.1697	0.1922	0.1949	0.1903	0.1607	0.1816
4–5	0.1670	0.1526	0.1437	0.1474	0.1435	0.1508
5–6	0.0999	0.0746	0.0726	0.0614	0.0987	0.0814
6–7	0.0410	0.0301	0.0240	0.0170	0.0441	0.0312
7–8	0.0130	0.0082	0.0072	0.0038	0.0171	0.0099
8–9	0.0048	0.0023	0.0021	0.0014	0.0068	0.0035
9–10	0.0014	0.0009	0.0007	0.0006	0.0015	0.0010
10–11	0.0002	0.0001	0.0003	0.0006	0.0014	0.0005
11–12	0	0	0	0	0.0002	0.0001
12–13	0	0	0	0	0	0
13–14	0	0	0	0	0	0
14–15	0	0	0	0	0	0

Table 3.9 Computed Lagrangian multipliers for the MEP distribution for Izmir, Turkey [10]

	α_0	α_1	α_2	α_3
1995	.7542272496	-.1416488238	-.3426759964	.3152539247
1996	.7881434168	-.3422200503	-.1119391018	.2473089644
1997	.9006720254	-.7708792993	.2551978888	.1626434861
1998	.6191240533	.1932578382	-.4700954520	.3050853139
1999	.6796604162	-.2133339307	.01213285454	.1634198531
5-year	.7624864361	-.3654586475	.00771480461	.1956133319

Table 3.10 Fitting criteria for the comparison of MEP and Weibull distributions with the wind speed data measured at Izmir, Turkey [10]

	1995		1996		1997	
Fitting criteria	MEP	Weibull	MEP	Weibull	MEP	Weibull
COD	0.9864	0.8955	0.9878	0.9266	0.9922	0.9328
RMSE	0.0085	0.0236	0.0086	0.0210	0.0070	0.0204
χ^2	0.0093	0.0931	0.0080	0.1538	0.0062	0.0602
	1998		1999		5-year	
Fitting criteria	MEP	Weibull	MEP	Weibull	MEP	Weibull
COD	0.9737	0.8777	0.9880	0.9161	0.9879	0.9151
RMSE	0.0133	0.0286	0.0079	0.0210	0.0084	0.0222
χ^2	0.0202	0.2224	0.0089	0.0619	0.0090	0.0789

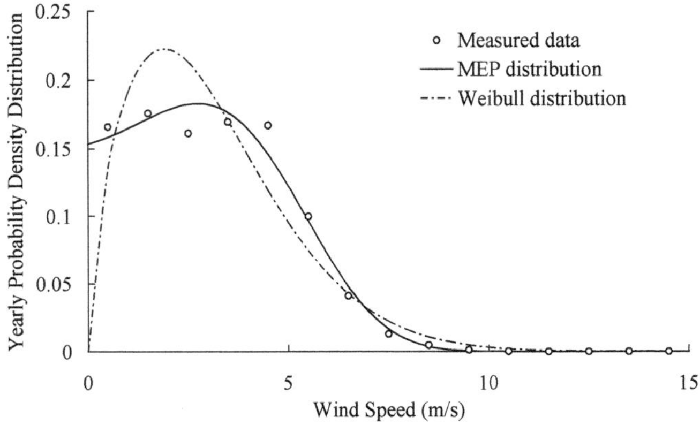

(a)

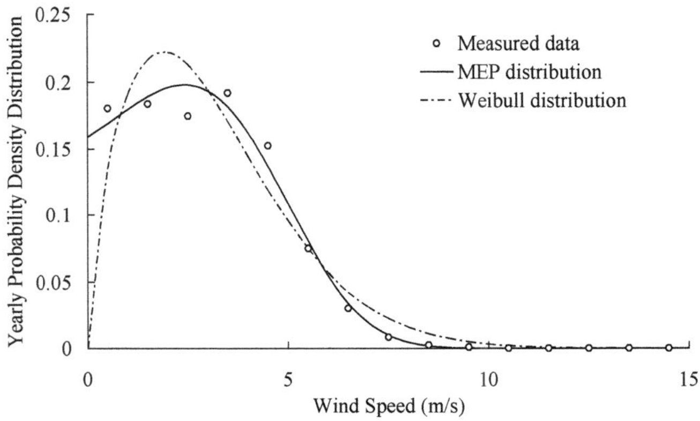

(b)

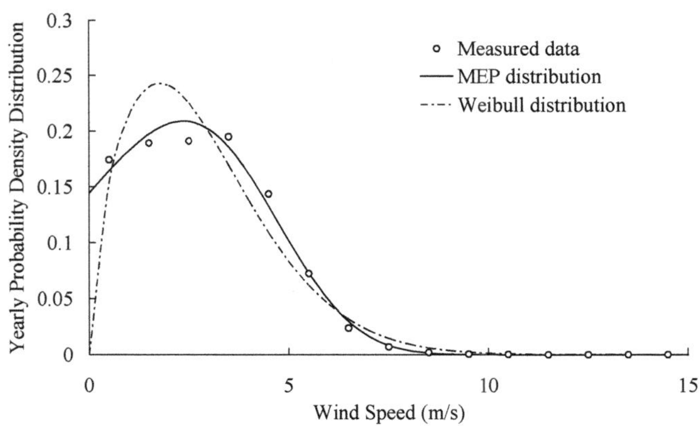

(c)

Figure 3.8 Comparison of the MEP distribution with the measured data [10] and the Weibull distribution for Izmir City, Turkey for the year **a** 1995; **b** 1996; **c** 1997; **d** 1998; **e** 1999; **f** 5-year

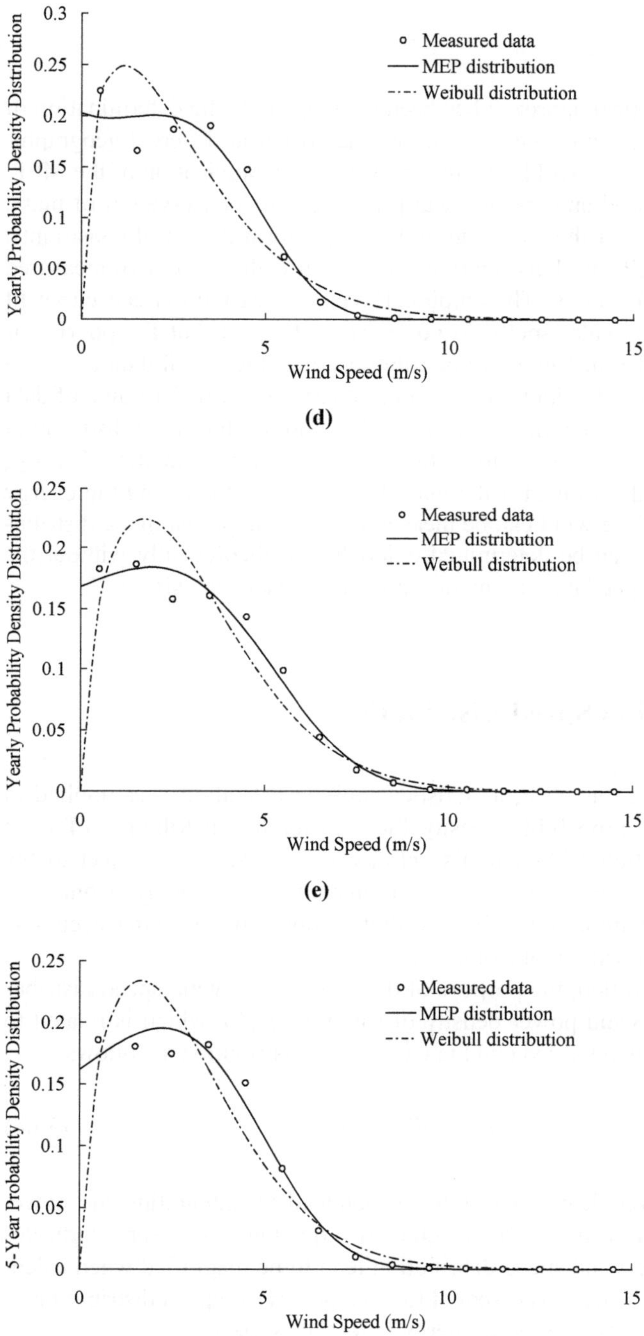

Figure 3.8 (continued)

3.5.4 Summary

In this section, a theoretical approach has been developed for the determination of the probabilistic distribution of wind speed that may occur at different geographical locations throughout the world. The theory is based on application of the MEP. Shannon's entropy is used and the physical principles, the conservation of mass, momentum, and energy for the air stream in the wind, form the set of constraints. It is shown that the MEP based distribution agrees well with a variety of measured data taken at various locations. The empirical Weibull distribution can curve fit the measured data when calm spells do not exist in the data, but fits poorly for other data sets. The MEP distribution not only describes the actual data more accurately than the Weibull distribution, but can also represent a wider range of data types, with or without the presence of calm spells. Further, the MEP distribution has a sound theoretical basis. In addition, the three mean wind velocities, V_{10}, V_{20}, and V_{30}, can easily be determined if the mass flow rate, the force, and the energy flow rate (or power) of the wind can be measured. Then the probabilistic distribution of the wind speed can be determined uniquely and theoretically without the need for lengthy wind speed measurements and tedious data analysis.

3.6 MEP-type Wind Speed Distribution

In the previous section, the MEP, a statistical inference method, was applied to theoretically determine probability density function for the distribution of wind speeds. The maximization of Shannon's entropy was carried out subject to the conservation principles for the wind mass, momentum, and energy. Compared with the Weibull distribution, the MEP distribution not only has better accuracy, but also can represent a wider range of data types.

In wind energy utilization, the purpose of determining the wind speed distribution, $f(V)$, is to obtain wind power density distribution, $f_P(V)$, which is related to the cubic power of the wind speed and the wind speed distribution as follows:

$$f_P(V) = \frac{1}{2}\rho V^3 f(V) \tag{3.68}$$

Then the wind power density can thus be obtained by integrating the power density distribution given in the above equation. Equation 3.68 clearly indicates that errors in predicting wind speed distribution tend to be magnified when calculating the wind power. Thus, it is essential to develop a wind speed distribution as accurate as possible for practical wind energy potential assessment.

In this section, the MEP distribution will be extended to improve further the accuracy of representing a variety of wind speed data. Semi-empirical distribution functions are proposed that combine the MEP and Weibull distribution in that an

MEP-type exponential family of functions is incorporated with the Weibull distribution by introducing a pre-exponential term to the theoretical MEP distribution. This MEP-type wind speed distribution is validated by comparing with a variety of measured data, as well as the corresponding MEP and Weibull distributions. The statistical analysis parameters for wind power density from Equation 3.68 are introduced as a suitability judgment of the wind speed distribution developed. It is shown that the MEP-type distribution not only fits better with the measured wind speed data, but can also yield better representations for the power density distribution.

3.6.1 Mathematical Formulation

It is observed that both the theoretical MEP distribution derived from maximization of Shannon's entropy given in the previous section and all the other empirical distributions used to describe the variation in wind speed are exponential functions. However, the empirical functions always have a pre-exponential term, which is a function of the wind speed to a non-negative power. Considering that this pre-exponential function of wind speed can lower the probabilities in the low speed range, the MEP-type exponential function is modified by adding wind speed to the rth power as the pre-exponential term to the theoretical MEP distribution given in Equation 3.60. Therefore, the semi-empirical MEP-type wind speed distribution can be written as:

$$f(V) = V^r \exp\left(-\alpha_0 - \alpha_1 V - \alpha_2 V^2 - \alpha_3 V^3\right) \tag{3.69}$$

where r may take non-negative values, and $r=0$ represents the MEP distribution as shown in Equation 3.60. The above equation with non-zero positive r values thus represents the MEP-type exponential family of distribution functions. Similar to the MEP distribution, the Lagrangian multipliers in Equation 3.69 are still determined with the four physical constraints, namely, the normalization condition and the conservation of mass, momentum, and energy, as follows:

$$\int_{V_{\min}}^{V_{\max}} V^r \exp\left\{-\alpha_0 - \alpha_1 V - \alpha_2 V^2 - \alpha_3 V^3\right\} dV = 1 \tag{3.70}$$

$$\int_{V_{\min}}^{V_{\max}} V^{r+1} \exp\left\{-\alpha_0 - \alpha_1 V - \alpha_2 V^2 - \alpha_3 V^3\right\} dV = V_{10} \tag{3.71}$$

$$\int_{V_{\min}}^{V_{\max}} V^{r+2} \exp\left\{-\alpha_0 - \alpha_1 V - \alpha_2 V^2 - \alpha_3 V^3\right\} dV = V_{20}^2 \tag{3.72}$$

$$\int_{V_{\min}}^{V_{\max}} V^{r+3} \exp\left\{-\alpha_0 - \alpha_1 V - \alpha_2 V^2 - \alpha_3 V^3\right\} dV = V_{30}^3 \tag{3.73}$$

The three mean speeds, V_{10}, V_{20}, and V_{30}, needed can be calculated from the measured wind speed distribution data, or alternatively these mean speeds can be relatively easily measured compared with the wind speed distribution data, as pointed out in the previous section. Similarly, the lower integration limit in Equations 3.70–3.73 can be set to $V_{min} = 0$ according to the measurement, and the upper limit, V_{max}, can be set to the maximum wind speed measured for the corresponding data set. Again, the modified second-order Newton–Raphson method [42] is used to solve this highly non-linear set of equations. All of these manipulations are the same as those employed for the theoretical MEP distribution described in the previous section. However, the value of r is taken as pre-determined and as a model input parameter. In this chapter, integer numbers are assumed and r values ranging from 0 to 5 are considered. The lower limit $r = 0$ corresponds to the theoretical MEP distribution and the upper limit $r = 5$ is determined from the present study when comparing with actual wind speed data, as shown later in Section 3.6.3.

3.6.2 Fitting Criteria for Comparison

Conventionally, the criteria for the suitability of a possible distribution function are made only based on wind speed distribution [43]. The criteria include the mean wind speed, standard deviation or the R^2 of the wind speed distribution and direct comparison of the distribution parameters. However, the wind speed distribution function will ultimately be used to determine wind power density in wind energy utilization applications or other applications, such as the effects of extreme wind conditions on structures. Estimating the mean wind speed (V_{10}) correctly does not necessarily mean the correct estimation of the wind energy (V_{30}). Therefore, it is essential that the criteria for the suitability of the distribution function should also be based on its ability to predict the wind power density. Celik [43] introduced standard deviation of the wind power density as the most appropriate characteristic parameter for the judgment. This idea is also adopted and applied in this section to judge the suitability of the MEP-type exponential family of the distribution functions and the Weibull distribution. The goodness-of-fit parameter, root mean square error (RMSE), is used as well for both wind speed distribution and power density distribution instead of the standard deviation. The coefficient of determination and RMSE are defined in Equations 3.65 and 3.67, respectively.

3.6.3 Comparison with the Measured Data and Weibull Distribution

To demonstrate the suitability of the MEP-type function developed in this section for the probabilistic distribution of wind speed, comparisons have been made with the Weibull distribution and a variety of measured wind speed data taken from different sources. The results are presented in this section. The wind speed data from three countries, the USA, Oman, and Turkey, used in Section 3.5.3 for the theoretical development of the MEP distribution function, are considered in this section, along with more data for further and extensive comparison.

The actual wind data for Dodge City and Kansas City in the USA for the year 1970 are tabulated in the forms of probability density function and cumulative distribution in [28]. It is repeated here that, due to limitations of the measurement technique used when the data were measured, the data points for the two cities are somewhat scattered and irregular. The shape parameters of the Weibull distribution for Dodge City and Kansas City are obtained as $k=2.110$ and $k=1.776$, respectively, and the scale parameters are $c=11.96$ kn and $c=7.65$ kn, respectively.

The results of the statistical analysis for the wind speed data measured at Dodge City are shown in Table 3.11. A comparison of the COD and RMSE values for the wind speed indicates that the MEP family of distributions with r values from 0 to 5 fit the measured wind speed data better than the Weibull distribution. As pointed out earlier, the distribution with $r=0$ corresponds to the theoretical distribution derived from the MEP (the so-called MEP distribution). On the other hand, the RMSE values for the power density show that the Weibull distribution fits the power density data slightly better than the MEP distribution and $r=1$ MEP-type distribution, but not as well as the other MEP-type distributions with higher values of exponent r. The MEP-type distribution with $r=4$ fits both the wind speed and power density data best among all the distributions investigated. This distribution ($r=4$) is plotted in Figure 3.9 along with the actual wind speed data, the theoretical MEP distribution ($r=0$) and the Weibull distribution. Clearly, the MEP-type distribution with $r=4$ fits the data better, especially for higher wind speeds.

The measured data for wind speeds at Kansas City have a large fluctuation for successive wind speed and are rather scattered. The actual data seem to have a peak in the low wind speed region and zigzag in the medium wind speed region. The second peak appears at a speed of 10 kn or so. The values of the fitting criteria for wind speed and power density are given in Table 3.12. It is seen that among these distributions, the theoretical MEP distribution fits best both the measured wind speed data and power density data. The MEP-type distributions with $r=1$ and $r=2$ fit the wind speed data better than the Weibull distribution, while only the one with $r=1$ is better than the Weibull distribution when fitting the power density data. This is because the power density distribution is heavily weighted towards the high wind speed range, as is evident in Equation 3.68, thus a good fit for the wind speed distribution does not necessarily show a good fit for the power density distribution. To demonstrate the comparison graphically, the measured data for Kansas City are also shown in Figure 3.10 with the Weibull distribution, the theoretical MEP distribution, and the MEP-type distribution with $r=1$. It is evident that a salient feature of the theoretical MEP distribution is that the probability density function starts at a non-zero value when the wind speed is zero, which is more consistent with the actual data points for some locations where calm spells exist. Calm spells represent the situation of zero wind speed or stationary air. However, similar to the Weibull distribution, the MEP-type distributions with $r>0$ always give a zero value for the probability density function whenever the wind speed is zero in any case, or it is unable to predict the existence of calm spells. The MEP-type distribution with $r=1$ here shows bi-modal shape, which is more consistent with the measured data than the other two distributions for this type of data set. It might be concluded that the MEP distribution is most appropriate for both wind speed distribution and wind power density distribution when calm spells exist in the particular set of wind speed data.

Table 3.11 Fitting criteria for the comparison of MEP-type and Weibull distributions with the wind speed data and power density measured at Dodge City, USA [28]

	Wind speed		Power density
	COD	RMSE	RMSE
Weibull	0.8281	0.01233	17.1028
MEP ($r=0$)	0.8406	0.01187	18.9617
$r=1$	0.8540	0.01136	17.7249
$r=2$	0.8610	0.01109	17.0918
$r=3$	0.8647	0.01094	16.7899
$r=4$	0.8663	0.01088	16.7148
$r=5$	0.8661	0.01088	16.8062

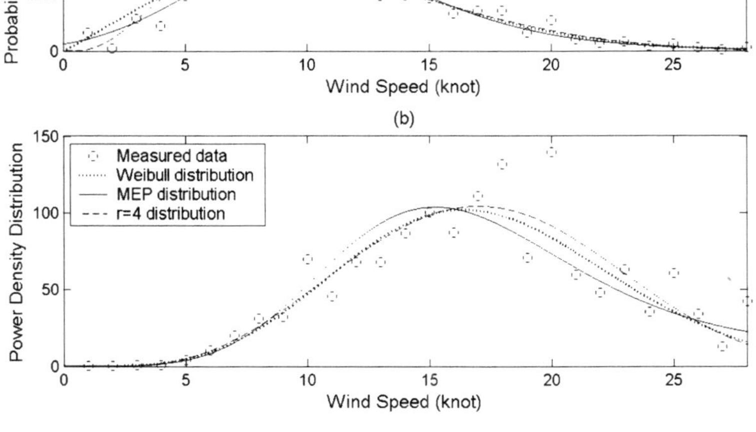

Figure 3.9 Comparison of MEP-type distributions with measured data [28] and the Weibull distribution for wind speeds measured at Dodge City, USA: **a** probability density distribution; **b** power density distribution

Some measured data for Oman are available in [42]. These data are summarized based on the monthly weather data from 1986 to 1998 for four stations in Marmul, Masirah, Sur, and Thumrait. The available data are tabulated in the form of cumulative distribution for the wind speed. The backward differencing scheme is used to transform the cumulative distribution into probability density distribution. Three methods were compared by Dorvlo [45] to estimate the Weibull distribution parameters, and the chi-square method was found to be the best in providing the numerical estimate of k and c for the measured data. Therefore, the Weibull distributions calculated with this method are cited in this section. Figure 3.11-1 shows the

Table 3.12 Fitting criteria for the comparison of MEP-type and Weibull distributions with the wind speed data and power density measured at Kansas City, USA [28]

	Wind speed		Power density
	COD	RMSE	RMSE
Weibull	0.5090	0.02784	16.5904
MEP ($r=0$)	0.6548	0.02334	15.6398
$r=1$	0.6264	0.02429	16.4243
$r=2$	0.5705	0.02604	17.2456
$r=3$	0.5027	0.02802	18.0346
$r=4$	0.4293	0.03002	18.7770
$r=5$	0.3544	0.03192	19.4718

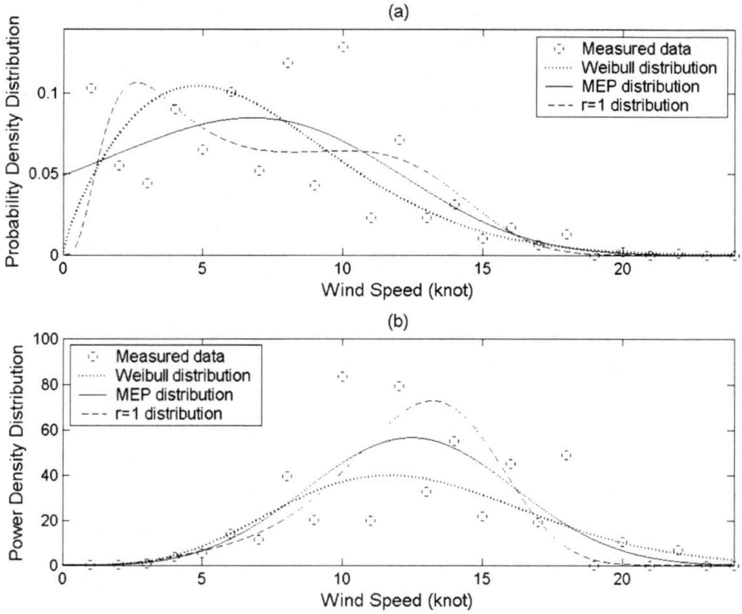

Figure 3.10 Comparison of MEP-type distributions with measured data [28] and the Weibull distribution for wind speeds measured at Kansas City, USA: **a** probability density distribution; **b** power density distribution

MEP distribution and the $r=5$ MEP-type distribution with the measured data and the corresponding Weibull distribution for the measurement taken at the Marmul station, Oman. The wind speed probability density distributions are presented in Figure 3.11-1a and the density power distribution in Figure 3.11-1b. It is seen from Figure 3.11-1a that all of these distributions have almost the same peak location, while the Weibull distribution has a higher peak value. It was concluded that the Weibull distribution fits the wind speed data equally well for this case as the MEP-type distributions in Section 3.5.3. However, the MEP-type distributions fit the higher speed regions much better than the Weibull distribution. This advantage is of great importance for the power density distribution. It can be seen from Figure 3.11-1b that the MEP-type distributions fit the power density data much better than the Weibull distribution, especially at the higher wind speed region.

The results of the statistical analysis for both the wind speed probability distribution and the power density distribution are given in Table 3.13. For the wind speeds measured at Marmul station, CODs for the wind speed distribution are 0.9043, 0.9084, 0.9199, 0.9297, 0.9384, 0.9461, and 0.9529 for the Weibull distribution, the theoretical MEP distribution, and the $r=1$ to $r=5$ MEP-type distributions, respectively. The corresponding RMSEs are 0.02652, 0.02595, 0.02426, 0.02272, 0.02128, 0.01990, and 0.01859. The RMSEs for the power density distribution are 6.6704, 3.9263, 3.6826, 3.4699, 3.2744, 3.0915, and 2.9190 for the Weibull distribution, the theoretical MEP distribution, and the $r=1$ to $r=5$ MEP-type distributions, respectively. These parameters indicate that all the MEP-type distributions fit better for the wind speed probability distribution data and power density data than the Weibull distribution for the data at the quite windy location with few calm spells. The higher the value of r, the better the fitting is. Compared with the Weibull distribution, the RMSEs for the power density decrease by 41 %, 45 %, 48 %, 51 %, 54 %, and 56 % for the theoretical MEP distribution and the $r=1$ to $r=5$ MEP-type distributions, respectively.

The statistical analysis for the three other stations is also tabulated in Table 3.13 and the measured data for the Masirah, Sur, and Thumrait stations are plotted in Figure 3.11-2, 3.11-3, and 3.11-4, respectively, with the corresponding three curves shown: the Weibull distribution, the theoretical MEP distribution, and the $r=5$ MEP-type distribution. Similar behaviors to the Marmul case are observed in the comparison of the MEP-type distributions with the Weibull distribution and the wind speed data measured. The Weibull distribution fits slightly better for the wind speed density data than the theoretical MEP distributions from the COD and the RMSE values, but not as good as other MEP-type distributions. However, the parameters for the power density distribution show that the MEP-type distributions fit much better than the Weibull distributions.

It is emphasized that the common features of the wind speed data measured at the four stations in Oman are that they have high long-term average wind speeds and almost no calm spells. For this type of data, the Weibull distribution function can fit the wind speed data quite well, as observed earlier, while it cannot fit the power density well due to the larger error at the high wind speed region compared with the MEP-type distributions.

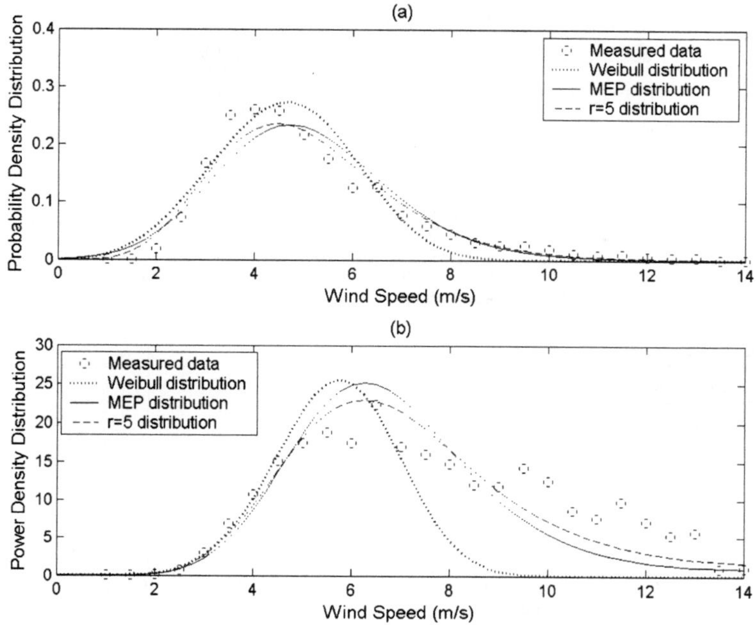

(1)

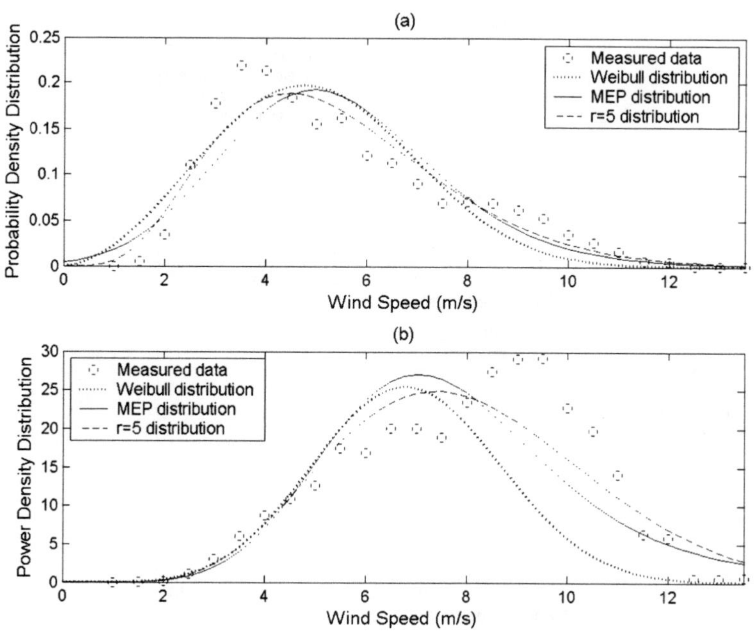

(2)

Figure 3.11 Comparison of the MEP-type distributions with the measured data [42] and the Weibull distribution for the wind speeds measured at the four stations in Oman: (1) Marmul; (2) Masirah; (3) Sur (4) Thumrait

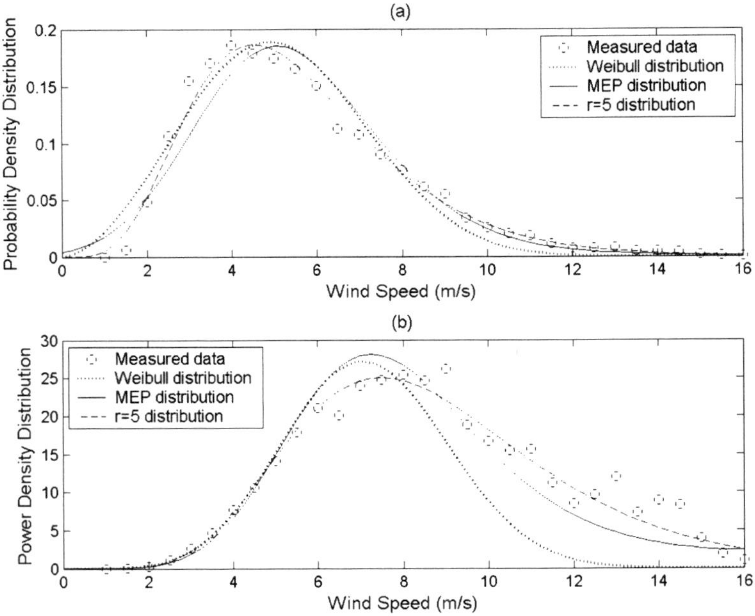

(3)

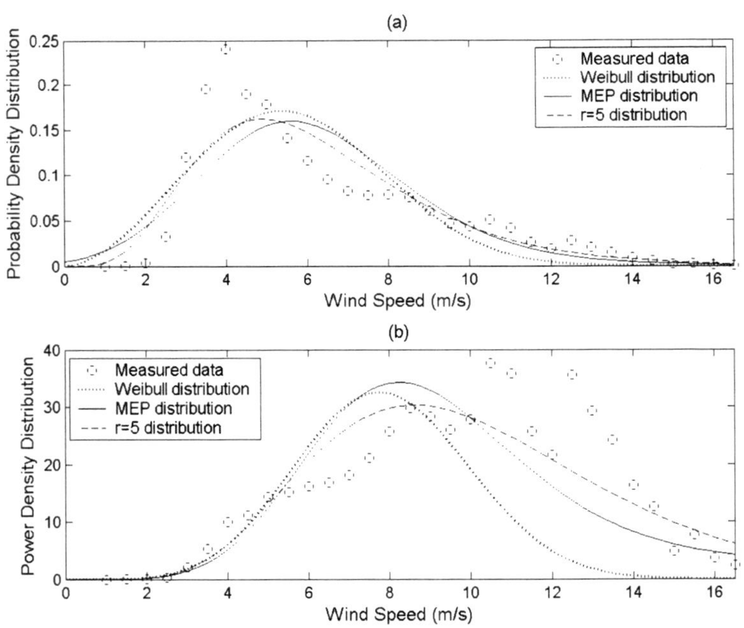

(4)

Figure 3.11 (continued)

Table 3.13 Fitting criteria for the comparison of MEP-type and Weibull distributions with the wind speed data and power density for the wind speeds measured at the four stations in Oman [42]

	Marmul			Masirah		
	Wind speed		Power density	Wind speed		Power density
	COD	RMSE	RMSE	COD	RMSE	RMSE
Weibull	0.9043	0.02652	6.6704	0.8362	0.02779	8.1044
MEP ($r=0$)	0.9084	0.02595	3.9263	0.8162	0.02944	5.4441
$r=1$	0.9199	0.02426	3.6826	0.8461	0.02694	5.0596
$r=2$	0.9297	0.02272	3.4699	0.8686	0.02489	4.7716
$r=3$	0.9384	0.02128	3.2744	0.8868	0.02311	4.5261
$r=4$	0.9461	0.01990	3.0915	0.9019	0.02150	4.3029
$r=5$	0.9529	0.01859	2.9190	0.9148	0.02005	4.0931
	Sur			Thumrait		
	Wind speed		Power density	Wind speed		Power density
	COD	RMSE	RMSE	COD	RMSE	RMSE
Weibull	0.9436	0.01534	5.9218	0.7154	0.03420	12.7970
MEP ($r=0$)	0.9336	0.01664	3.0497	0.6888	0.03576	8.1812
$r=1$	0.9521	0.01413	2.6019	0.7270	0.03349	7.4972
$r=2$	0.9649	0.01210	2.2795	0.7571	0.03159	6.9854
$r=3$	0.9742	0.01038	2.0384	0.7825	0.02989	6.5619
$r=4$	0.9809	0.00892	1.8682	0.8045	0.02834	6.1967
$r=5$	0.9857	0.00772	1.7655	0.8239	0.02690	5.8736

On the other hand, the wind speed data measured at Izmir, Turkey has significant periods of calm winds and the low speed range accounts for a relatively large percentage in the measurements over the 5-year period from 1995 to 1999 [46], as shown in Table 3.8. The annual average, 3 m/s or so, is much lower than the values measured in Oman. Table 3.14 presents the statistical analysis for wind speed and power density for the available six sets of data. Similar trend of variation is observed, that is, the theoretical MEP distributions fit best both wind speed and power density data among all the seven distributions. The CODs decrease while RMSEs increase with increasing r values for the MEP-type distributions. For 1995 data, when r reaches 5, the MEP-type distribution is even slightly worse than the Weibull distributions, although the power density distribution is still slightly better. The RMSEs for the power density decrease by 63 %, 48 %, 34 %, 22 %, and 11 % for the MEP-type distributions for r values from zero to 4 compared with the Weibull distribution. The results for the other years shown in Table 3.14 are very similar. For example, the RMSEs for 1998 power density decrease by 72 %, 66 %, 59 %, 53 %, 46 %, and 40 % for r from zero to 5. Figure 3.12-1–3.12-6 shows the curves of the theoretical MEP distribution, the MEP-type distribution with $r=2$, and the Weibull distribution with the corresponding measured data for the six cases. Figure 3.12-1–3.12-5 are for the years 1995 to 1999, respectively; and Figure 3.12-6 is for the 5-year data. It can be seen from the figures that the theoretical MEP distributions closely match the measured data points, much better than the corresponding Weibull distributions curve fitted from the data for all the speed range, especially for the high wind speed range that is important for wind energy utilization applications. The finite probability of zero wind speed is predicted by the theoretical MEP distribution for the significant number of calm spells measured, whereas the Weibull distribution and the MEP-type distributions cannot fit the existence of calm spells. However, the MEP-type distributions do express the existence of many calm spells with the bi-modal shape wind speed distributions.

Table 3.14 Fitting criteria for the comparison of MEP-type and Weibull distributions with the wind speed and power density for the wind speeds measured at Izmir, Turkey [10]

	1995			1996		
	Wind speed		Power density	Wind speed		Power density
	COD	RMSE	RMSE	COD	RMSE	RMSE
Weibull	0.8955	0.02363	1.5204	0.9068	0.02367	1.1141
MEP ($r=0$)	0.9864	0.00851	0.5562	0.9878	0.00856	0.3719
$r=1$	0.9819	0.00984	0.7854	0.9781	0.01147	0.6150
$r=2$	0.9644	0.01380	1.0006	0.9568	0.01612	0.8302
$r=3$	0.9441	0.01728	1.1907	0.9356	0.01967	1.0148
$r=4$	0.9171	0.02105	1.3604	0.9116	0.02304	1.1771
$r=5$	0.8797	0.02535	1.5133	0.8810	0.02674	1.3237
	1997			1998		
	Wind speed		Power density	Wind speed		Power density
	COD	RMSE	RMSE	COD	RMSE	RMSE
Weibull	0.9328	0.02045	1.0450	0.8777	0.02864	1.9247
MEP ($r=0$)	0.9922	0.00698	0.3451	0.9737	0.01328	0.5296
$r=1$	0.9819	0.01061	0.5315	0.9644	0.01545	0.6462
$r=2$	0.9618	0.01542	0.7091	0.9358	0.02075	0.7811
$r=3$	0.9407	0.01921	0.8643	0.9105	0.02451	0.9130
$r=4$	0.9159	0.02287	1.0023	0.8891	0.02728	1.0389
$r=5$	0.8843	0.02683	1.1280	0.8677	0.02979	1.1578
	1999			5-year		
	Wind speed		Power density	Wind speed		Power density
	COD	RMSE	RMSE	COD	RMSE	RMSE
Weibull	0.9161	0.02095	1.2768	0.9151	0.02224	1.3134
MEP ($r=0$)	0.9920	0.00649	0.4723	0.9879	0.00840	0.4683
$r=1$	0.9871	0.00822	0.7913	0.9765	0.01170	0.7019
$r=2$	0.9681	0.01292	1.0641	0.9526	0.01662	0.9076
$r=3$	0.9490	0.01632	1.2927	0.9273	0.02057	1.0854
$r=4$	0.9246	0.01985	1.4910	0.8971	0.02447	1.2421
$r=5$	0.8433	0.02862	1.6738	0.8582	0.02873	1.3824

Another typical set of wind speed data is given by Seguro and Lambert [44]. In their effort to estimate the parameters of the Weibull wind speed distribution, they modified the maximum likelihood method and compared it with the classical maximum likelihood method and a graphical method. From their results they recommended the modified maximum likelihood method for use with wind data in frequency distribution format. Therefore, the parameters calculated from this method, $k=2.99$ and $c=5.77$ m/s, are also used here to compare with the MEP-type distribution functions. The statistical analysis parameters for the seven distributions considered, namely the Weibull distribution, the theoretical MEP distribution, and the MEP-type distributions with r up to 5, are given in Table 3.15. It can be seen that for this set of data, the MEP-type exponential family of distribution functions always yields better agreement than the Weibull distribution for both wind speed distribution and power density distribution. The MEP-type distribution with $r=2$ is the best one for estimating wind speed, whereas for power density, slightly better agreement with the data is obtained for the MEP-type distributions with higher values of r, although the value of r does not have a significant effect on the power density distribution. This can also be observed from the wind speed probability density distribution shown in Figure 3.13a and the power density distribution shown in Figure 3.13b. In these figures, the theoretical MEP distribution and the MEP-type distributions with $r=2$ and $r=5$ are shown along with the measured data and the corresponding Weibull distribution. The advantage of the MEP-type distributions over the Weibull is clearly shown, while only small variations are observed for the different MEP-type distributions.

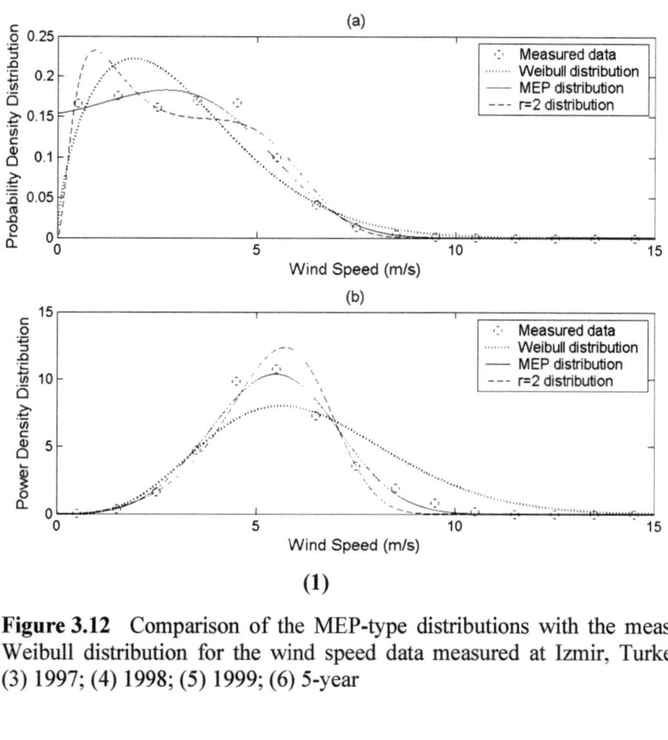

(1)

Figure 3.12 Comparison of the MEP-type distributions with the measured data [10] and the Weibull distribution for the wind speed data measured at Izmir, Turkey: (1) 1995; (2) 1996; (3) 1997; (4) 1998; (5) 1999; (6) 5-year

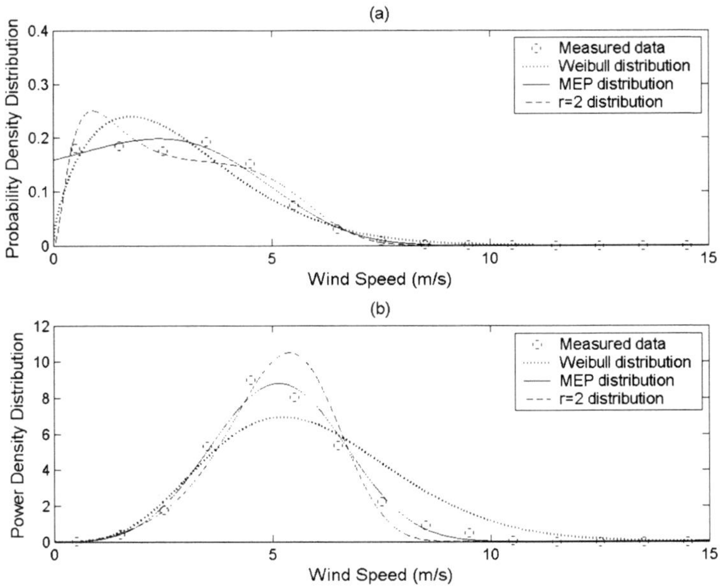

(2)

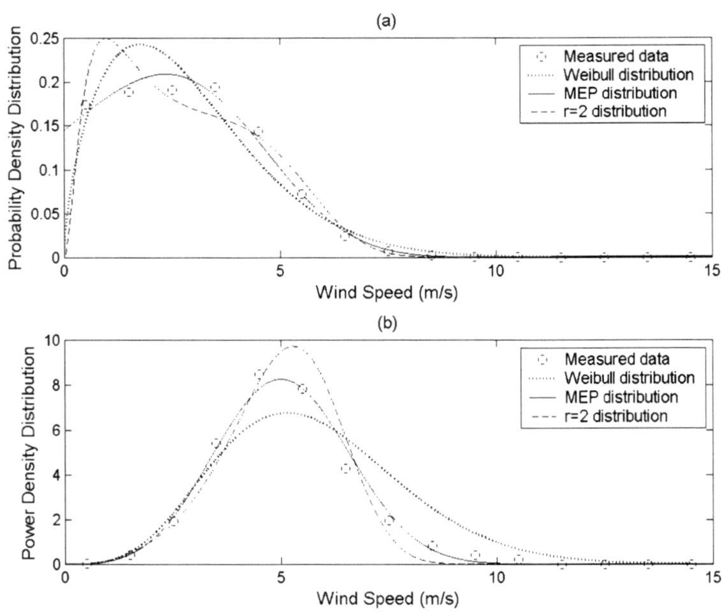

(3)

Figure 3.12 (continued)

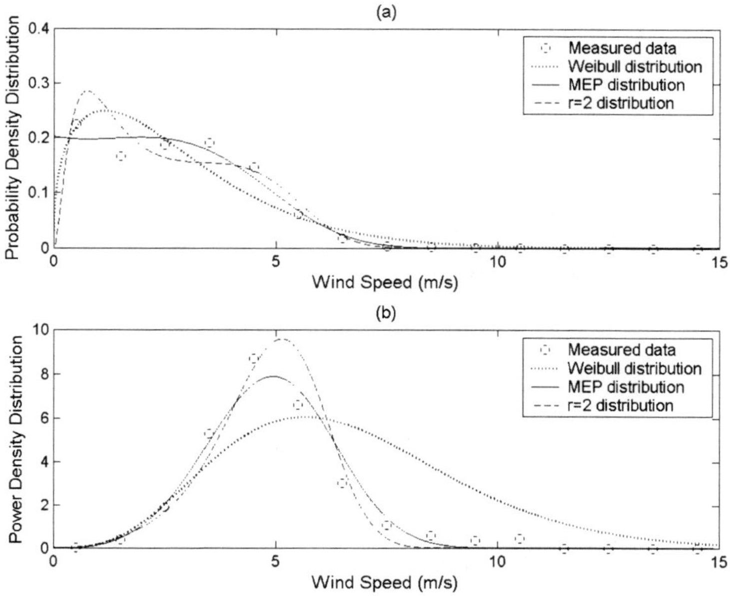

(4)

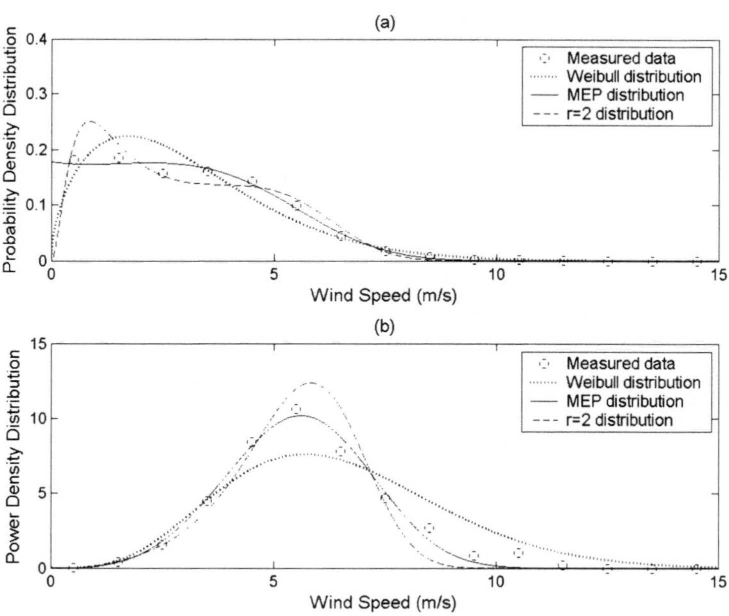

(5)

Figure 3.12 (continued)

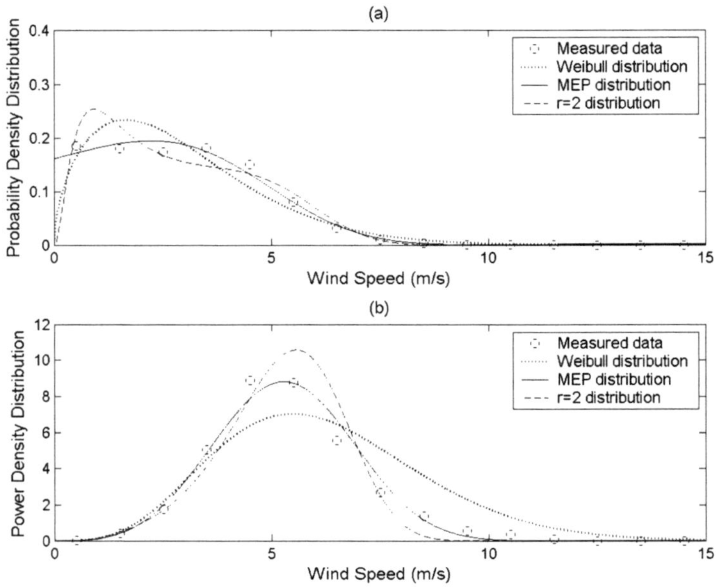

(6)

Figure 3.12 (continued)

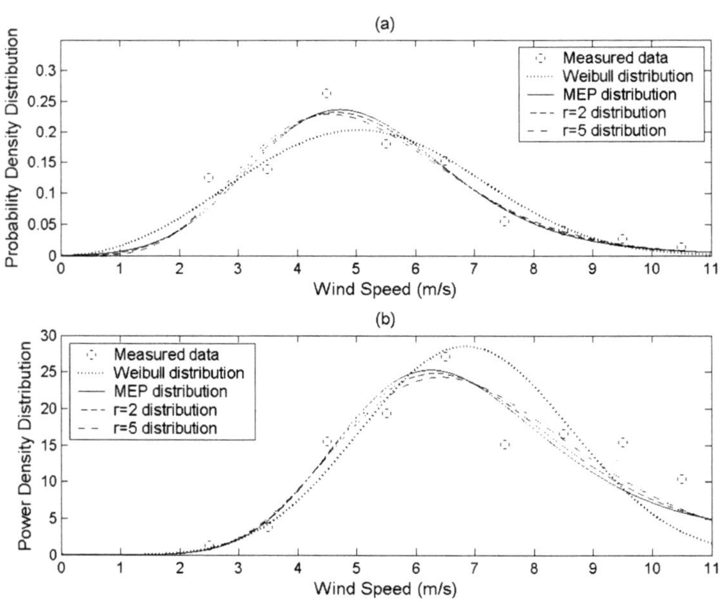

Figure 3.13 Comparison of the MEP-type distributions with the measured data [44] and the Weibull distribution for a sample data set: **a** probability density distribution; **b** power density distribution

Table 3.15 Fitting criteria for the comparison of MEP-type and Weibull distributions with the wind speed data and power density for the measured data given in [44]

	Wind speed		Power density
	COD	RMSE	RMSE
Weibull	0.8281	0.03037	5.2380
MEP ($r=0$)	0.8742	0.02597	3.6137
$r=1$	0.8766	0.02572	3.5258
$r=2$	0.8773	0.02565	3.4597
$r=3$	0.8769	0.02570	3.4086
$r=4$	0.8756	0.02583	3.3697
$r=5$	0.8736	0.02604	3.3415

Table 3.16 Fitting criteria for the comparison of MEP-type and Weibull distribution with the wind speed data and power density for the data measured at Elazig, Turkey [45]

	1998			1999		
	Wind speed		Power density	Wind speed		Power density
	COD	RMSE	RMSE	COD	RMSE	RMSE
Weibull	0.7988	0.04632	2.0450	0.7879	0.04778	2.1722
MEP ($r=0$)	0.8608	0.03853	1.8402	0.8373	0.04186	1.9738
$r=1$	0.8958	0.03333	1.5384	0.8759	0.03654	1.6657
$r=2$	0.9243	0.02841	1.2753	0.9081	0.03144	1.4045
$r=3$	0.9484	0.02346	1.0386	0.9357	0.02631	1.1768
$r=4$	0.9673	0.01868	0.8372	0.9578	0.02130	0.9853
$r=5$	0.9805	0.01443	0.6848	0.9742	0.01668	0.8260
	2000			2001		
	Wind speed		Power density	Wind speed		Power density
	COD	RMSE	RMSE	COD	RMSE	RMSE
Weibull	0.8136	0.04385	1.9581	0.7962	0.04280	2.0600
MEP ($r=0$)	0.8769	0.03564	1.7413	0.8390	0.03803	1.9941
$r=1$	0.9101	0.03045	1.4395	0.8883	0.03169	1.5667
$r=2$	0.9363	0.02564	1.1769	0.9236	0.02620	1.2728
$r=3$	0.9576	0.02091	0.9410	0.9499	0.02122	1.0673
$r=4$	0.9736	0.01652	0.7379	0.9678	0.01702	0.8951
$r=5$	0.9837	0.01295	0.5708	0.9792	0.01367	0.7432
	2002			5-year		
	Wind speed		Power density	Wind speed		Power density
	COD	RMSE	RMSE	COD	RMSE	RMSE
Weibull	0.8035	0.04482	2.4033	0.7972	0.04538	2.1527
MEP ($r=0$)	0.8585	0.03803	2.2286	0.8559	0.03825	1.9389
$r=1$	0.8912	0.03336	1.9018	0.8937	0.03285	1.5961
$r=2$	0.9176	0.02902	1.6153	0.9236	0.02785	1.3088
$r=3$	0.9394	0.02490	1.3566	0.9481	0.02295	1.0625
$r=4$	0.9556	0.02131	1.1297	0.9667	0.01840	0.8577
$r=5$	0.9657	0.01873	0.9360	0.9789	0.01463	0.6836

Akpinar and Akpinar [45] presented the wind speed data over a 5-year period from 1998 to 2002 in their work to analyze the wind energy potential at Elazig, Turkey. These data are tabulated in the form of cumulative distribution. Therefore, the backward differencing scheme is used to transform the cumulative distribution into probability density distribution. The Weibull parameters for all the six sets can be found in tabulated form [45]. The goodness of the fit among the distributions considered and the measured wind speed data is given in Table 3.16. It can be seen that for all the data sets, the MEP-type distributions fit both the wind speed data and the power density data much better than the Weibull distribution, and the agreement becomes better with increasing r value. Compared with the RMSEs for the Weibull distributions, the RMSEs for the $r=5$ MEP distributions decrease by 67 %, 62 %, 71 %, 64 %, 61 %, and 68 % for 1998 to 2002 and the 5-year data, respectively. The best fitting curves ($r=5$) for the six sets of data are plotted in Figure 3.14 with the corresponding measured data, the theoretical MEP distribution and the Weibull Distribution. Figure 3.14-1–3.14-5 are for the years 1998 to 2000, respectively, while Figure 3.14-6 is for the 5-year data. It can be seen that the $r=5$ MEP-type distributions closely match both the measured wind speed and the power density data points. This clearly demonstrates the superiority of the present proposed MEP-type distribution functions over the empirical Weibull distribution for both wind speed distribution and wind power density distribution.

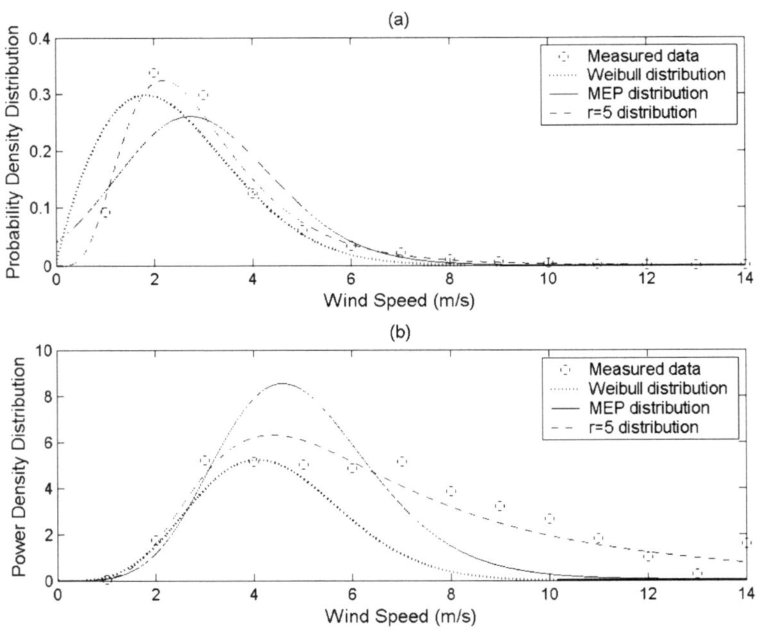

(1)

Figure 3.14 Comparison of the MEP-type distributions with the measured data [45] and the Weibull distribution for the wind speeds measured at Elazig, Turkey: (1) 1998; (2) 1999; (3) 2000; (4) 2001; (5) 2002; (6) 5-year

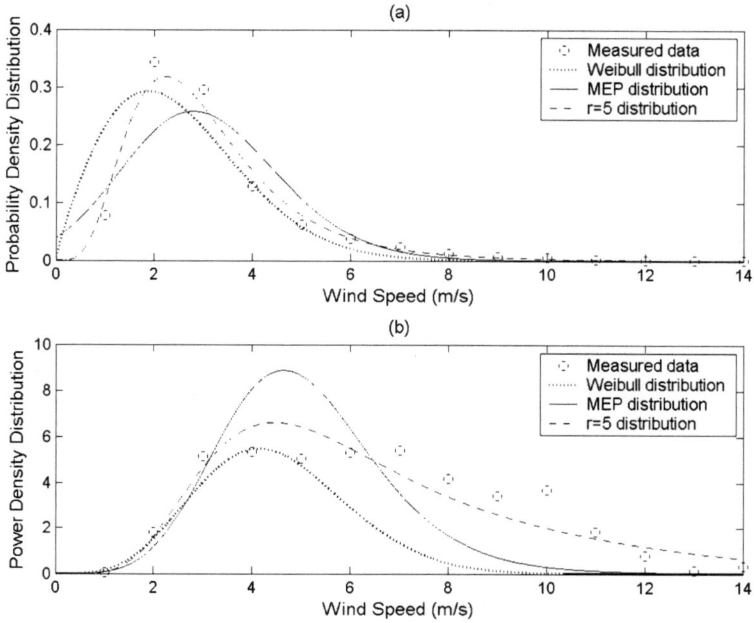

(2)

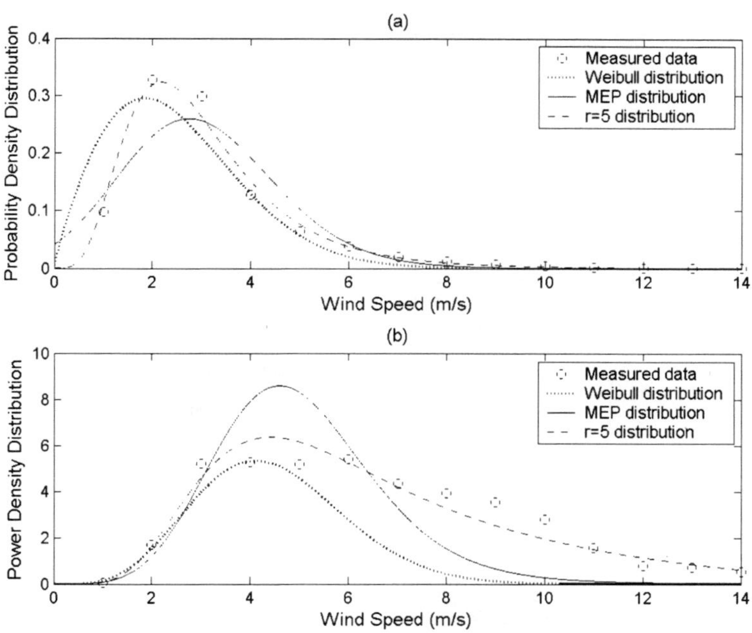

(3)

Figure 3.14 (continued)

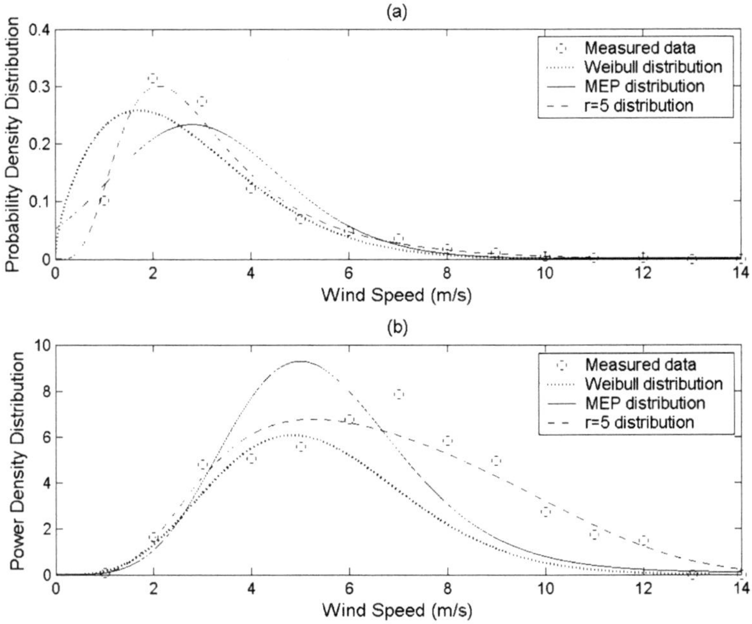

(4)

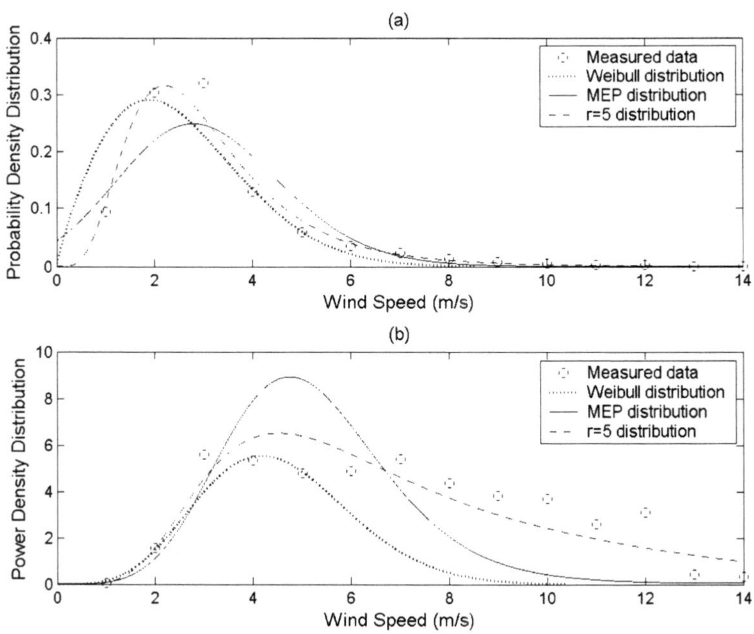

(5)

Figure 3.14 (continued)

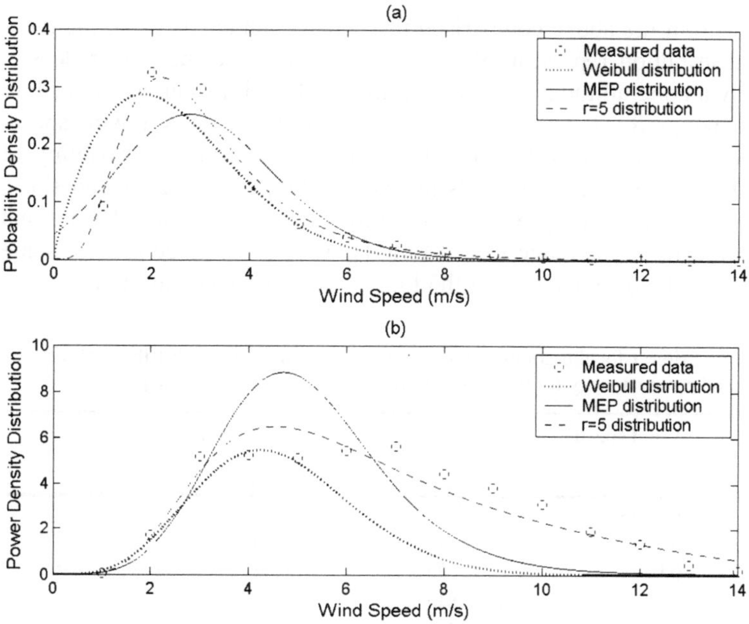

(6)

Figure 3.14 (continued)

The last set of wind speed data considered in this section is from Celik's work in his estimate of the wind energy output for small-scale wind power generators using Weibull-representative wind data [6]. The data are arranged in frequency distribution format based on the measured data in time series for January, 1996 at Cardiff, UK. The results for the statistical analysis are shown in Table 3.17. The COD. and RMSE for the Weibull distribution indicate that this empirical distribution is quite reasonable for the estimation of the wind speed and the power density data measured. However, the theoretical MEP distribution provides much better agreement with the measured data for both the wind speed and the power density; and the RMSEs for the wind speed and the power density distribution decrease by 48 % and 7 %, respectively. The $r=1$ MEP-type distribution fits the wind speed data better than the Weibull distribution, although the RMSE for the power density is slightly larger. The other MEP-type distributions are not as good as the Weibull distribution, as shown in Table 3.17. The measured data for this case are shown in Figure 3.15 along with the Weibull distribution, the theoretical MEP distribution, and the $r=1$ MEP-type distribution. It is also seen from the figure that the theoretical MEP distribution predicts the existence of physically realistic calm spells, while all other distributions fail to do so, as observed earlier for this type of wind speed data.

It is evident from the above comparison that when calm spells exist, the theoretical MEP distribution represents the wind speed data and the power density most accurately among all the distributions considered; and the empirical Weibull distribution cannot predict the existence of a finite probability for zero wind speed. For other types of wind speed data that do not have a significant probability of calm spells, an optimal value of r exists for the MEP-type distribution functions, which can fit both the wind speed and the power density best; and the optimal value of r is typically in the range from 1 to 5. However, it is noted that for some types of data, *e.g.*, the data taken from the four stations in Oman and Elazig, Tur-

Table 3.17 Fitting criteria for the comparison of MEP-type and Weibull distribution with the wind speed data and power density for the wind speeds measured at Cardiff, UK [6]

	Wind speed		Power density
	C.O.D	RMSE	RMSE
Weibull	0.9538	0.01659	1.3341
MEP	0.9876	0.00861	1.2401
$r=1$	0.9727	0.01273	1.3491
$r=2$	0.9467	0.01781	1.5247
$r=3$	0.9128	0.02278	1.7127
$r=4$	0.8721	0.02758	1.8966
$r=5$	0.8263	0.03215	2.0707

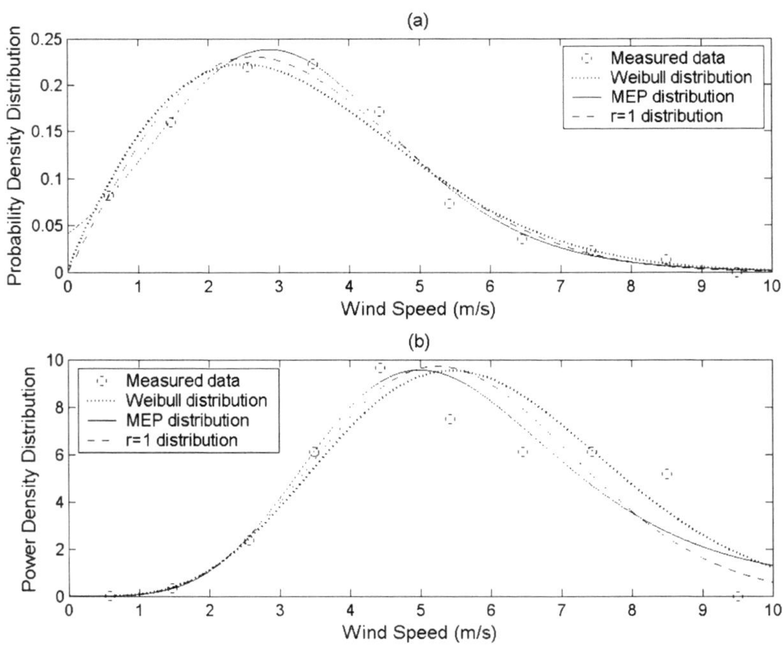

Figure 3.15 Comparison of the MEP-type distributions with the measured data and the Weibull distribution for the wind speeds measured at Cardiff, UK [6]

key, optimal r value can be larger than 5. Two sets of this type of data, one taken from the Sur station, Oman and the other is 1998 data for Elazig in Turkey, are examined for $r=6$. Comparison for the two curves with $r=5$ and $r=6$ for each data set is shown in Figure 3.16. The two curves for the Sur station almost coincide with each other and are almost indistinguishable. For the 1998 Elazig data, the peak of the $r=6$ MEP-type distribution is slightly higher for the wind speed distribution and is slightly lower for the power density distribution when compared with the $r=5$ distribution, and the $r=6$ distribution clearly agrees better with the measured data for both wind speed and power density distributions, especially at high wind speeds for the power density distribution. The results of the statistical analysis are given in Table 3.18 for the fitting criteria. When r is increased from 5 to 6, the RMSE for the power density decreases by about 2 % for the Sur station data (see Tables 3.13 and 3.18) and 14 % for the 1998 Elazig data (see Tables 3.16 and 3.18). Clearly, the $r=6$ MEP type distribution is much better than the corresponding Weibull distribution for both wind speed and power density distributions. However, a larger value of r is more computationally demanding for determination of the Lagrangian multipliers in the MEP type distribution, a decision might have to be made as to the value of r for this type of data, based on the balance between the accuracy in the wind speed/power representation and the computational power required for the calculations.

In summary, the Weibull distribution does reasonably fit the wind speed distribution for a wide range of data, although the accuracy deteriorates for the wind power density distribution. However, the MEP-type distributions can always agree with both the wind speed data and wind power density data better than the corresponding empirical Weibull distribution for a variety of wind speed data taken from different sources measured at different geographical locations in the world. When the wind speed distribution is ultimately used for the wind energy utilization applications, the suitability judgment of a particular distribution based on the statistical analysis on the wind power density is of importance. A suitable distribution function should not only have the ability to predict accurately the wind speed, but the wind power density as well. From the above comparison and discussion, it is evident that MEP-type distribution functions provide much better fitting for the power density than the Weibull distribution. It can thus be concluded that the MEP-type distribution is better suited to the statistical description of wind speed distribution and can be a better alternative to the Weibull distribution, especially in the field of wind energy utilization applications.

Table 3.18 Fitting criteria for the comparison of MEP-type and Weibull distributions with the wind speed data and power density when $r=6$ for the wind speed data measured at (1) Sur, Oman [45]; (2) Elazig, Turkey in 1998 [45]

		Wind speed		Power density
		COD	RMSE	RMSE
$r=6$	Sur	0.9889	0.006811	1.7269
	Elazig	0.9882	0.01121	0.5857

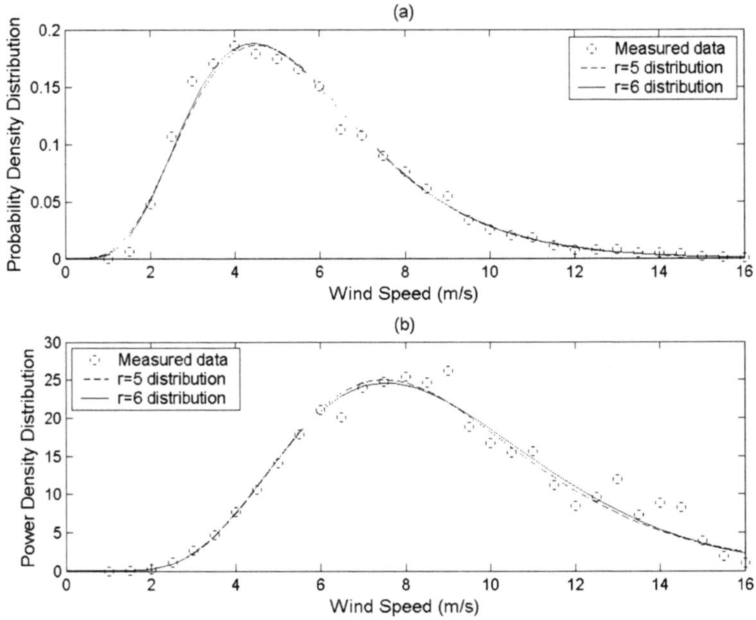

(1)

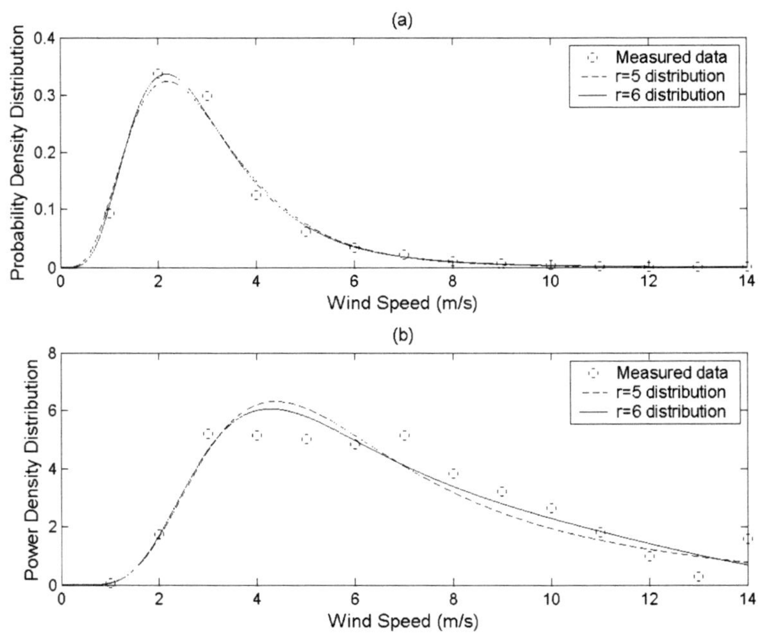

(2)

Figure 3.16 Comparison of the MEP-type distributions with $r=5$ and $r=6$ for the wind speed data measured at (1) Sur, Oman [19]; (2) Elazig, Turkey in 1998 [45]

3.6.4 Summary

In this section, the MEP distribution is extended to the MEP-type exponential family of distribution functions for the description of the probabilistic distribution of wind speed, and comparison is made with the wind speed data taken from different sources and measured at different geographical locations in the world. These MEP-type distributions are developed by introducing a pre-exponential term to the theoretical MEP distribution that is derived from the maximization of the Shannon's entropy based on the MEP. The conservation of mass, momentum, and energy for the air stream in the wind, form the set of constraints to determine the Lagrangian multipliers involved in these distributions. The statistical analysis parameters based on wind power density are introduced as the fitting criteria for the judgement of the suitability of the distribution functions. It is shown that the MEP-type distributions not only agree better with a variety of the measured wind speed data than the conventionally used empirical Weibull distribution, but represent a wider range of data types as well. The MEP-type distributions describe the wind power density more accurately than the Weibull distribution. Therefore, the MEP-type distributions are more suitable for the assessment of the wind energy potential and the performance of wind energy conversion systems.

3.7 Summary and Outlook

Owing to the increased need for environmental protection, mitigation of global climate change, energy security, and sustainable economic and social development, green and renewable energy sources and technologies are being developed and deployed at an accelerated rate. The cost of electricity generated from wind turbines has been reduced significantly and it has become comparable with the electricity from conventional power plants. As a result, wind energy utilization has become the fastest growing among all alternative and renewable energies. In order to install wind turbines at sites appropraite to the local wind energy resources as well as to determine the wind energy eonomics (such as investment payback time), wind energy potential at the selected sites must be aessessed, and as such the key factor is to determine an accurate wind speed or power density distribution for such calculations. An accurate wind speed distribution function is also important in other applications, such as in the determination of wind forces exerted on man-made structures for structure design or retrofit.

In this chapter, we first discuss the representation of the measured, but discrete, wind speed data as well as the various characteristic parameters for wind speed distribution. Then, commonly used empirical distribution functions, Weibull and Rayleigh distributions, are described, and their characteristics evaluated. Next the maximum entropy principle (MEP) is introduced, and is applied to derive the theoretical MEP distribution function. The MEP distribution is compared with the Weibull distribution under a variety of different types of wind speeds at different geographical locations. It is shown that the MEP distribution represents the measured wind speed data much more accurately than the empirical Weibull distribution. Finally, the MEP distribution is extended to the MEP-type exponential family of distribution functions for the description of the probabilistic distribution of wind speed. The statistical analysis parameters based on wind power density are introduced as the fitting criteria for judgement of the suitability of the distribution functions. It is shown that the MEP-type distributions not only agree better with a variety of measured wind speed data than the conventionally used empirical Weibull distribution, but represent a wider range of data types as well. The MEP-type distributions describe the wind power density more accurately than the Weibull distribution. Therefore, the MEP-type distributions are more suitable for the assessment of wind energy potential and the performance of wind energy conversion systems.

Literature is abundant in the use of the empirical Weibull or other distributions for the assessment of wind energy potential for various specific sites around the world, and it is recognized that the empirical distributions are limited in the accuracy of representation over the actual wind energy data. However, literature is scarce in developing theoretical wind speed distribution from first principles. Extensive future research effort is required to transition the current wind energy potential assessment beyond empiricism.

References

[1] Muljadi E., Pierce K., Migliore P. Soft-stall control for variable-speed stall-regulated wind turbines. J Wind Eng Ind Aerodynam 2000; 85:277–291.

[2] Chang TJ, Wu YT, Hsu HY, Chu CR, Liao CM. Assessment of wind characteristics and wind turbine characteristics in Taiwan. Renewable Energy 2003; 28:851–871.

[3] Ackermann T, Soder L. An overview of wind energy-status 2002. Renewable Sustainable Energy Rev 2002; 6:67–128.

[4] Wind energy comes of age: global wind energy sector celebrates success, tackles challenges. AWEA News Release. March 25, 2004.

[5] Elliott DL, Schwartz MN. Wind energy potential in the United States, September 1993. PNL–SA–23109. Richland, WA: Pacific Northwest Laboratory. NTIS no. DE94001667.

[6] Celik AN. Energy Output estimation for small–scale wind power generators using Weibull-representative wind data. J Wind Eng Ind Aerodynam 2003; 91:693–707.

[7] Patel MR. Wind and Solar Power Systems. New York: CRC Press, 1999.

[8] Mungwena W. The distribution and potential utilizability of Zimbabwe's wind energy resource. Renewable Energy 2002; 26:363–77.

[9] Weisser D. A wind energy analysis of Grenada: an estimation using the 'Weibull' density function. Renewable Energy 2003; 28:1803–1812.

[10] Ulgen K, Hepbasli A. Determination of Weibull parameters for wind energy analysis of Izmir, Turkey. Int J Energy Res 2002; 26:495–506.

[11] Atsu S, Dorvlo S. Estimating wind speed distribution. Energy Conversion Manage 2002; 43:2311–2318.

[12] Li M, Li X. On the probabilistic distribution of wind speeds: theoretical development and comparison with data. Int J Exergy 2004; 1:237–255.

[13] Li M, Li X. MEP–type distribution function: a better alternative to Weibull function for wind speed distributions. Renewable Energy 2005; 30:1221–1240.
[14] Naif MA, Shafiqur R. Wind speed and wind power characteristic for Gassim, Saudi Arabia. Int J Green Energy 2009; 6(2).
[15] Corotis RB, Sigl AB, Klein J. Probability models of wind velocity magnitude and persistence. Solar Energy 1978; 20:483–493.
[16] Deaves D M, Lines IG. On the fitting of low mean windspeed data to the Weibull distribution. J Wind Eng Ind Aerodynam 1997; 66:169–178.
[17] Garcia A, Torres JL, Prieto E, De Francisco A. Fitting wind speed distributions: a case study. Solar Energy 1998; 62:139–144.
[18] Milborrow D. Economics of wind power and comparisons with conventional thermal plant. Renewable Energy 1994; 5:692–699.
[19] Turksoy F. Investigation of wind power potential at Bozcaada, Turkey. Renewable Energy 1995; 6:917–923.
[20] Mayhoub AB, Azzam A. A Survey on the assessment of wind energy potential in Egypt. Renewable Energy 1997; 11:235–247.
[21] Algifri AH. Wind energy potential in Aden–Yemen Renewable Energy 1998; 13: 255–260.
[22] Persaud S, Flynn D, Fox B. Potential for wind generation on the Guyana Coastlands. Renewable Energy 1999; 18:175–189.
[23] Ilinca A, McCarthy E, Chaumel JL, Retiveau JL. (2003), Wind potential assessment of Quebec province. Renewable Energy 2003; 28:1881–1897.
[24] Weibull W. A statistical theory of the strength of materials. Ingeniors Vetenskaps Akademiens Handlingar (Royal Swedish Institute for Engineering) 1939; 151:1–45.
[25] Weibull W. A statistical distribution function of wide applicability. J Appl Mech – Trans ASME 1951; 18:292–297.
[26] Abernethy RB. The New Weibull Handbook, 4th edn. North Palm Beach, FL: self published 2002.
[27] Ahmed Shata AS, Hanitsch R. Evaluation of wind energy potential and electricity generation on the coast of Mediterranean Sea in Egypt. Renewable Energy 2006; 31:1183–1202.
[28] Johnson GL. Wind Energy Systems, Englewood Cliffs, NJ: Prentice–Hall, 1985.
[29] Takle ES, Brown JM. Note on the use of Weibull statistics to characterize wind speed data. J Applied Meteorol 1978; 17:556–559.
[30] Tuller SE, Brett AC. The characteristics of wind velocity that favor the fitting of a Weibull distribution in wind speed analysis. J Climate Appl Meteorol 1984; 23:124–134.
[31] Justus CG, Hargraves WR, Mikhail A, Graber G. Methods for estimating wind speed frequency distributions. J Applied Meteorol 1978; 17:350–353.
[32] Benard A, Bos–Levenbach EC. Het uitzetten van waarnemingen op waarschijnlijkdeids–papier (The plotting of observations on probability paper). Statististica Neerlandica 1953; 7:163–173.
[33] Johnson NL, Kotz S. Continuous Univariate Distributions, Vol. 2. Houghton Mifflin, 1970; 306.
[34] Christofferson RD, Gillette DA. A simple estimator of the shape factor of the two–parameter Weibull distribution. J Climate Appl Meteorol 1987; 26:323–325.
[35] Bagiorgas HS, Mihalakakou G, Matthopoulos D. A statistical analysis of wind speed distributions in the area of Western Greece. Int J Green Energy 2008; 5:120–137.
[36] Mathew S, Pandey KP, Kumar AV. 2002. Analysis of wind regimes for energy estimation. Renewable Energy 2002; 25:381–399.
[37] Kapur JN, Kesavan HK. *G*eneralized Maximum Entropy Principle (with Applications). The Stanford Educational Press, 1987.
[38] Shannon CE. (1948), A mathematical theory of communication. Bell System Tech J 1948; 27:379–623.
[39] Jaynes ET. Information theory and statistical mechanics. Phys. Rev. 1959; 106:620–630.
[40] Li M., Li X. A second-order Newton–Raphson method for improved numerical stability in the determination of droplet size distributions in sprays 2006; 16:71–82.

[41] Holman JP, Gajda WJ Jr. Experimental Methods for Engineers, 4th edn. New York: McGraw-Hill, 1984.
[42] Dorvlo ASS. (2002), Estimating wind speed distribution. Energy Conversion Manage 2002; 43:2311–2318.
[43] Celik AN. Assessing the suitability of wind speed probability distribution functions based on wind power density. Renewable Energy 2003; 28:1563–1574.
[44] Seguro JV, Lambert TW. Modern estimation of the parameters of the Weibull wind speed distribution for wind energy analysis. J Wind Eng 2000; 85:75–84.
[45] Akpinar EK, Akpinar S. An analysis of the wind energy potential of Elazig, Turkey. Int J Green Energy 2004; 1:193–207.

Chapter 4
Co-combustion and Gasification of Coal and Cattle Biomass: a Review of Research and Experimentation

Nicholas T. Carlin, Kalyan Annamalai, Hyukjin Oh, Gerardo Gordillo Ariza, Ben Lawrence, Udayasarathy Arcot V, John M. Sweeten, Kevin Heflin, and Wyatte L. Harman

4.1 Introduction

Large confined animal feeding operations (CAFOs), including cattle feedlots and dairies, have been cited as major point sources for air, soil, and water pollution. Subsequent land application of manure solids and the usage of liquid manure (> 90 % moisture) as irrigation water on nearby crop fields have been sited as non-

Nicholas T. Carlin
Department of Mechanical Engineering, Texas A&M University, College Station, Texas, USA

Kalyan Annamalai
Department of Mechanical Engineering, Texas A&M University, College Station, Texas, USA, e-mail: kannamalai@tamu.edu

Hyukjin Oh
Department of Mechanical Engineering, Texas A&M University, College Station, Texas, USA

Gerardo Gordillo Ariza
Department of Mechanical Engineering, Texas A&M University, College Station, Texas, USA

Ben Lawrence
Department of Mechanical Engineering, Texas A&M University, College Station, Texas, USA

Udayasarathy Arcot V.
Department of Mechanical Engineering, Texas A&M University, College Station, Texas, USA

John M. Sweeten
Texas Agricultural Experiment Station, Texas A&M University System, Agricultural Research and Extension Center, Amarillo, Texas, USA

Kevin Heflin
Texas Agricultural Experiment Station, Texas A&M University System, Agricultural Research and Extension Center, Amarillo, Texas, USA

Wyatte L. Harman
Blackland Research and Extension Center, Texas A&M University System Temple, Texas, USA

point sources for soil and water pollution, particularly with respect to phosphorus overloading in local watersheds and streams [1]. For example, annual average phosphorus concentrations in the North Bosque River Watershed, in Texas, USA are limited to between 15 and 50 μg/L [2]. American Agriculture has become an increasingly mechanized and specialized system capable of growing more food on less land. Consequently, many smaller, family oriented farms that tended to grow a variety of crops and keep a diverse population of animals have disappeared. In order to stay competitive, most farmers have had to specialize in one or two crops or keep many of one type of animal. Growing single monocultures and keeping one animal type allows farms to produce more food, more efficiently [3–5].

The volume and concentration of animals, and hence manure, makes proper waste disposal and environmental stewardship challenging and expensive for many American farmers. For farmers who house dairy cows, beef cattle, hogs, chickens, and other traditional farm animals, the amount of manure produced from the hundreds, sometimes thousands, of animals on the farm is a significant undertaking, because manure disposal systems are typically less developed or sophisticated than other focal aspects of the agricultural production system, such as feeding, milking, slaughtering, *etc.* [3]. Large cattle feedlots and dairy operations are a cornerstone of the agricultural economy in Texas and neighboring states in the Southern Great Plains; the manure from these two types of operations will be discussed here. Since animal wastes are similar in chemical composition to the agricultural ration (animal feed: grain, corn, *etc.*) the animal wastes will be termed as biomass. CAFOs show the potential for pollution, yet the concentration of the manure makes this low-energy feedstock a viable source of fuel for combustion and emission control systems.

The main contention of the present study is that electrical and thermal energy production facilities, particularly coal-fired power plants, can benefit from animal wastes as bio-fuel feedstock. The recent increased concern over CO_2 emissions and global warming is only the latest in an unending call to reduce the amount of fossil fuels used for heat and energy and the resulting emissions from fossil fuel power plants. Nitrogen oxides (NO_x), sulfur oxides (SO_x), mercury (Hg), and particulates have all been regulated emissions from coal-fired power plants, and restrictions on these products of combustion will probably continue to increase. Animal waste-based biomass may supplement recent improvements in NO_x emission reductions, as well as reduce non-renewable carbon emissions and Hg emissions.

Theoretically, cattle manure (cattle biomass, CB), like most other solid fuels, including coal, can undergo numerous conversion processes with various reactants, different heat inputs and outputs, different temperature ranges, and a variety of end products such as gaseous bio-fuels, heat energy, thermal commodities (*e.g.*, steam), or electricity. Some of these thermal conversion processes and their end products are illustrated in the triangular diagram in Figure 4.1. Each apex of the triangle represents a chemical element present in the solid fuel. Depending on the process and the reactants (*e.g.*, air, pure oxygen, steam, or hydrogen), the fuel's basic chemical elements will be converted to char, gaseous fuels, or products of combustion. In general these processes include pyrolysis, gasification, and combustion.

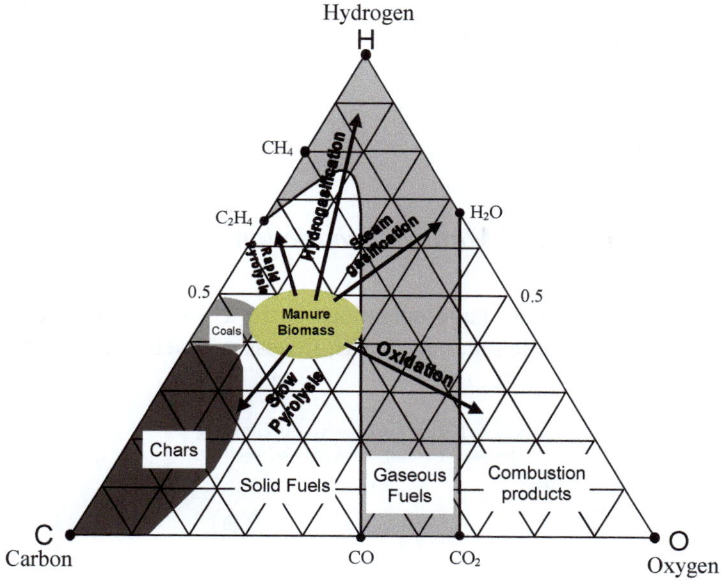

Figure 4.1 Thermal conversion of manure biomass (adapted from Probstein *et al.* [6])

Pyrolysis usually occurs in the absence of oxygen at temperatures below 600 °C (1100 °F) to make gaseous and liquid products as well as carbonaceous chars. Gasification occurs at higher temperatures between 800 and 1100 °C (1470 and 2000 °F) in the presence of one or more reactants such as hydrogen, pure oxygen, air, or air-steam mixtures. Almost all of the char in biomass is converted into gas, leaving gaseous products, ash, and sometimes small quantities of tar. The chemical makeup and heat value of the gaseous products depend on the reactants and the temperature of the process. These gases may be utilized as synthetic gaseous fuels to replace fossil fuels (*e.g.*, natural gas) or product gas for direct firing in heat furnaces. Combustion or oxidation processes occur at temperatures over 1500 °C (2700 °F) [6, 7]. The products generally include CO_2, H_2O, and heat. In complete combustion processes all carbon in the biomass fuel is converted to CO_2. The heat from combustion is typically used to generate thermal commodities such as high temperature gases for gas turbines or steam for thermal energy or, with steam-turbine generators, electricity.

However, CB may not be suitable for all of these thermal conversion processes due to its low heat content, high moisture, and high ash content. The limits of usefulness of CB have presented many challenges during research and experiments, yet, as will be seen in the present review, manure has a tremendous upside as a supplementary fuel and as an emissions controller. This review will cover five general paths that animal waste-based biomass may undergo to generate heat and electrical power while reducing emissions from coal. An exhaustive review of experiments and modeling work will be presented. Moreover, a brief discussion of economic modeling of animal biomass systems will follow. However, first it is necessary to discuss the available quantity of CB in the USA, its fuel properties, and the emissions that may be affected by its use in coal-fired furnaces.

4.2 Background Information

4.2.1 Cattle Populations and Manure Production

The population of feedlot cattle on large operations in the USA is illustrated in Figure 4.2a. The three largest cattle states are Texas, Kansas and Nebraska, respec-

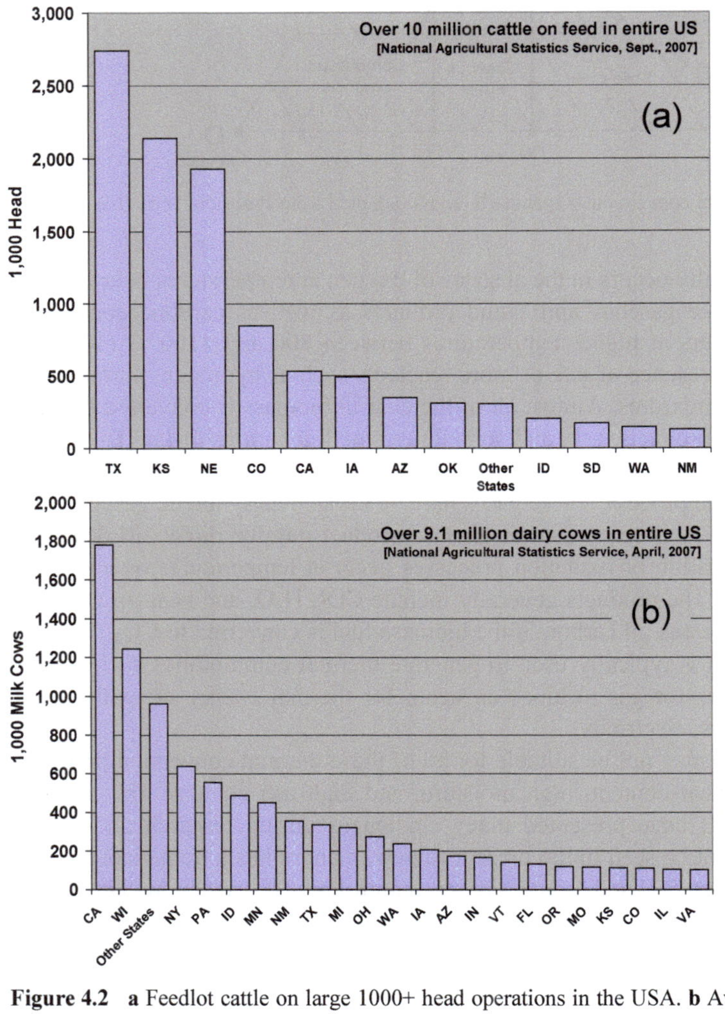

Figure 4.2 **a** Feedlot cattle on large 1000+ head operations in the USA. **b** Average numbers of dairy cows in the USA (not including heifers) for 2006 [8, 13]

tively. These three states produce more feedlot cattle than the other 47 states combined. Most of the Texas feedlots are concentrated in the Panhandle region of the state [8]. Feedlots in the Texas and Oklahoma Panhandle regions can range between 5000 and 75,000 head [9]. Moreover, feedlot cattle can produce 5 to 6 % of their body weight in manure each day; roughly 5.5 dry kg (12 lb) per animal per day [10]. Thus, nearly 20 billion dry kg (22 million tons) of cattle manure per year comes from large feedlot CAFOs. Texas alone produces over 27 % of this annual total.

US dairy operations are not quite as congested as feedlots, although there are certain areas of the country, such as the Bosque River Watershed near Waco, Texas and many parts of California that contain dozens of large dairy operations, each with over 500 milking cows (Figure 4.2b). The dairy cows in the Bosque River Watershed make up about 25 % of the total number of dairy cows in Texas. The Californian counties of Tulare (26 %), Merced (14 %), and Stanislaus (10 %), house about 50 % of the 1.74 million dairy cows in California [11]. Full-grown milking cows can produce 7 to 8 % of their body weight in manure per day; roughly 7.3 dry kg (16 lb) per animal per day [12]. About 24 billion dry kg (26.7 million tons) of dairy manure is produced per year in the USA. Texas dairy cows produce about 890 million dry kg (980,000 tons) of manure per year. The term "cattle biomass (CB)" will refer to both feedlot and dairy manure in general. Manure from feedlots will be termed feedlot biomass (FB) and manure from dairies will be termed dairy biomass (DB).

Figure 4.3 illustrates some of the most common ways DB and FB is disposed of in most animal feeding operations. The possible seepage of manure nutrients to surface and ground water sources has been a major concern. The high cost of transporting manure solids from feeding operations to composting sites and application fields, together with relatively shallow top soil and high intensity rainfall, limit the

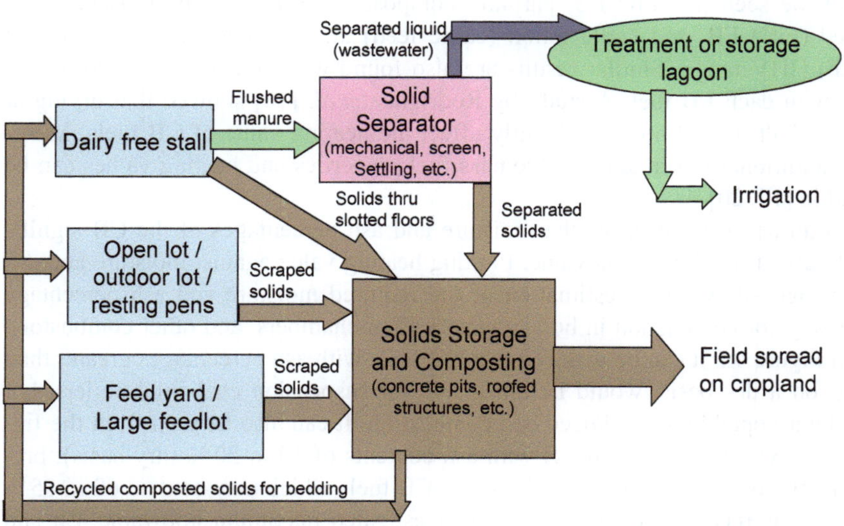

Figure 4.3 Current dairy and feedlot manure disposal (adapted from Schmidt *et al.* [12])

ability to properly distribute the manure. Moreover, the amount of manure to be applied is usually determined by the amount of nitrogen contained in the solids. Sometimes this can lead to an overloading of phosphorus on the land. Only recently have farms begun to switch to P-based land application and composting [1, 4, 14].

Moreover, when the manure gets very dry (<20 % moisture), the cattle's feet grind the dry manure, creating a dust problem. Particulate matter (PM) or dust from feedlot ranges from 8.5 to 12 microns. The total suspended particles (TSP) in feedlot dust can range from 150 μg/m^3 to 400 μg/m^3. Average units have values exceeding 1000±400 μg/m^3 [15]. The PM 10 regulation requires concentration of particles less than 10 μm should be less than 150 μg/m^3. Moreover, when wet and composting manure streams decompose or anaerobically digest in relatively uncontrolled settings, such as poorly maintained manure storage lagoons, methane (CH_4) and malodorous odors can form, reducing the quality of life near the farm [16]. Methane is also a very strong greenhouse gas; about 24 times more harmful than CO_2.

4.2.2 Fuel Properties

Results of the proximate, ultimate, and heat value analyses of the biomass fuels and coals used are presented in Table 4.1. The CB fuels are higher in ash (on a dry basis), lower in heat content, higher in moisture, and higher in nitrogen and sulfur compared to lignite and sub-bituminous coals. Raw manure can be composted in order to generate decomposed, relatively pathogen free, matter for use as a soil conditioner, fertilizer, cattle feed amendment, or free stall bedding. Generally, the higher heating value (HHV) of CB on a dry ash free basis tends to be between 18,000 and 22,000 kJ/kg depending on the animal's feed ration [17]. In Figure 4.4, it may be seen that raw FB, partially composted (PC) FB, fully/finished composted (FiC) FB, and cattle ration (cattle feed) all fall under this dry, ash-free (DAF) HHV range. Similar results are also found when blending 5 % crop residues with each FB fuel. A study by Rodriguez *et al.* [18] showed that drying at 400 °C (750 °F) did not significantly affect the heating value of CB fuels. Moreover, additional information of biomass fuel properties and heating values can be found in [19, 20].

On an as received basis, the moisture and ash percentages of the CB significantly affect the fuel's heat value. Plotting heating value against moisture and ash percentage allows some estimation of the required moisture and ash percentage necessary for combustion in boilers, gasification chambers, and other combustors. From Figure 4.5 it can be deduced that CB fuels with ash percentages greater than 40 %, on a dry basis, would be unsuitable for suspension coal-fired boilers, but may be acceptable for fluidized bed combustion. It can also be seen from the figure that, even for lower ash CB with ash contents of 10 to 20 % (dry basis); predrying processes may be required for raw CB fuels which may contain 60 to 85 % moisture as received. Carlin [21] also investigated maximum allowable ash and moisture contents of DB for small-scale, on-the-farm combustion schemes.

As part of determining coal and biomass fuel properties, a thermal gravimetric analyzer/differential scanning calorimeter (TGA/DSC) instrument provided information on the moisture and combustible volatile matter released during pyrolysis and early gasification processes. The tests were performed using a TA Q600 TGA/DSC. The purge gases typically used are N_2 for pure pyrolysis and air for oxidation reactions. The instrument monitored temperature, particle mass, and heat flow versus time. The pyrolysis temperature was found to be approximately 350 °C (650 °F) and the ignition temperature about 320 °C (620 °F) for Texas lignite. Pyrolysis temperatures were generally 60 to 70 °C less for pure CB fuels.

Table 4.1 Fuel compositions of DB, FB, Texas lignite, and Wyoming sub-bituminous on an as received and dry, ash-free basis

	DB separated solids[1]	DB vacuumed solids[2]	Low-ash FB[1]	High-ash FB[1]	Texas Lignite[3]	WY Sub-bituminous[3]
	As received					
%Moisture[4]	25.26	83.63	29.25	27.31	38.34	32.88
%Ash	14.93	3.57	9.61	32.88	11.46	5.64
%Fixed Carbon	13.00	--	12.94	7.31	25.41	32.99
%Volatile Matter	46.88	--	47.97	32.50	24.79	28.49
%Carbon	35.20	6.80	35.11	23.51	37.18	46.52
%Hydrogen	3.12	0.83	4.17	2.80	2.12	2.73
%Nitrogen	1.93	0.42	2.37	1.68	0.68	0.66
%Oxygen	19.15	4.68	19.11	11.51	9.61	11.29
%Sulfur	0.43	0.07	0.38	0.31	0.61	0.27
HHV (kJ/kg)	12,817	2,733	13,195	8,172	14,290	18,194
	Dry ash free basis					
%Moisture	0.00	0.00	0.00	0.00	0.00	0.00
%Ash	0.00	0.00	0.00	0.00	0.00	0.00
%Fixed Carbon	21.74	--	21.16	18.36	50.62	53.66
%Volatile Matter	78.38	--	78.46	81.64	49.38	46.34
%Carbon	58.85	53.13	57.43	59.06	74.06	75.67
%Hydrogen	5.22	6.48	6.82	7.03	4.22	4.44
%Nitrogen	3.23	3.28	3.88	4.22	1.35	1.07
%Oxygen	32.02	36.56	31.26	28.91	19.14	18.37
%Sulfur	0.72	0.55	0.62	0.78	1.22	0.44
HHV (kJ/kg)	21,429	21,352	21,581	20,529	28,467	29,594
Emperical formulae	$CH_{1.06}N_{0.05}O_{0.41}S_{0.003}$	$CH_{1.44}N_{0.05}O_{0.51}S_{0.004}$	$CH_{1.42}N_{0.06}O_{0.41}S_{0.003}$	$CH_{1.42}N_{0.06}O_{0.37}S_{0.005}$	$CH_{0.68}N_{0.02}O_{0.19}S_{0.006}$	$CH_{0.70}N_{0.01}O_{0.18}S_{0.003}$

[1]Adopted from Sweeten *et al.* [88]; compost bedding used in the dairy freestalls; low-ash FB from fly-ash paved feed pens; high-ash FB from soil surface feed pens.

[2]Obtained from Mr. Barry Goodrich, Dept. of Bio. and Ag. Engineering, Texas A&M University; compost bedding used in the dairy freestalls.

[3]Adopted from TAMU Coal and Biomass Energy Lab website [25].

[4]Moisture in DB separated solids is low due to solar drying prior to fuel analysis, typically 80 % moisture before drying

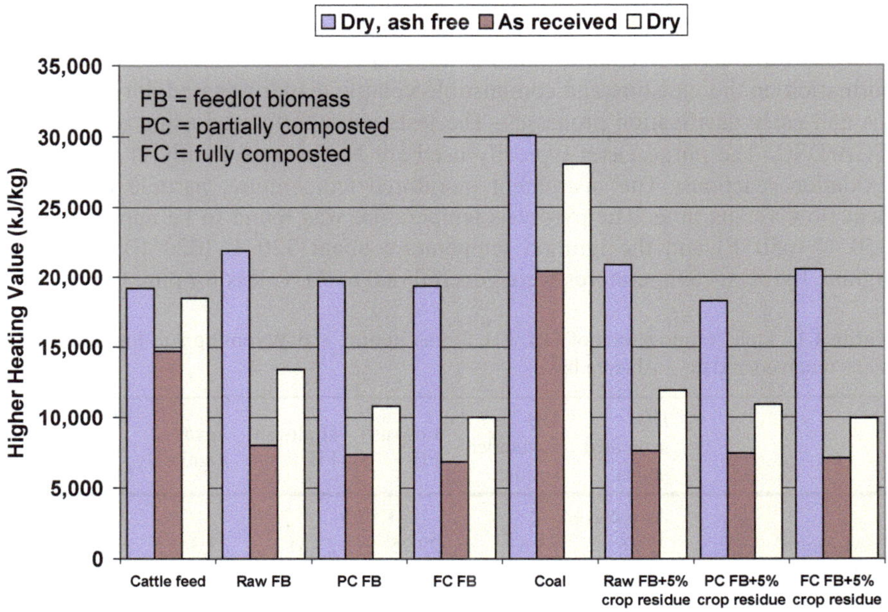

Figure 4.4 Higher heating values for cattle ration, raw FB, partially composted FB, finished composted FB, coal, and respective FB+5 % crop residue blends (adopted from [17])

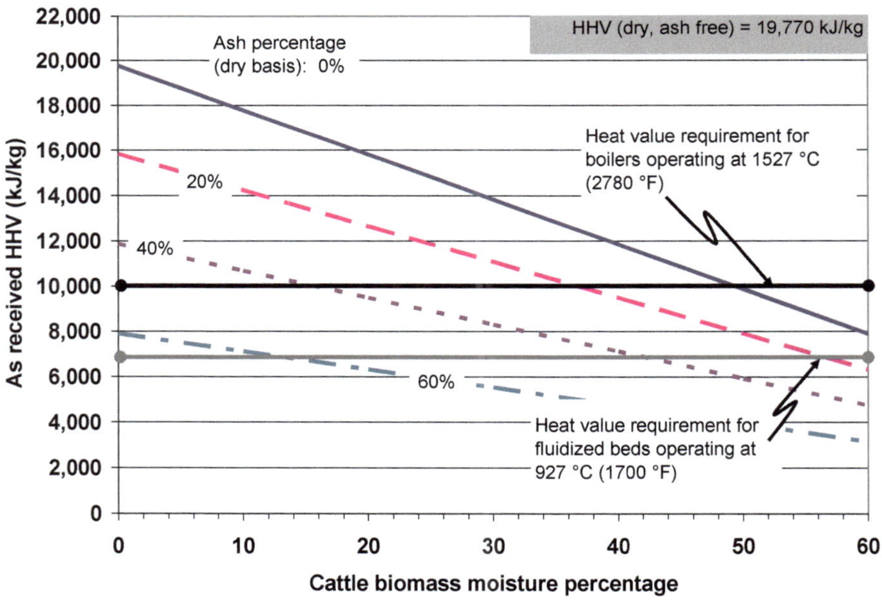

Figure 4.5 Higher heating value of cattle biomass versus moisture and ash percentage (assuming a dry, ash-free HHV of 19,770 kJ/kg)

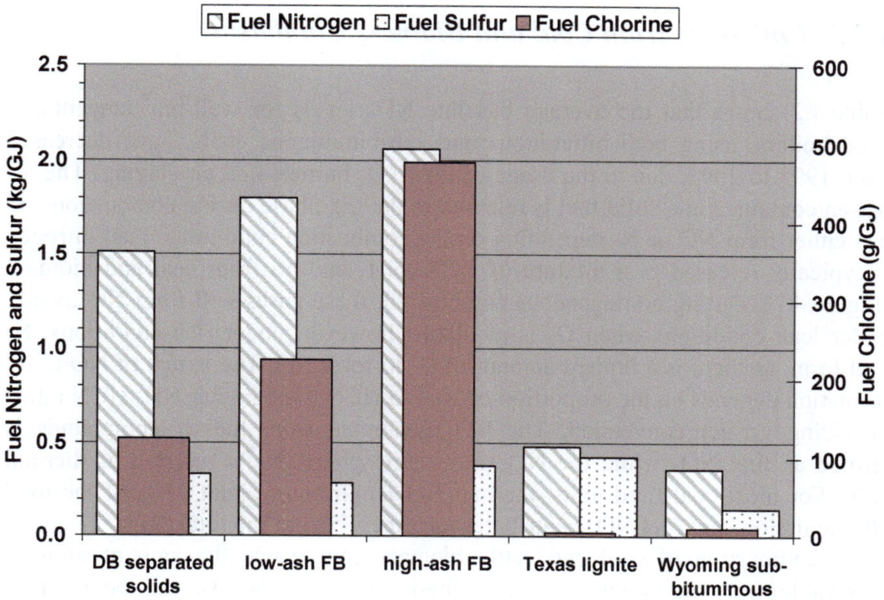

Figure 4.6 Nitrogen, sulfur, and chlorine contents of DB, FB, Texas lignite, and Wyoming sub-bituminous (adapted from TAMU [25] and Arcot Vijayasarathy [26])

Previous TGA experiments by Martin *et al.* [22] found similar results for FB. Pure biomass samples had an average ignition temperature of 474 °C (885 °F), while biomass blends with Texas lignite coal had an average ignition temperature of 292 °C (560 °F). The lower ignition temperatures in coal-CB blends are generally due to the high amount of fixed carbon in coal that is not present in pure biomass fuels. Ignition temperature did not vary appreciably between high ash FB and low ash FB. Nor did it vary significantly with average particle size or coal:FB blend ratio. Additional TGA analysis of FB pyrolysis is also provided by [23].

Finally, in Figure 4.6, the nitrogen, sulfur, and chlorine contents of DB, FB, lignite, and sub-bituminous coal are compared on an energy basis. Nitrogen contents of DB and FB are about 2 to 3 times those of lignite and sub-bituminous coal, which suggest higher NO_x emissions if the biomass fuels are burned under fuel lean conditions. Sulfur contents of DB and FB seem to be slightly lower than that of lignite; however, Wyoming sub-bituminous has the lowest sulfur content of all. Lower SO_2 emissions allowances, under the American Clean Air Interstate Rule, have recently increased the demand for this low sulfur sub-bituminous coal (mined mostly out of the Powder River Basin) [24]. Moreover, chlorine content of DB is much greater than either lignite or sub-bituminous coal, which suggests the possibility of higher Hg oxidation during coal and biomass co-combustion.

4.2.3 Emissions from Coal and Biomass Combustion

Table 4.2 shows that the average baseline NO_x levels for wall and tangentially fired boilers, using both bituminous and sub-bituminous coals, have decreased from 1995 to 2003, due to the usage of low-NO_x burners and air staging. The nitrogen contained in a solid fuel is released to the gas phase during combustion and can either form NO or N_2 depending on the combustion conditions. Fuel nitrogen is typically released as a mixture of HCN, NH_3 and N_2 from coal and biomass (Figure 4.7). During homogeneous combustion, these gases will form NO mostly under lean conditions when O_2 is available; however, under rich conditions, N_2 will form, as there is a limited amount of O_2 to form NO. The extent of NO to N_2 formation depends on the proportion of NH_3 to HCN, with higher NH_3/HCN ratios providing greater conversion. The NO_x generated from fuel N compounds is termed as fuel NO_x while the NO_x from atmospheric N_2 is referred as thermal NO_x. For most coal-fired units, thermal NO_x contributes about 25 % of the total NO_x emission, and fuel NO_x contributes the other 75 % of the total [27].

Co-combustion of coal and cattle biomass can reduce the amount of non-renewable CO_2 emissions from coal-fired power plants. As can be seen in Figure 4.8, coal combustion accounted for 36 % of the roughly 5.9 billion metric tons of CO_2 released by anthropogenic sources in the USA during 2006. Of course, burning coal with cattle biomass will not reduce CO_2 emissions to accept-

Table 4.2 NOx reduction performance of primary control technology applications on coal-fired boilers (adopted from Srivastava *et al.* [28])

Boiler type*	Coal type	Primary control technology	2003 Average controlled NO_x emission (g/GJ)	Average NO_x reduction efficiency from 1995 levels (%)	Range of NO_x Reduction efficiencies (%)	Number of boilers
Wall-fired	Bituminous	LNB	177	39.2	8.6–70.1	62
Wall-fired	Bituminous	LNBO	151	53.3	32.7–71.9	16
Wall-fired	Sub-bituminous	LNB	121	45.5	19.4–80.3	16
Wall-fired	Sub-bituminous	LNBO	60	63.4	40.0–80.9	4
Tangential-fired	Bituminous	LNC1	168	35.0	17.2–65.4	26
Tangential-fired	Bituminous	LNC2	134	36.6	23.3–70.8	15
Tangential-fired	Bituminous	LNC3	108	54.9	38.1–72.2	19
Tangential-fired	Sub-bituminous	LNC1	90	45.4	11.3–74.4	18
Tangential-fired	Sub-bituminous	LNC2	99	45.6	33.9–65.4	3
Tangential-fired	Sub-bituminous	LNC3	60	60.5	48.2–77.2	23

Notes: LNB = low-NO_x burner; LNBO = LNB with over fire air; LNC1 = LNB with close-coupled OFA; LNC2 = LNB with separated OFA; and LNC3 = LNB with both close-coupled and separated OFA.

*All boilers are dry-bottom type.

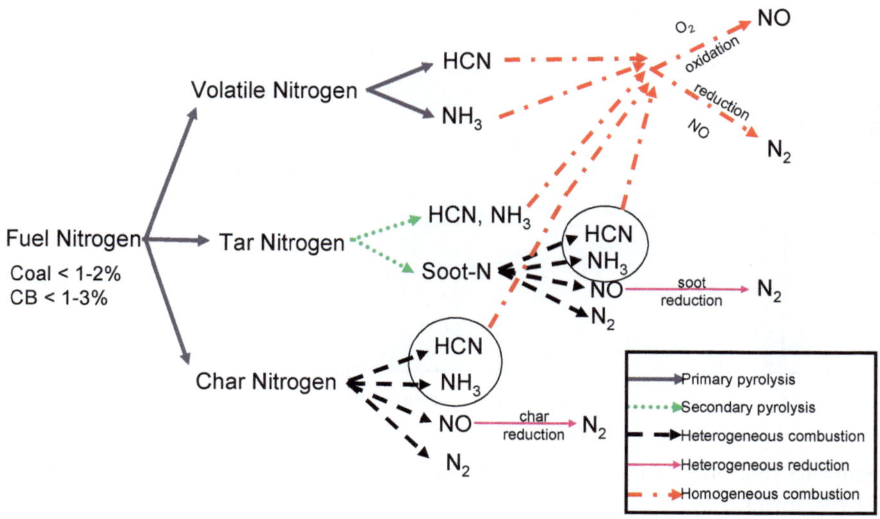

Figure 4.7 Fuel nitrogen paths to NO and N2 (adapted from Di Nola [30])

able levels alone; however, as discussed by Pacala *et al.* [29], biomass combustion can be one of many wedges of development in alternative technologies that can create an energy economy capable of sustaining our climate and our way of life (Figure 4.9). Most of these technologies, such as nuclear power, solar energy, and bio-fuel combustion, are already well understood, but they still need to be implemented into our current energy production systems.

Carbon emissions from a fuel are directly related to the hydrogen-carbon (H/C) ratio and the oxygen–carbon (O/C) ratio, as can be seen in Figure 4.10. Note that this figure shows total CO_2 emission (both for renewable and non-renewable fuels). Plotting CO_2 emissions in this way can provide an estimation of carbon released during synthetic gas (*e.g.*, methane) and value added liquid fuel (*e.g.*, ethanol) processing. For instance, CO_2 released by the generation of ethanol from corn

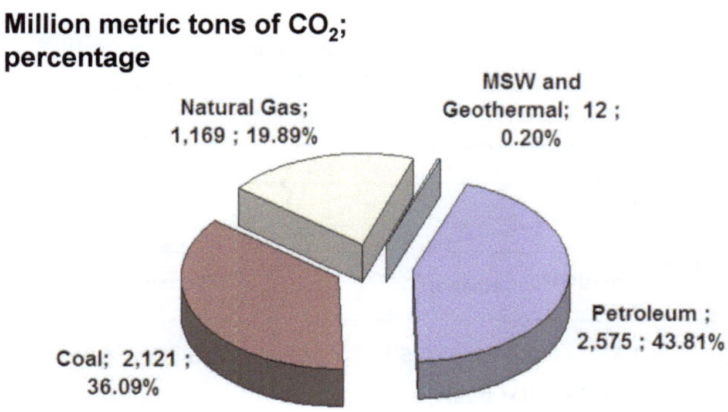

Figure 4.8 CO_2 emissions from various fuels in the USA (estimated 2006 emissions) [31]

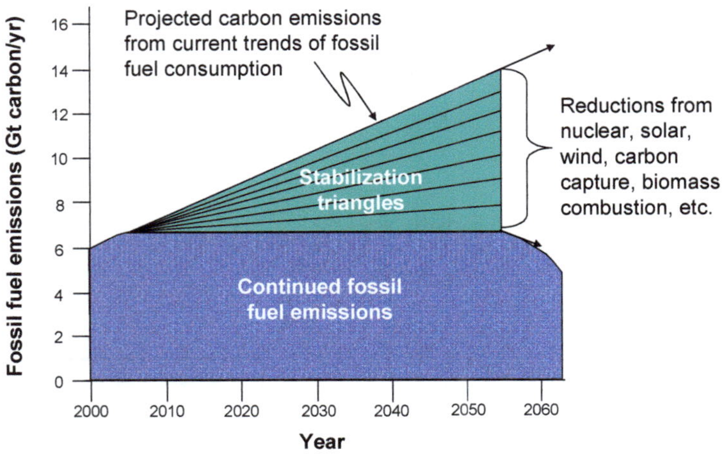

Figure 4.9 Stabilization triangle of avoided emissions (top) and allowed emissions (bottom) (adapted from Pacala *et al.*, 2004 [29])

or grain through fermentation is indicated on Figure 4.10. This difference of CO_2 emission between corn (approximately the same as coal) and ethanol must be accounted for when determining carbon savings and footprints. Ideally, this difference in CO_2 emission should also come from a renewable source when generating value added liquids and gases.

The drive for cleaner air has also caused an increased concern for control of toxic metal emissions from coal combustion systems. In particular, mercury has

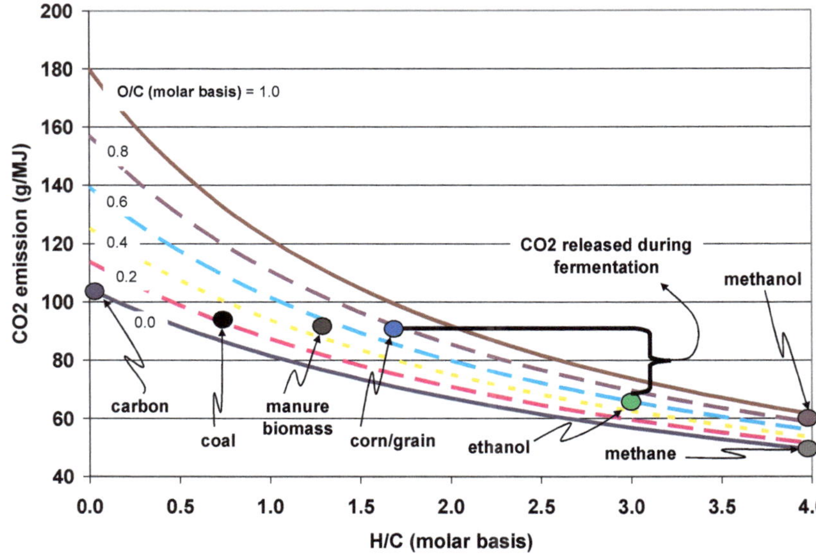

Figure 4.10 CO_2 emission versus O/C and H/C ratios, with various fuels indicated

been targeted for control; due to its unique characteristics such as high volatility, bio-accumulation and other toxic properties. To date, there are no post-combustion treatments that can effectively capture elemental mercury vapor. The Environmental Protection Agency (EPA) has released a Clean Air Mercury Rule, which caps the mercury emission from coal-fired power plants from a current rate of 158 tons per year to 15 tons per year by 2018. Elemental mercury (Hg^0), due to its volatile nature, exists in vapor phase in the flue gases which escapes into the atmosphere without being captured in any environmental emission capturing devices currently available, while an oxidized form of mercury (Hg^{2+}, *e.g.*, $HgCl_2$) can be captured in commonly used flue gas desulphurization (FGD) units, since oxidized mercury is soluble in water. Moreover, particulate mercury (Hg_p), which is found in fly ash, can be captured in bag houses and electrostatic precipitators (ESP) (Figure 4.11). The objective of this aspect of the research is to use cattle biomass and coal/biomass blends as fuel to effectively convert Hg^0 to its oxidized form which can be captured more easily when using traditional environmental devices. The high chlorine content of CB, particularly DB, may allow more Hg^0 to be converted to Hg^{2+}.

Yet, due to the higher amount of fuel-bound nitrogen in CB fuels, it is critical to determine what happens to this nitrogen during combustion. Otherwise, any reductions in CO_2 and Hg may be overruled by higher NO_x emissions. In following sections of this chapter, CB co-firing, reburning, and gasification experiments will be reviewed along with an overview on the economics of utilizing CB in existing coal-fired power plants. This review includes most of the work conducted at the Coal and Biomass Laboratory of Texas A&M University as well as limited literature review. There will also be a brief review of biological conversion processes for high moisture manure and small scale, onsite combustion systems for direct manure-waste disposal.

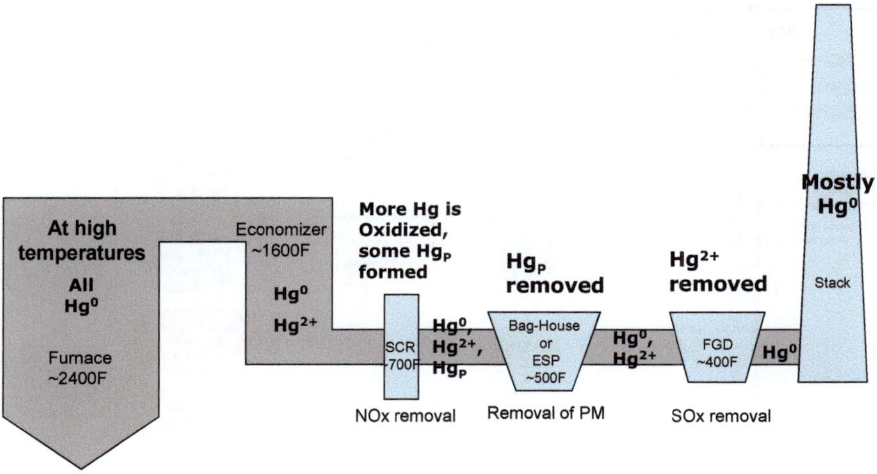

Figure 4.11 Mercury reduction co-benefits from secondary combustion controls (adapted from Arcot Vijayasarathy [26])

4.3 Energy Conversion of Cattle Biomass

Five different paths for energy production from CB are illustrated in Figure 4.12. From the high moisture CB exiting dairy free stalls, open lots, and feed yards, the remaining liquid stream from mechanical (screen) separation can be sent through an anaerobic digestion system, which will be referred to as Path 1. Anaerobic digestion allows for methane capture from natural biological decomposition of manure wastes. The resulting methane/CO_2 gas (usually termed biogas) can then be burned in gas turbines, IC engines, or heat furnaces for energy production.

The remaining manure solids from the mechanical separation can be dried, ground, and either gasified or fully combusted. Path 2 in the present discussion will be thermal gasification of CB either with air or air-steam mixtures to produce low to medium-calorific value gases which can then be utilized in a variety of different combustion processes. Moreover, gasification allows the usage of coarser, higher ash, and higher moisture CB solids. Path 3 will be burning coal in a primary burn region and then reburning the coal with CB in a secondary or re-burn zone of an existing coal-fired power plant. The primary purpose of reburning is to reduce NO_x emissions. Path 4 will be co-firing coal with CB in the primary burn region of a coal-fired power plant. In Paths 3 and 4 CB is utilized as a supplementation to coal, which can reduce the amount of non-renewable CO_2 emissions from large coal plants; however, both of these applications generally require drying and fine grinding of both coal and CB fuels. Finally, Path 5 is the direct firing or incineration of manure wastes on the farm. The primary purpose of Path 5

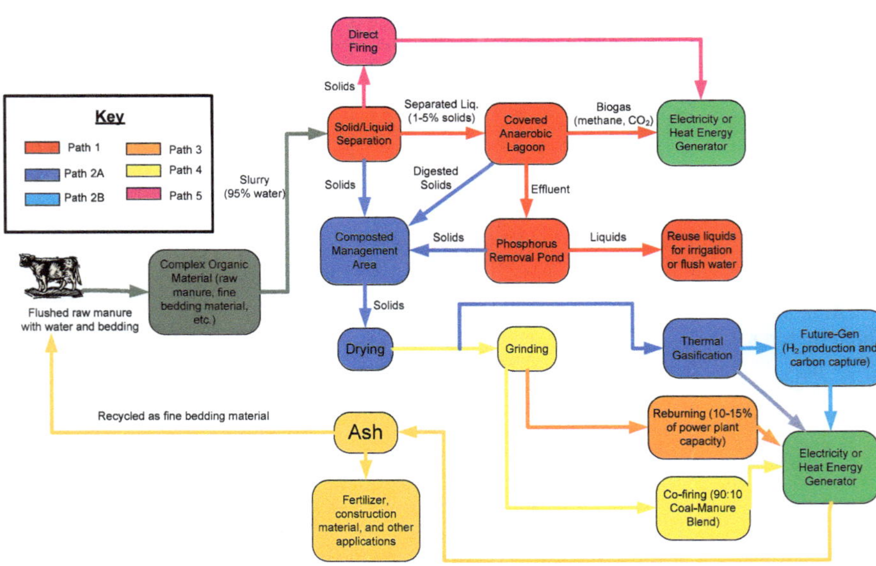

Figure 4.12 Five paths to heat and electrical energy production from CB (adapted from Annamalai *et al.* [32])

is to dispose of manure while reducing the need of transporting large quantities of raw manure off the farm or land application of unprocessed solids as fertilizer. Electrical or heat energy is an additional benefit of Path 5 systems.

In this section, there will first be a brief discussion of anaerobic digestion. Anaerobic digestion is the most commercially available energy conversion system involving cattle biomass. Investigators at Texas A&M have studied Paths 2 through 5 extensively for research and development purposes. Results from experimental research on gasification, reburning, and co-firing (Paths 2, 3, and 4, respectively) are presented in order to show how CB can reduce coal consumption, NO_x emissions, and Hg emissions. The economic viability of installing a co-combustion system on an existing coal plant will also be discussed. Finally, computational models and system designs for Path 5 (on-site, direct firing) will also be reviewed.

4.3.1 Biological Gasification of Cattle Biomass Through Anaerobic Digestion

To date, most of the research on energy conversion systems that involve high moisture and/or high ash animal biomass have dealt with capturing methane (CH_4) or biogas (mixture of CH_4, CO_2, and other trace gases) biologically produced from anaerobic digesters. The biological conversion of CB to biogas can occur anaerobically (absence of oxygen) in three different steps. First hydrogen producing acetogenic bacteria consume organic acids to produce hydrogen, CO_2, formate, and acetate. Secondly, homoacetogenic bacteria form more acetate from H_2, CO_2, and formate. Finally, methanogenic bacteria produce CH_4. There may be an initial hydrolysis step to break down lignocellulosic material, but this step is primarily used when digesting crop residues. In contrast to plant or crop-based biomass, manure typically does not contain enough lignin to make an initial hydrolysis step cost effective [6].

Ideally all three steps should occur in separate reactors; however, as shown in Figure 4.13, manure is usually first sent to an equalization pond so that a mixture of biomass, water and nutrients can form a homogeneous substrate before entering a digester. Sometimes the digester will be divided into two stages, with mechanical mixing occurring in the first stage, although mixing may be found not to be energy efficient. Wen *et al.* [33] found that treating the manure before sending it to the digester, and thereby increasing its organic strength, had more of an effect on biogas production and organic matter removal than mixing. Plug flow reactors and completely mixed reactors performed very similarly when the manure was treated before entering the digester. It was also found that liquid manure from storage lagoons did not produce as much biogas as liquid manure created from as-excreted manure plus fresh tap water (1:4 and 1:2, by weight, mixtures of manure and tap water were investigated). Reduction of total solids in the manure varied between 27 and 48 % and reduction of volatile solids varied between 26 and 47 %, with higher reductions being found for the 1:2 mixture of excreted manure to tap water.

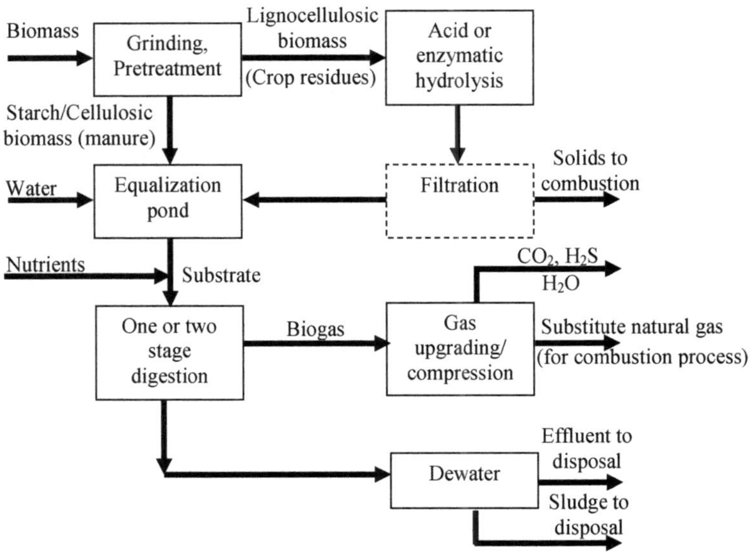

Figure 4.13 Simplified anaerobic digestion flow diagram (adopted from Probstein *et al.* [6])

Depending on whether the desired end product is a low, medium, or high calorific value gas, the biogas can be upgraded by removing undesirable contaminants, such as hydrogen sulfide (H_2S), and neutral gases, such as CO_2 and water vapor. Generally, both of these products must be removed to various extents in order to provide an adequate natural gas substitute. Hydrogen sulfide and CO_2 are usually removed from a product gas through liquid absorption [34]. However, if the biogas is to be burned in a conventional, industrial IC engine for electric power, on or very near the animal feeding operation, usually only particulate matter and water vapor removal are required.

The bacteria involved in anaerobic digestion can also be divided into the temperature range in which they thrive. Psychophilic bacteria thrive at near ambient temperatures (25 °C or 77 °F), mesophyllic dominate at about 35 °C (95 °F) and thermophyllic bacteria dominate at higher temperatures of 57 °C (135 °F). The selection of operating temperature determines which bacteria group will thrive, and hence also determines the percentage of CH_4 in the biogas, the conversion rates, the residence time of the substrate in the digester, and the overall cost of the system [6]. Typically biogas can contain between 55 and 70 % CH_4 and 30 to 45 % CO_2; although, there have been reports of biogases with as much as 90 % CH_4, even without upgrading [35].

Nutrients such as nitrogen, phosphorus, and alkali metals must be present for anaerobic bacteria to survive, thus manure, which contains all of these, is an ideal feedstock for these systems. Yet, very high ammonia (NH_3) concentrations may be toxic. Therefore, the carbon to nitrogen ratio in animal biomass is critical. Dairy manure has higher than average C/N ratios, approximately 20, as can be derived from data in Table 4.1. Higher C/N ratios make retention times relatively short.

However, in batch processes, there is a limit to C/N ratio in which anaerobic digestion will stop. Nitrogen concentrations are usually higher in the effluent after digestion, making the effluent an even more valuable fertilizer [6, 33, 36].

The percentage of CH_4 may be reasonably predicted using atom conservation equations for the reaction between digestible solids and H_2O. Krich *et al.* [35] and Probstein *et al.* [6] presented this atom balance generalized as cellulose as follows:

$$(-C_6H_{10}O_5-)+H_2O\xrightarrow{bacteria}3CO_2+3CH_4 \qquad (4.1)$$

From this equation it can be deduced that an ideal methane production of 0.3 kg CH_4 per kg of biomass may be expected. Krich *et al.* [35] also conducted atom balances for wastes containing proteins ($C_{10}H_{20}O_6N_2$, which produces a CH_4:CO_2 ratio of about 55:45) and fats or triglycerides ($C_{54}H_{106}O_6$, which produces a ratio of about 70:30). The actual CH_4:CO_2 ratio produced from manure, however, depends on a number of other factors such as temperature in the digester, residence time, pretreatment of the substrate, *etc.*

Moreover, Carlin [21] and Annamalai *et al.* [19] conducted atom balances from ultimate analysis of manure biomass with the following chemical balance equation:

$$\begin{aligned}&\mathbf{CH}_h\mathbf{N}_n\mathbf{O}_o\mathbf{S}_s+N'_{H2O}H_2O_{(l)}\\&\xrightarrow{bacteria}N'_{CH4}CH_{4(g)}+N'_{CO2}CO_{2(g)}+N_nS_{s(solid)}\end{aligned} \qquad (4.2)$$

From here, methane concentrations and higher heating values of the resultant biogas can be estimated in terms of the C/N and C/O ratios (see Figure 4.14a and b). Also, note that the water in these reaction equations is very small compared to the amount of water entering the digester. The effluent contains nearly all of the water that was present in the substrate. The remaining material in the effluent can sometimes be further processed to make fertilizer. Effluent can also be recycled back to make more substrate, however, this is limited because toxins may build up in the system and harm the methane producing bacteria. The pH level of the digester should be maintained between 6.6 and 7.6 [6, 36, 37]. An overall mass and energy balance for CB digestion was presented by Probstein *et al.* [6], based on results found by Chen, *et al.* [38], and is summarized in Table 4.3.

However, despite being the most commercial energy conversion systems for manure biomass, digestion systems installed on animal feeding operations are not very common in the USA. There were only 41 operational systems in this country as of November 2007; 29 of those systems were installed on dairy farms, 10 were installed on swine farms, one on a duck farm and one on a chicken farm [39]. However, near Stephenville, Texas, the largest manure-to-natural gas plant in the USA has recently been completed. The plant obtains manure from local dairies in Erath County and mixes it with restaurant grease and other wastes to produce biogas. The biogas is then upgraded to industry standards as a natural gas replacement fuel. It is expected that the natural gas from this plant can produce enough energy to power 11,000 homes [40].

Goodrich *et al.* [41] studied the overall biogas production and economics of a digester system installed on an 800-cow dairy in Princeton, Minnesota. In this example, the dairy has profited greatly from the production of biogas, using it to meet most of the electrical energy needs of the farm and either selling unused electricity to the utility or flaring excess biogas produced in the digester that cannot be consumed by the engine. Goodrich *et al.* [41] do admit that most dairies may not be able to profit on the installation of an anaerobic digester, as the installation and operation & maintenance costs may be too overwhelming and feasibility is very "site specific". Long operation times and biodegradable bedding (in-

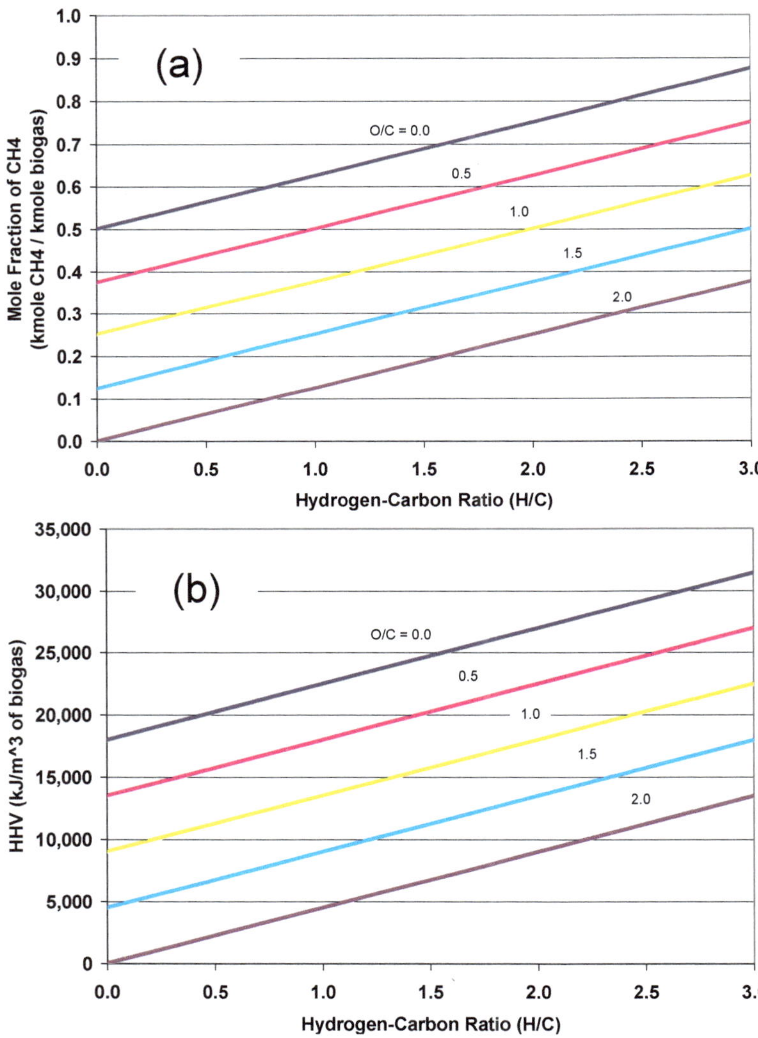

Figure 4.14 **a** Mole fraction of methane in biogas versus H/C and O/C ratios in flushed DB. **b** HHV of biogas versus H/C and O/C ratios in flushed DB (adopted from Carlin [21])

stead of sand) are also necessities to make manure digesters profitable. However, some digesters are not installed to obtain an economic profit. Rather, they are installed to control odors and to reduce pathogens from manure storage. Relatively inexpensive digesters can be installed by simply covering a storage lagoon with a tarp and placing a flare to burn the biogas. With these systems, no economic return is expected from electricity sales; although small boilers may be installed with the flare to provide heat to buildings at or near the farm. The USEPA [39] also provides estimated installation costs for all current, on-the-farm operating systems, which vary from $ 15,000 to $ 1.4 million, depending on what kind of digester is installed and whether or not an engine-generator set is purchased.

Further discussions of biological energy conversion of manure-based biomass solids can be found in the literature. These include, but are not limited to: Meyer [42], who discussed a digester on a 700-cow dairy in Iowa; Chang [43], who compared the efficiencies and sustainability of anaerobic digesters to non-biological, thermal gasification; Simons [44], who reviewed some of the initiatives for digester installations in California; Ghaly [37], who compared the anaerobic diges-

Table 4.3 Material and thermal balance for anaerobic digestion of cattle manure (adopted from Probstein *et al.* [6])

	Mass, kg	Heat Content, MJ/kg	Heat, MJ	Heat, % of total
IN				
Cattle manure (dry)	100*	13.4	1340	84.7
Dilution water (10 % solids)	900	--	--	--
Substrate heating**	--	--	157	9.9
Mixing energy	--	--	55	3.5
Gas scrubbing	--	--	21	1.3
Methane compression (to 1 MPa)	--	--	9	0.6
Total	1000		1582	100.0
OUT				
Methane (21 m^3)	15	55.6	834	52.7
Carbon dioxide	41	--	--	--
Moisture in gas***	6	--	--	--
Effluent water	888	--	--	--
Sludge (dry) and losses (by difference)	50	--	748	47.3
Total	1000		1582	100.0

$$\textit{Thermal efficiency} = \frac{834}{1582} * 100 = 53\%$$

*About 22 bovine-days worth.
**Assuming 50 % heat recovered from the effluent.
***Assuming gas saturated at 55 deg C and 101 kPa.

tion of cheese whey and dairy manure in a two-stage reactor; and Monnet [36], who provides a general review of digester types and operating conditions.

4.3.2 Non-biological Gasification of Cattle Biomass

Extensive literature exists on technologies used to gasify coals. These systems are very briefly summarized by Hotchkiss [45], including fixed-bed, fluidized bed, and entrained gasifiers. More detailed discussions of coal gasification are also available by Howard-Smith [46], Nowacki [47], and Probstein *et al.* [7]. Moreover, a review of common gasification technologies for wood and crop-based biomass is provided by Quaak *et al.* [48].

However, there have been limited studies on non-biological, thermal gasification of high moisture, high ash manure biomass. An advanced gasification system discussed by Young *et al.* [49] proposed the use of separated DB solids pressed to 70 % moisture from an auger press in a high temperature, air blown gasifier to produce synthetic gas. Gasification conversion efficiencies can range between 65 and 81 %, and the gasification temperature was assumed to be about 1300 °C (2370 °F). The product gas composition was found to be approximately 30.2 % (molar) CO, 5.5 % CO_2, 25.7 % H_2, and 38.6 % N_2, with a heating value of 7,140 kJ/kg (3,076 Btu/lb). The gas could then be fired in an IC engine to generate electrical power. The dairy would be able to produce twice its electrical energy requirement from the synthetic gas.

There is also a prototype system developed by Skill Associates, Inc. called Eli-manure™ that can eliminate both the liquid and solids of any animal manure. Waste manure up to 95 % moisture enters large drying units and is mixed by large augers with hot air. The temperature in the drying units reaches 71 C (160 F) and the manure is dried to about 40 % moisture. The water vapor is ventilated out of the drying unit, while the 40 % moisture solid manure is sent to a thermal gasification boiler where it is burned at 1090 °C (2000 °F). The boiler generates steam which runs turbines to generate electricity. During the first two hours of operation, the system uses propane or some other fuel to start up, but after that, the dried manure can sustain the process. Besides water vapor from the drying process, the only byproduct is a grey powdery ash which contains the inorganic or non-combustible material in the manure. The facility was constructed in Greenleaf, Wisconsin in 2005 [50].

Experiments are being performed by investigators at Texas A&M University on gasification of coal and CB. Furthermore, modeling results on gasification of FB with air and air-steam mixtures as oxidizing agents have also been generated [51, 52].

There has also been extensive work and application on thermal gasification of CB in fluidized bed combustors. Sweeten *et al.* [53] conducted several experiments on a 0.305 m (1.0 ft) diameter pilot plant fluidized bed combustor. It was found that when bed and vapor space temperatures exceeded 788 °C and 927 °C (1450 °F and 1700 °F), respectively, severe ash agglomeration and plugging occurred in the vapor space transition duct and hot cyclone. However, ash agglomeration, plugging and slagging could be avoided if the combustor was operated on a re-circulating bed mode, which lowered vapor space temperature to 649 °C (1200 °F). Annamalai *et al.* [54] found that on a 0.15 m (0.5 ft) diameter fluidized bed combustor with bed temperatures ranging from 600 °C to 800 °C (1112 °F to 1472 °F), gasification efficiencies ranged from 90 to 98 % and combustion efficiencies ranged from 45 to 85 %. Moreover, Raman *et al.* [55] developed a mathematical model to describe FB gasification in fluidized bed combustors. It was assumed that initial volatilization of the biomass fuel was nearly instantaneous and only secondary reactions including char gasification and water–gas shift reaction were modeled. The resulting calculations suggested that the water–gas shift reaction was the most dominant reaction. However, when comparing computational results to experimental data, it was found that reactions involving volatiles from devolatilization (pyrolysis) should be included.

The applications of CB gasification have proven to be beneficial as a renewable source of energy. For example, Panda Ethanol, Inc., based out of Dallas, Texas, has recently completed a bubbling fluidized bed gasification plant near Hereford, Texas, which will convert FB from local feedlot operations to synthetic gas. The gases can then be burned in a combustor to produce steam, which is necessary to process corn into ethanol (see Figure 4.15 for an overview of this

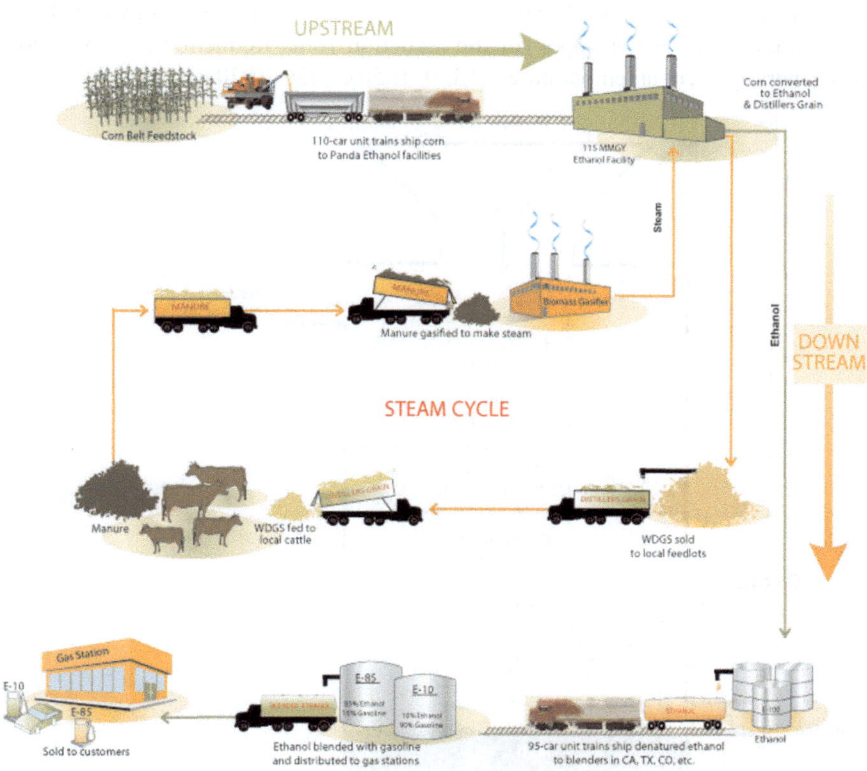

Figure 4.15 Cattle manure gasification for corn ethanol production [56]

system). A by-product of the ethanol production process is wet-distiller grains, which can be used as an additive in cattle feed ration. The bubbling sand bed is operated at low temperatures to avoid ash agglomeration and fouling [56]. The disposal of ash from manure gasification has proven to be one of the more challenging aspects because unlike ash from coal combustion, manure ash is not suitable as a replacement for Portland cement or as aggregate in Portland cement concrete. However, there are alternative uses for manure ash such as road base, flowable fills, and soil amendments [57].

4.3.2.1 Modeling Gasification

For fixed bed gasification, the products from thermal gasification are emitted at different zones in the gasifier's chamber. The different zones that the CB, or any solid fuel, encounters during the gasification process, as well as the main expected product gas species from coal and biomass gasification, are listed in Figure 4.16. In the drying zone, with temperatures of 25–130 C (78–265 F), the fuel's moisture is vaporized. In the pyrolysis section, 130–330 C (265–630 F), the fuel is broken down into volatile gases and solid char. The char, carbon dioxide, and water vapor undergo reactions in the reduction zone in which carbon monoxide (CO) and hydrogen (H_2) are produced. The remaining char is combusted in the oxidation zone providing heat, carbon dioxide, and water vapor for the reduction zone. Reduction and oxidation zones generally occur at temperatures of 330–1050 C (630–1920 F).

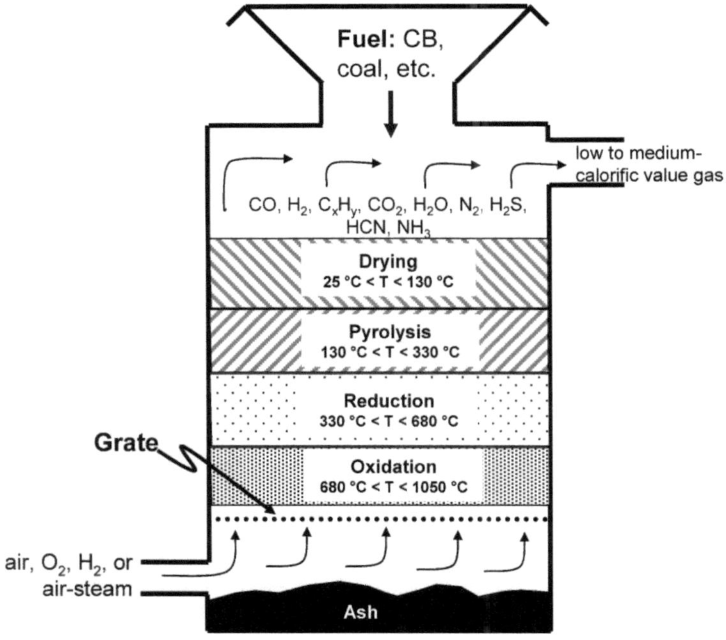

Figure 4.16 Different zones in an updraft, fixed-bed gasifier (adapted from [58])

The significant global reactions typically used to simplify modeling of gasification are summarized below:

$$C+\frac{1}{2}O_2 \rightarrow CO, \qquad \Delta H_R = -9205\ kJ/kg \tag{4.3}$$

$$C+O_2 \rightarrow CO_2, \qquad \Delta H_R = -32{,}765\ kJ/kg \tag{4.4}$$

$$C+CO_2 \rightarrow 2CO, \qquad \Delta H_R = -14{,}360\ kJ/kg \tag{4.5}$$

$$CO+\frac{1}{2}O_2 \rightarrow CO_2, \qquad \Delta H_R = -10{,}105\ kJ/kg \tag{4.6}$$

$$C+H_2O \rightarrow CO+H_2, \qquad \Delta H_R = 10{,}930\ kJ/kg \tag{4.7}$$

$$CO+H_2O \rightarrow CO_2+H_2, \qquad \Delta H_R = -6250\ kJ/kg \tag{4.8}$$

$$C+2H_2 \rightarrow CH_4, \qquad \Delta H_R = -6230\ kJ/kg \tag{4.9}$$

where the negative enthalpies of reaction indicate exothermic reactions and positive values indicate endothermic reactions.

The stoichiometric reaction for CB gasification remains the same as normal air-oxidation processes, even when considering air-steam reactants.

$$CH_hO_oN_nS_s + iO_2 \rightarrow jCO_2 + kH_2O + lSO_2 \tag{4.10}$$

The composition of the principal products can be predicted using mass and energy conservation equations and through atom balances. The principal products (dry basis) obtained from ideal gasification of CB, are CO_2, CO, CH_4, H_2, N_2, H_2S, HCN, and NH_3.

$$\begin{aligned} &CH_hO_oN_nS_s + a\left(O_2+3.76N_2\right)+bH_2O \\ &\quad \rightarrow cCO_2 + dCO + eCH_4 + fH_2S + gN_2 + hH_2 + mHCN + pNH_3 \end{aligned} \tag{4.11}$$

Other compounds can be considered as traces. Chemical equilibrium equations may be used to predict molar compositions or the number of species can be produced. Figure 4.17 shows the gas composition of gasified FB obtained with an adiabatic equilibrium model, as a function of equivalence ratio (ER) and a constant air/steam ratio (ASR, on a molar basis) of 0.25. The equilibrium model used for this study was the Chemical Equilibrium with Applications (CEA) computer model developed by NASA [59]. For gasification studies, ER and ASR are defined as:

$$ER = \frac{stoichiometric\ O\ atoms}{external\ total\ O\ atoms} = \frac{2i}{2a+b} \tag{4.12}$$

$$ASR = \frac{O\ atoms\ from\ air}{external\ total\ O\ atoms} = \frac{2a}{2a+b} = \frac{a}{i} * ER \tag{4.13}$$

Figure 4.17 Adiabatic equilibrium gas composition of FB versus equivalence ratio, with ASR at 0.25 (adopted from Gordillo *et al.*, [60])

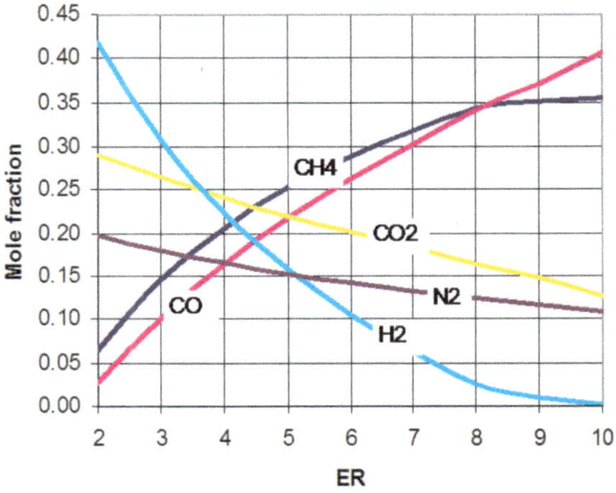

Note, that for gasification, the actual amount of oxygen in the reaction (denominator of the ER) includes the oxygen in the steam. The minimum external oxygen mass required for gasification with air-steam reactants may be found with the following analysis:

$$C + a'O_2 + b'H_2O \rightarrow c'CO + h'H_2 \tag{4.14}$$

With an atom balance, for C, O, and H, it can be determined that $c' = 1$, $h' = b'/2$, and $a' = (1 - b')/2$. The minimum ER is thus,

$$ER_{min} = \frac{2i}{2a' + b'} = 2i \tag{4.15}$$

While the minimum ASR is,

$$ASR_{min} = \frac{a'}{i} ER_{min} = \frac{(1 - b')}{2i} * 2i = (1 - b') \tag{4.16}$$

When the amount of steam in the reactants approaches zero (*i.e.*, $b' \rightarrow 0$): $ER_{min} = 2i$ and $ASR_{min} \rightarrow 1$. But when $b' \rightarrow 1$, then air is not required (*i.e.*, $a' \rightarrow 0$), and $ASR_{min} \rightarrow 0$. Moreover, during pure pyrolysis, both a' and b' tend to zero, in which case $ER \rightarrow \infty$, and only volatile matter is released from the fuel.

When $2i < ER < 1.0$, there could be oxidation of volatile matter to CO and CO_2 and/or oxidation of CO and H_2 produced by gasification of fixed carbon. In general, as can be verified by Figure 4.17, when the ER is increased, the availability of oxygen in the gasifier to reach complete combustion decreases; thus, the production of CO_2 and H_2 decreases, while the production of CO and CH_4 increases. The results from the equilibrium model show that for FB gasification at constant ASR and ER <9, increasing ER diminishes the production of H_2 and CO_2 and increases the concentration of CO and CH_4. For ERs > 9, the concentration of CH_4 tends to be constant and the concentration of H_2 tends to be zero.

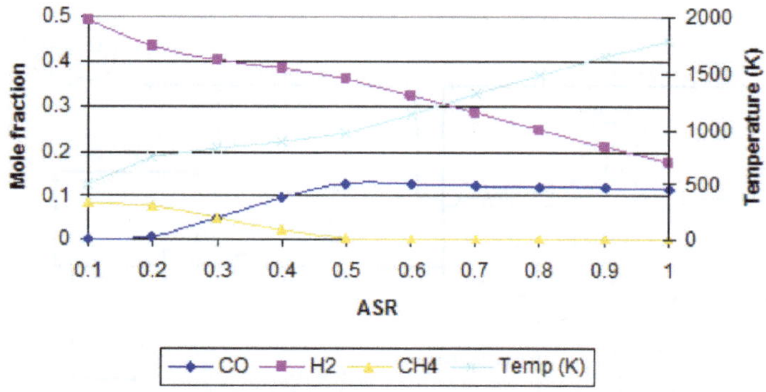

Figure 4.18 Adiabatic equilibrium gas composition of FB and temperature versus ASR with ER at 2

It is also apparent from Figure 4.17 that the concentration of H_2 could be increased to almost 45 % under ideal conditions when the ER approaches 2. On the other hand, under richer conditions (greater oxygen deficiency, higher ERs), FB gasification would tend to produce poor hydrogen mixtures. It seems that the key to producing H_2-rich gases is to maintain a low ER of about 2 and as low of an ASR as possible (*i.e.*, having more oxygen atoms in the form of H_2O). However, as can be seen in Figure 4.18, lowering ASR for a fixed ER also reduces the equilibrium reaction temperature, since the H_2O reaction with fuel is endothermic. For example, if ASR was reduced to 0.1, the mole fraction of H_2 would reach 0.5, but the temperature would only be 500 K. Adiabatic gasification would not be possible at this condition since the reaction rate with H_2O is extremely slow. Hence, ASR should be maintained above 0.25 or 0.3 to produce H_2-rich gases and maintain a reasonable reaction temperature (see [60] for more discussion on modeling CB gasification using air-steam mixtures as oxidizing agents).

4.3.2.2 Experimental Gasification Results

Gasification experiments were conducted and discussed by Priyadarsan [61], and Priyadarsan *et al.* [52, 58, 62]. In these experiments FB, chicken litter biomass (LB), blends of LB with FB, and blends of LB and FB with coal were each used as fuel for an updraft, fixed-bed gasifier. The experiments were performed on a 10 kW (30,000 Btu/h) counter-current fixed bed gasifier using air as the oxidizing agent. The experimental setup in Figure 4.19 is similar to that used by Priyadarsan *et al.*, except that steam is added to the reactant air in current experiments.

Figure 4.20 shows the species compositions measured at different distances over the grate, and the bed temperature profile from FB gasification experiments obtained by Priyadarsan *et al.* [61] in a study of fixed gasification at normal pressure. Here, results for experiments performed at 45 SCFM (76.5 m^3/h) are presented. The figure shows that the maximum production of CO (31 %) is reached at half an inch (12.7 mm) above the grate and starts to stabilize through the grate at

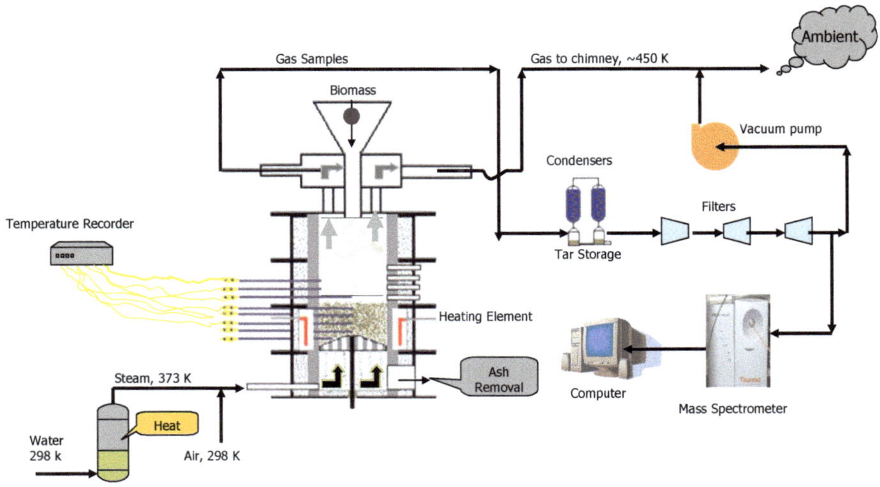

Figure 4.19 Schematic of 10 kW (30,000 Btu/h) fixed-bed counter-flow gasifier

about 27 %. On the other hand, the H_2, CO_2, and CH_4 concentrations increase continuously, indicating the presence of pyrolysis throughout the bed. However, the increased rate of the production of CO_2 and H_2 is higher than the production of CH_4. Also, the experimental results show that the maximum production of H_2 (8 %) and CO_2 (10 %) are reached at the top of the bed (7 inch, 178 mm). Beyond two and a quarter inches (57 mm) above the grate, there is a rapid increase in the production of H_2 and CO_2 marking the drying and devolatilization zone in the bed. It should be noted that, although there seems to be devolatilization taking place over the entire bed, the oxidation and gasification reaction's maximum contribution to the product gas species is near the base of the bed. The bed temperature data curve shows a peak at two and a quarter inches (57 mm) indicating that, be-

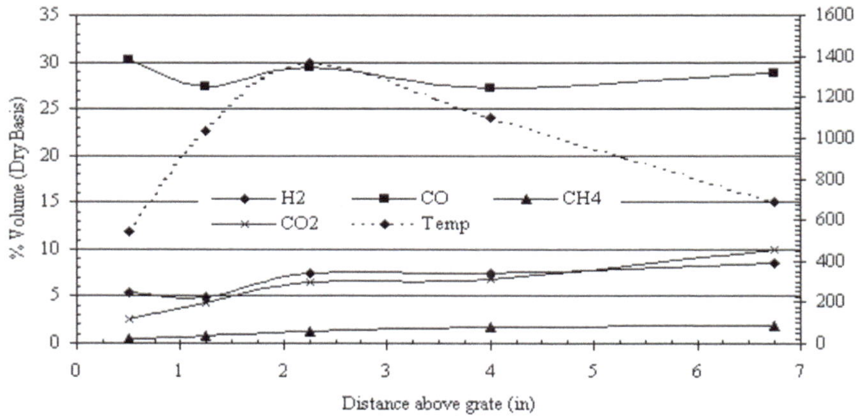

Figure 4.20 Gas species profiles for FB at air flow rate of 45 SCFH (adopted from Priyadarsan [61])

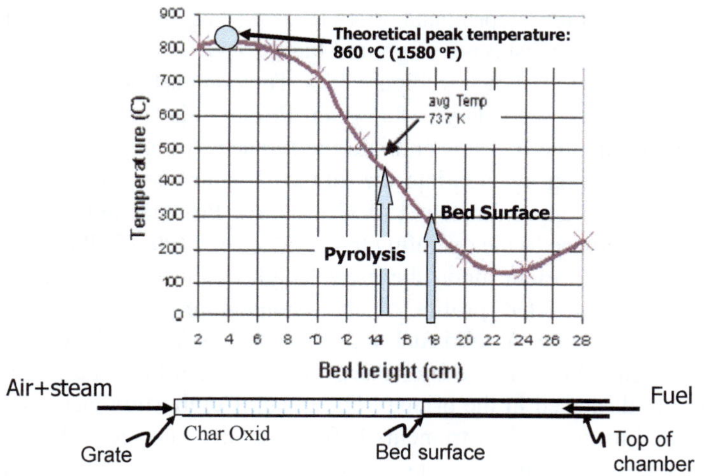

Figure 4.21 Experimental bed-temperature profile for DB at a time interval of 140 minutes with equivalence ratio of 1.8 and ASR of 0.38

fore this point, oxidation and gasification occur and after this point, drying and devolatilization processes occur. Typically, the gas composition data is correlated with the peak temperature in fixed bed gasification. Additional information on animal biomass gasification and co-gasification of animal biomass fuels with coals can be found in [58, 62].

Experiments on the gasification of CB with air and air-steam mixtures as oxidizing agents are currently underway. In Figure 4.21, a temperature profile of a DB gasification experiment with ER of 1.8 and ASR of 0.38 is presented. The peak temperature of 860 °C occurs near the grate.

4.3.3 Co-firing Coal and Cattle Biomass in Primary Burn Zones

Cattle biomass can also be used as a fuel by mixing it with coal and firing it in the primary burn zone of an existing coal suspension fired combustion system. This technique is known as co-firing. The high temperatures produced by the coal allow the biomass to be completely combusted. Previous work concerned with co-firing FB with coal may be found in [17, 32, 63, 64]. Di Nola [30] conducted co-firing experiments for coal-chicken litter biomass blends. Most co-firing experimental results obtained when using coal and finely ground CB seem to indicate that NO_x emissions did not increase and sometimes even decreased when co-firing coal and animal biomass.

Co-firing experiments have been conducted with two primary objectives: (1) access the feasibility of using CB to reduce NO_x and (2) measure the mercury emissions when co-firing coal with CB. All co-firing experiments were conducted using a 30 kWth (100,000 Btu/h, approximately 15 lb or 6.80 kg of coal/h) small scale furnace capable of firing most types of ground fuels.

A schematic of the furnace is shown in Figure 4.22. The combustion air is split between primary (~20 %) and secondary (~80 %) air. Primary air is necessary to propel the solid fuel through the fuel line and into the furnace. A blower provides the secondary air. Before being injected into the furnace, the air passes through a pre-heater to heat the air to approximately 150 °C (300 °F). Propane and natural gas are used to heat the furnace to the operating temperature of 1100 °C (2000 °F). Type-K thermocouples are present along the axial length of the furnace. These thermocouples provide a detailed profile of the temperature of the furnace throughout the combustion zone. A solid fuel hopper feeds coal and coal/biomass blends during experiments. Solid fuel comes out of the solid fuel line as a finely ground powder lightly dispersed in the primary air stream. Combustion gas compositions are measured using an E-Instruments 8000 Flue Gas Analyzer. The instrument measures the dry volume percentages of CO, CO_2, O_2, SO_2, NO, NO_2, and combined NO_x. Another method for detecting NO_x with laser sensors is discussed by Thomas *et al.* [67]. At the furnace's final port, a probe is used to sample the flue gases. Just after the final port, exhaust gases pass through a water cooling spray to significantly lower the temperature of the gases. This water is pumped out

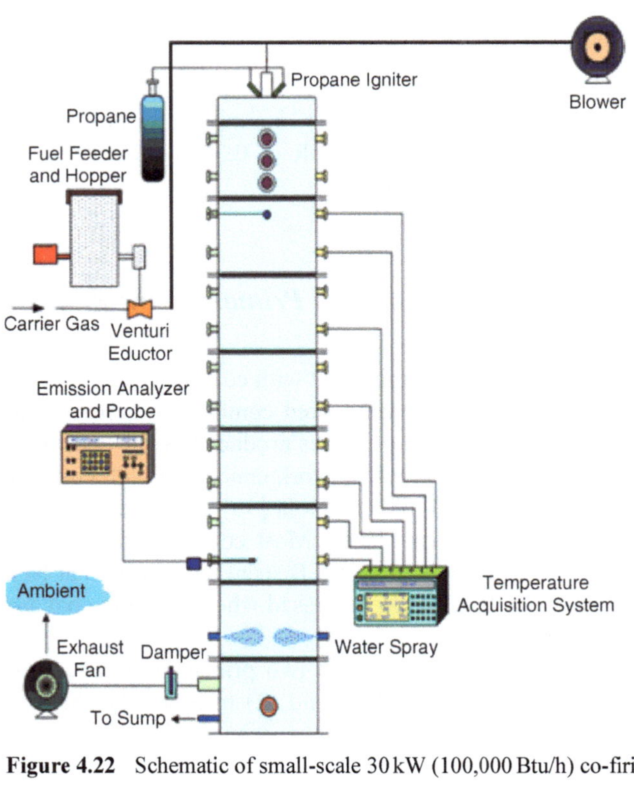

Figure 4.22 Schematic of small-scale 30 kW (100,000 Btu/h) co-firing experimental setup

of the furnace by a sump pump. More details are provided by Annamalai *et al.* [32, 65, 66]. A study on fouling during co-firing may be found in [68].

4.3.3.1 NO_x Reduction While Co-firing Coal With Cattle Biomass

Results from the study by Annamalai *et al.* [66] are shown in Figure 4.23. These co-firing experiments were conducted on the same boiler described above. The figure shows that even for fuel lean (excess oxygen) conditions, NO_x emissions for a 90:10 coal to FB blend were lower than those for burning pure coal. However, similar experiments conducted by Arumugam *et al.* [69] (see Figure 4.24) seem to suggest that lower NO_x emissions are not seen until 20% (by mass) biomass is blended with coal. Residence time, fuel particle size, and other factors may have caused the slight difference between the two results.

Present experiments continue the work begun by Annamalai *et al.* Experiments in which pure Wyoming Powder River Basin coal (a sub-bituminous coal), Texas Lignite coal, and blends of each coal with partially composted separated DB solids were burned have been conducted. Each coal was blended in 95:5, 90:10, and 80:20 (by mass) blends with DB. Experiments were performed for equivalence ratios of 0.8, 0.9, 1.0 (stoichiometric), 1.1, and 1.2. The results of these experiments may be found in a study by Lawrence [70].

In principle, it is difficult to hypothesize whether or not NO_x emissions, particularly during lean (low ER) combustion, will be lower for coal-CB blends. Since CB has more fuel-bound nitrogen, one would think that more NO_x would be formed if oxygen is available. However, since CB has a higher volatile content that is released quicker than the volatiles in coal, the oxygen might react more quickly to the nitrogen in the volatile matter, creating local fuel rich areas where NO reduction can take place, even during an overall lean combustion process.

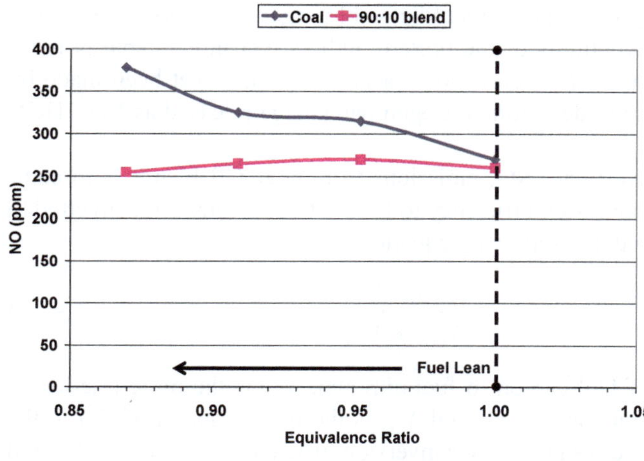

Figure 4.23 NO emission from coal and 90:10 coal–FB blends (adopted from Annamalai *et al.* [66])

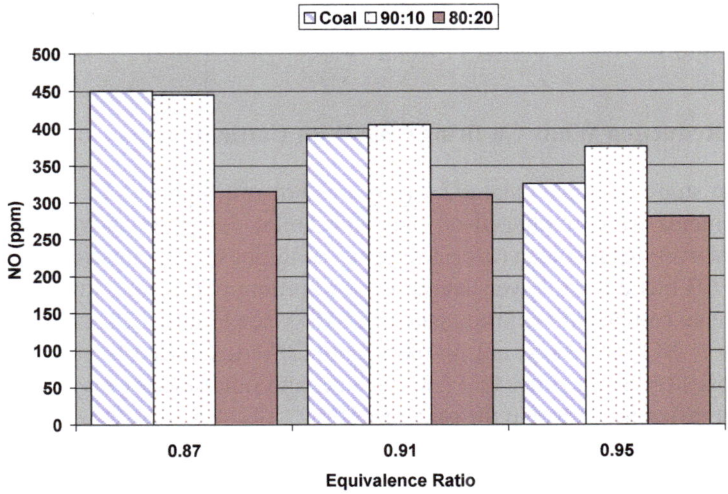

Figure 4.24 NO emission for coal, 90:10, and 80:20 blends of coal and FB versus equivalence ratio (adapted from Arumugam *et al.* [69])

This reduction would be intensified if the fuel N in the CB was released as NH_3. Moreover, the size of the FB particles is significant. Smaller FB particles can release volatile matter during combustion more quickly than larger particles. In both studies by Annamalai *et al.* [66, 69], the fuels were ground to similar sizes, such that 70 % of the FB particles in each study passed through a 170 μm sieve.

The manure contains two forms of nitrogen (N): (1) less stable inorganic (ammonium) N present as urea in urine (50 % of total N); and (2) more stable organic N present in the feces, which is released more slowly [71]. The organic N is a mixture of proteins, peptides, *etc.* of undigested feed. While the N slightly decreases during composting, the heat value decreases much faster and as such N per GJ increases with composting [17] since the stable organic N decomposes very slowly. Thus it is the hypothesis of the present authors that more stable N is released at high temperatures. However, no measurements have yet been made by the authors of this article to determine nitrogen percentage released as NH_3, HCN and other forms.

One other way to investigate NO_x emissions from coal–CB blends is to compute the fuel nitrogen conversion efficiency to NO_x. The fuel nitrogen converted to NO_x can be estimated with the following equation:

$$N_{CONV} = \frac{(c/n) * X_{NO}}{X_{CO2} + X_{CO}} \tag{4.17}$$

where c/n is the ratio of fuel carbon to fuel nitrogen, X_{NO} is the mole fraction of NO, X_{CO2} is the mole fraction of CO_2 and X_{CO} is the mole fraction of CO in the combustion products. The results for N conversion efficiency for coal, 90:10, and 80:20 blends of coal and FB can be found in Figure 4.25. In general, N conversion

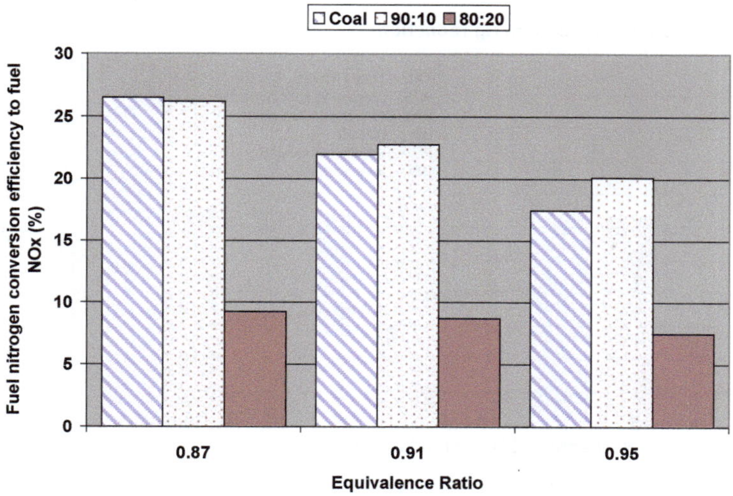

Figure 4.25 Fuel nitrogen conversion efficiency to fuel NO_x for coal, 90:10, and 80:20 blends of coal and FB (computed from data found by Arumugam *et al.* [64])

efficiency to NO_x was lowest for the 80:20 blend of coal and FB. However, conversion efficiency for 90:10 blends was higher than pure coal combustion except for leaner combustion at equivalence ratio of 0.87. Yet for all operating conditions, the conversion efficiency of fuel-N to NO_x was never significantly higher than efficiencies found when burning pure coal. The greatest difference between pure coal combustion and 90:10 combustion was only a 3 % increase seen at ER of 0.95. These results, at least ease some concern over producing significantly more NO_x when blending FB (a higher nitrogen containing fuel) with coal. In fact, the 80:20 results imply that if FB had a similar nitrogen content to that of coal, it may produce less NO_x than pure coal. For more discussion on fuel nitrogen conversion efficiency please refer to work by Annamalai *et al.* [66] and Lawrence [70].

4.3.3.2 Mercury Measurements During Co-firing Experiments

During the recent co-firing experiments with Wyoming sub-bituminous coal–FB blends, mercury measurements were also taken. Elemental mercury measurements were made using a Mercury Instrument VM 3000 based on the cold vapor atomic absorption principle. The oxidized mercury is measured using the wet chemistry method. To better understand the speciation of mercury compounds, this wet chemistry based flue gas conditioning system was developed which is based on a modified Ontario Hydro Method for online detection. The results presented in Figure 4.26 are promising, as they indicate greater reductions of elemental Hg when more DB is blended with both Texas lignite and Wyoming subbituminous. For a full discussion of mercury emissions during co-firing experiments, please refer to the study by Arcot Vijayasarathy [26].

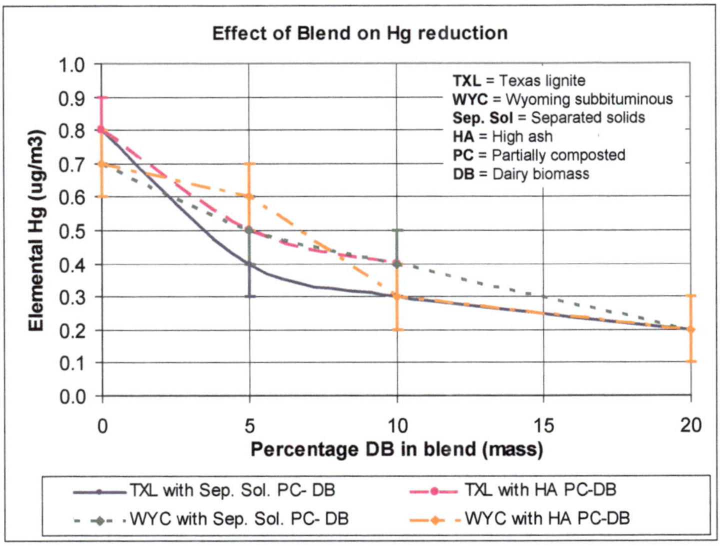

Figure 4.26 Elemental Hg reductions while co-firing coal with cattle biomass (adopted from Arcot Vijayasarathy [26])

4.3.3.3 Pilot-Scale Co-firing Tests

Pilot-scale co-firing tests of coal–FB blends performed at a 150 kW DOE-NETL facility in Pittsburgh, Pennsylvania revealed that due to higher volatile contents of FB, longer flames moved closer to the burner when firing coal:FB blends. Moreover, the NO_x emissions increased only slightly even though the nitrogen content in the coal:FB blended fuel increased by 25 %, and the Coal:FB blend resulted in ash deposits on heat exchanger tubes that were more difficult to remove than baseline coal ash deposits. The increase of slagging and fouling behavior was essentially due to the higher ash loading and ash composition of FB. The detailed results of this pilot-scale study are currently under patent process.

4.3.3.4 Economics of Cofiring Coal with Cattle Biomass

The economics of co-firing coal with CB in an existing power plant is currently being modeled, and should be similar to the economics of CB reburning discussed in the next section of this paper. However, in general, NO_x reductions and hence revenue from NO_x credits, will probably be less of a factor or non-existent for co-firing. Whereas the primary function of reburning with CB is to reduce NO_x, from small-scale experimental results discussed above, co-firing coal with CB in primary burn zones may not change NO_x emissions significantly. Hence, a greater dependence on savings from lower non-renewable CO_2 emissions and possible lower fueling costs can be expected for co-firing economics.

The economics spreadsheet computer model for reburning, discussed later, is currently being modified to simulate the economics of CB co-firing applications. The greatest challenge will be to determine the most realistic capital cost of installing CB co-firing capabilities on an existing coal plant. Robinson *et al.* [72] reported that for wood-based biomass, a mode capital cost of \$ 60 per kW_{th} supplied from biomass could be expected for co-firing rates of less than 2 % biomass on an energy basis. If more than 2 % biomass is to be fired, then capital costs of \$ 200/kW_{th} biomass can be expected because separate hoppers, feeders and injection systems would need to be installed to handle the higher mass fueling rates. Since CB's heating value is significantly lower (especially high-ash CB) than that of coal, capital costs could reach \$ 300 to \$ 400/kW_{th} biomass.

4.3.4 Reburning Coal with Cattle Biomass

It was seen during co-firing experiments that NO_x was reduced slightly, when coal was co-fired with FB at many operating conditions, possibly due to significant release of fuel-bound nitrogen in the form of NH_3. Therefore, experiments have also been performed to determine FB's effectiveness as a reburn fuel for NO_x reduction in coal-fired power plants. The basic reburn technology in coal-fired plants uses two separated combustion zones: the primary combustion zone where primary fuel is fired (*e.g.*, coal producing NO_x) and the reburn combustion zone where the additional fuel (typically natural gas, NG) is fired in order to reduce NO_x produced in the main burner. In the primary zone, coal is fired under normal to low excess air conditions with 70 to 90 % of the total heat input.

In the case of NG reburn fuels that do not contain fuel-bound nitrogen, 10 to 30 % less NO_x is produced simply because 10 to 30 % (by heat) of the primary coal fuel is displaced by the reburn fuel. Moreover, in the reburn zone, while operating in a fuel-rich regime, the NG molecules break down to hydrocarbon fragments which react with NO_x to form hydrogen cyanide (HCN) and NH_3. These nitrogen compounds then react with other nitrogen-containing species to form N_2. Similar processes also occur when reburning with nitrogen containing fuels such as coal, oil, and biomass (see Figure 4.27 for a general illustration of the reburning process). The optimal ER in the reburn zone is 1.18 to 1.05 (stoichiometric ratio of 0.85 to 0.95, respectively) [27]. The reburn fuel produces 10 to 30 % of the total heat input. Overfire air (OFA) is typically used downstream of the reburn zone to create a burnout zone for complete combustion.

Nitrogen oxide (NO_x) reduction while reburning with OFA using a down-fired pilot-scale (300 kW) combustion facility was examined for the effect of metal-containing compounds by Lissianski *et al.* [73]. Natural gas was used as the main fuel and the reburn fuel. Metal-containing compounds such as sodium carbonate, potassium carbonate, calcium acetate, and fly ash were injected with the main fuel or the reburn fuel. Reburning resulted in a 50 % reduction of SO_2. The baseline

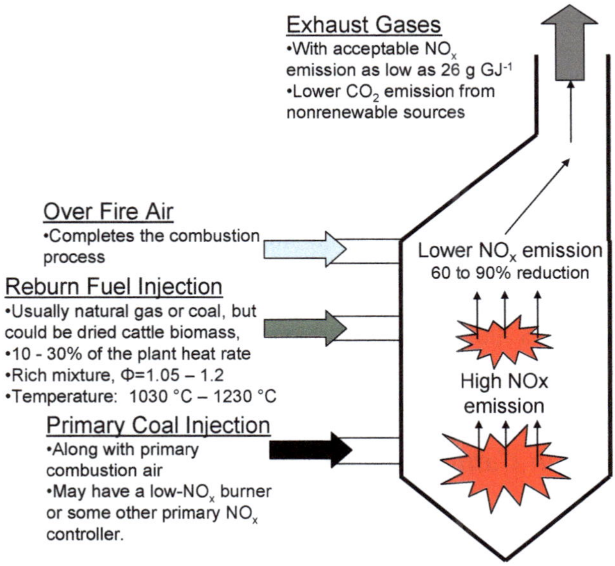

Figure 4.27 Simplified schematic of the reburn process

NO_x level was 600 ppm. Reburning with NG alone provided a 66 % NO_x reduction, while injection of 100 ppm of the metal compounds provided a 4–7 % additional reduction in NO_x.

A study by Yang *et al.* [74], investigated NO reduction when coal was reburned with more coal in the reburn zone. Eight different bituminous coals were used as reburn fuel. A maximum NO_x reduction of 65 % was found, as well as an optimum residence time of 450 ms. Base NO_x levels of 600 ppm from the primary burn zone were simulated by burning a mixture of propane and NH_3. Reburn zone stoichiometry and the volatile matter content of the reburn fuel were found to influence the reburn performance the most. However, fuel nitrogen content of the reburn fuel was found not to be as important as most of the other operating conditions of the experiments. Yet the nitrogen content of the reburn fuel may become significant in cases where the NO concentration from the primary zone is very low, such as cases when low-NO_x burners and air staging are installed. Yang *et al.* [74] suggests that at some point (about 150–170 ppm) NO-forming mechanisms over take NO-reducing reactions, causing a net increase in NO_x. This places a limitation on the practicality of reburning with nitrogen containing fuels such as coal, oil and CB.

Table 4.4 is a summary of various reburn experiments and existing systems and their bottom line NO_x reduction performances. Some descriptions of the experimental parameters are also indicated; however, please refer to each source for a full explanation of the tested systems.

Table 4.4 Reburn systems and experimental results with various reburn fuels

Primary Fuel	Reburn Fuel	NO_x Reduction Results	Source	Comments
Coal	Straw, miscanthus (type of grass), natural gas, and wood	Measured NO_x levels between 125 to 150 ppm	Rudiger *et al.* [86]	300 kW_{th} total heat input. Reduction zone residence time varied from 1.0 to 1.6 seconds. Under same conditions coal reburning only achieved a NO_x level of 250 ppm.
Coal	Pyrolysis gases from coal and miscanthus (with and without tars)	Measured NO_x levels as low as 100 ppm	Rudiger *et al.* [86]	Best results were from coal pyrolysis gases with tars produced at 800 °C. Gas composition, stoichiometry, and residence time in the reburn zone all influenced the NO_x reduction results.
Propane + NH_3	Eight different bituminous coals	Up to a 65 % reduction from a base level of 600 ppm	Yang *et al.* [74]	NH_3 in primary fuel is used to simulate base NO_x level. Reburn fraction was varied between 10 and 35 %. Reburn stoichiometry was found to be the most significant factor for NO_x reduction. Optimum reburn zone residence time was found to be 450 ms. Smaller particles with higher volatile matter improve reduction.
Natural gas and coal	Fir lumber wood waste, coal, coal pond fines, carbonized refuse derived fuel, orimulsion (bitumen-based fuel)	Up to 70 % reduction from basic reburning (20 % heat input), 85 to 95 % reduction from advanced reburning (10 % heat input)	Maly *et al.* [89]	Base NO_x levels were varied between 200 and 1300 ppm. Fuel N content, volatile content, and ash content significantly affected reductions. Most effective promoters during advanced reburning were alkalis, such as sodium compounds.
Coal	Coal and micronized coal	25 to 55 % reduction	DOE [27], Srivastava *et al.* [28]	Survey of coal reburning applications on US coal-fired boilers. Four applications in total. Three applications were still operational as of 2005.
Coal	Natural gas	40 to 68 % reduction	DOE [27], Srivastava *et al.* [28]	Survey of gas reburn applications on US coal-fired boilers. 22 operations in total. All operations, as of 2005, were decommissioned or not operating, except for one application at the Somerset plant in New York that also had a selective non-catalytic reduction system.

Table 4.4 (continued)

Primary Fuel	Reburn Fuel	NO_x Reduction Results	Source	Comments
Natural gas and Illinois and Ohio coals	Natural gas, coal, furniture waste, willow wood and walnut shells	60 % reduction from gas and biomass reburning, 50 % reduction from coal reburning	Zamansky *et al.* [81]	Base NO_x levels were varied from 400 to 900 ppm. Fibrous willow wood was the most difficult biomass fuel to prepare for combustion. Economics was also discussed.
Natural gas and Illinois and Ohio coals	Advanced reburning with above fuels and nitrogen agent injection	83 to 90 % reductions	Zamansky *et al.* [81]	Three advanced reburn technologies tested: injection of nitrogen agent with over fire air, injection of nitrogen agent in rich reburn zone, and SNCR reagent injected downstream of over fire air.
Natural Gas+NH_3	Natural Gas	66 % reduction from a base level of 600 ppm	Lissianski *et al.* [73]	Reburn fuel was 25 % of total heat rate. NH_3 in primary fuel is used to simulate base NO_x level.
Natural Gas+NH_3	Natural Gas+100 ppm Na, K, or Ca	70–73 % reduction from base level of 600 ppm	Lissianski *et al.* [73]	Reburn fuel was 25 % of total heat rate. 15–20 % reduction when Ca injected alone without natural gas.
Propane + NH_3	Coal, feedlot cattle manure biomass, 90:10 and 50:50 blends of coal and biomass	Up to a 40 % reduction with coal, up to 80 % reduction with pure biomass from a base level of 600 ppm	Annamalai *et al.* [69, 75]	It was speculated that most of the nitrogen in FB exists as NH_3, and volatile matter of FB (little fixed carbon) is twice that of coal and hence FB serves as a better reburn fuel in controlling the NO_x emissions.
Natural gas+NH_3	Low-nitrogen switch grass and hig-nitrogen alfalfa	Up to 70 % reduction from a base level of 500 ppm	Sweterlitsch *et al.* [90]	Initial O_2 percentage varied between 1 and 6 % and reburn fraction varied from 0 to 23 % (heat). Two reburn fuel carrier gases were investigated: N_2 and steam.
Propane + NH_3	Coal, feedlot cattle manure biomass, 90:10 and 50:50 blends of coal and biomass	Up to 14 % reduction with coal, up to a 62 % reduction with pure biomass from a base level of 600 ppm	Arumugam [77]	Equivalence ratio varied from 1.0 to 1.2. Circular jet and flat spray nozzel injectors were used.

4.3.4.1 NO_x Reduction from Reburning with Cattle Biomass

A study by Annamalai *et al.* [69, 75] showed that 80 % NO_x reductions were possible when reburning with pure FB (Figure 4.28). Again, base NO_x levels from the primary zone were about 600 ppm. Greater reductions were seen with higher (more fuel rich) equivalence ratios, except for pure FB reburning which did not vary as significantly with ER. The reductions from pure FB reburn fuels were found to be greater than those for coal reburning or reburning with blends of coal and FB.

The small scale facility used for reburn experiments in the present study is essentially the same as that used by Annamalai *et al.* [75]. The small-scale facility has a capacity of 30 kW_{th} (100,000 Btu/h). The facility, illustrated in Figure 4.29, also simulated primary NO_x levels by burning a mixture of propane and NH_3. Pulverized reburn fuels were injected into the reburn zone through a hopper. Temperature distribution along the boiler was measured, as were emissions. For reburn experiments of the present study, results for reburning with Texas Lignite (TXL) coal and Wyoming coal are used for the base case. NO_x emissions from reburning with pure TXL, TXL:low-ash partially composted FB blends, TXL:high ash partially composted FB blends, pure WY coal, and WY coal:low ash partially composted FB blends were obtained. Some of these results are presented in the graph in Figure 4.30. Experiments with pure high ash partially composted FB were not performed due to severe amounts of slag built up in the area near the reburn injectors in the furnace. A study by Oh *et al.* [76] investigated ash fouling from reburning with CB. The level of NO_x emissions in the exhaust decreased with increased equivalence ratio and CO percentage in the exhaust gases. With increased equivalence ratio, the oxygen in the combustion zone was depleted quickly and hence

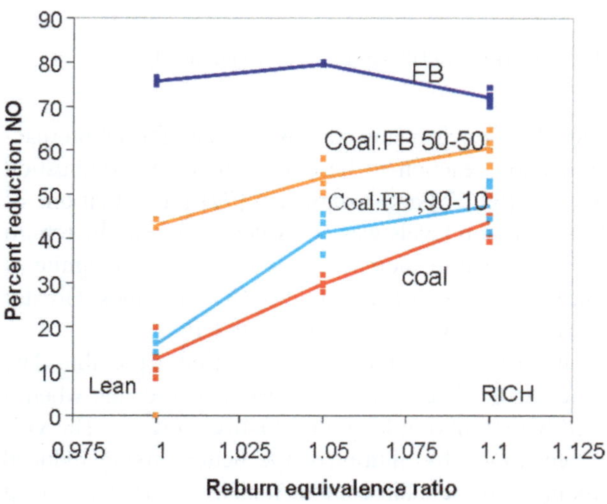

Figure 4.28 NO reduction percentage with coal, feedlot biomass, and blends of coal and FB from a base level of 600 ppm NO (adopted from Annamalai *et al.* [69, 75])

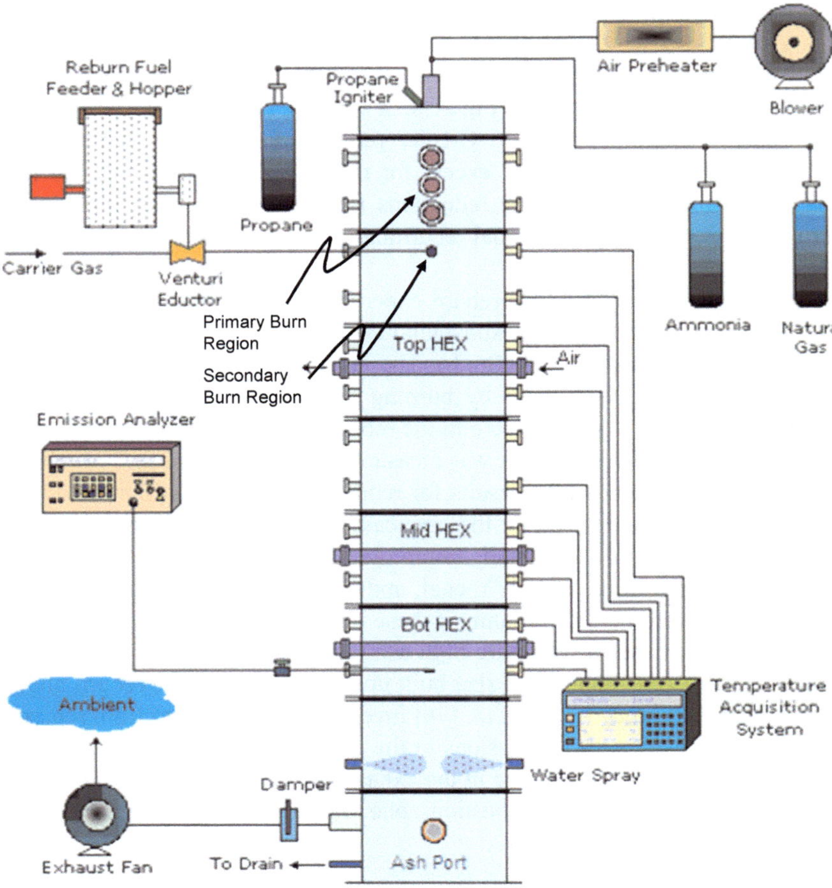

Figure 4.29 Schematic of small-scale 30 kW (100,000 Btu/h) coal reburn facility

more CO was formed. Low levels of oxygen slowed down the NO formation reaction and allowed the NO reduction reaction to be dominant in the combustion zone. Studies by Annamalai *et al.* [69, 75] and Arumugam [77] found that one of the greatest influences on NO_x emission levels was equivalence ratio in the reburn zone. Greater NO_x reduction from cattle biomass fuels compared to lignite is clearly shown in Figure 4.30, but at the expense of higher CO emissions. Greater amounts of burn-out air may be required when reburning with CB.

With a longer residence and reaction time, more NO_x reduction is possible. The estimation of the mixing time for the lateral injection was 0.32 seconds when a linear mixing model was used with a mixing length of 30.48 cm (12 in). The vitiated air reduced oxygen concentration by dilution while better mixing reduced oxygen concentration by mixing with the main combustion stream. Better mixing also caused the fuel to combust faster and thus reduced the oxygen levels. The reduced oxygen levels inhibited NO_x formation mechanisms.

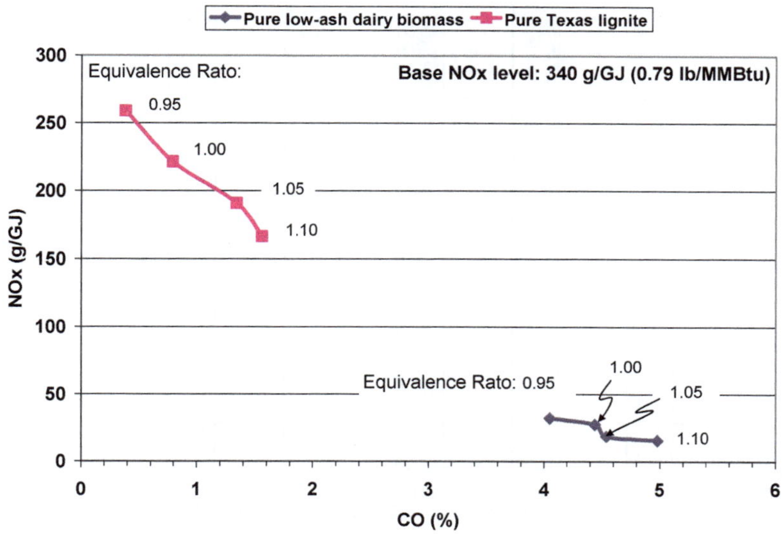

Figure 4.30 NOx versus CO for two reburn fuels from a base NOx level of 340 g/GJ (0.79 lb/MMBtu)

4.3.4.2 Mercury Oxidation During Reburning

Oxidized mercury (Hg^{2+}) measurements for CB reburning have not yet been conducted on the small-scale experiments discussed above. However, Hg^{2+} emissions were numerically simulated by Colmegna *et al.* [78]. The simulation accounted for the high amounts of Cl in cattle biomass, allowing for very high mercury oxidation. This may be seen in Figure 4.31, which is a plot of oxidized mercury percentage of total mercury emitted during combustion. Similar oxidation was also predicted for coal–CB blended reburn fuels. It can be deduced that CB helps the oxidation of mercury, which in turn, signifies that mercury can be captured more effectively at the exhaust by FGD-type environmental controls.

4.3.4.3 Economics of Reburning Coal with Cattle Biomass

Although reburning coal with CB may decrease NO_x emissions in power plants while minimizing the impact of large concentrated animal feeding operations (CAFOs) on the environment, it must also prove to be economically viable for energy producers. For this reason, the economics of reburning with CB has been modeled with a deterministic computer spreadsheet program that simulates a single coal-fired unit that utilizes CB from nearby dairies, feedlots or other animal feeding operation as a reburn fuel. The analysis includes quantifying investment (capital) costs and operation and maintenance (O&M) costs, as well as revenues from NO_x credits, non-renewable carbon allowances, and possible fuel savings.

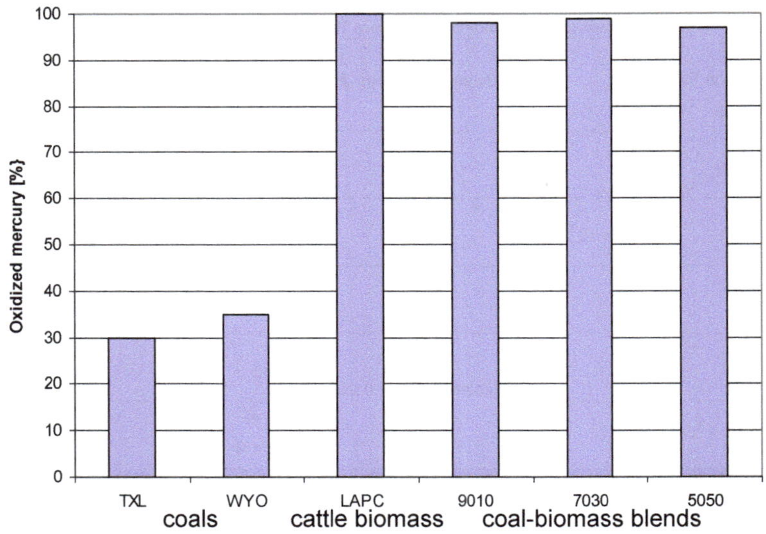

Figure 4.31 Numerical simulation of Hg oxidation from reburning coal with various fuels (adopted from Colmegna, *et al.* [78])

The following steps were taken to model the economics of a CB reburning facility:

- Obtained capital and O&M estimates of all possible aspects of transportation, preparation, and combustion of CB from literature review, manufacture estimates and government reports.
- Computed overall capital and annual O&M costs for CB reburning, as well as other competing technologies such as selective catalytic reduction (SCR), selective non-catalytic reduction (SNCR), and low NO_x burners with air-staging.
- Estimated the specific cost of NO_x reduction ($ per ton of NO_x reduced or cent per kWh) of a CB reburn system retrofitted on an existing coal-fired power plant, and subsequently, compared to other possible common NO_x control retrofits.

The economic modeling equations for low-NO_x burners, SCR and SNCR were constructed in a similar fashion to baseline costs and scaling factors for capital, fixed O&M, and variable O&M costs presented by the USEPA [79] for the Integrated Planning Model (IPM). The IPM is a deterministic linear programming model of the US electric power sector. It is used by the EPA to compare energy policy scenarios and governmental mandates concerning electric capacity expansion, electricity dispatch and emission control strategies. For the present study only equations for NO_x emission control technologies of the IPM were necessary.

However, reburning technologies are not covered by the latest version of the IPM, because very few coal plants in the USA consistently reburn coal, mostly due to higher natural gas prices and the recent effective installation and use of low-NO_x burners [28]. Although, in an earlier USEPA report [80], released when many pilot-scale reburn tests were being conducted on existing coal plants in the 1990s, reburning with natural gas and coal in cyclone boilers was modeled. Capital costs of reburning with wood-based biomass were found by Zamansky *et al.* [81] to be very similar to the capital costs of coal reburning facilities. This was an underlying assumption for CB reburning in the economics analysis presented here. Fixed O&M cost equations for coal reburning were scaled to account for CB's poorer heat value to coal when computing fixed O&M costs for CB reburning. Moreover, variable O&M costs, unique to CB reburning, such as transportation, drying, and higher ash disposal, were computed in a more itemized fashion. These more detailed variable O&M calculations allowed for easier study of the overall system cost's sensitivity to variables such as transportation distance, CB moisture content, and CB ash content.

There are over 90 inputs to the CB reburn model, covering all aspects of the system, including boiler type, primary NO_x control-type, biomass transportation distance, transportation vehicles, biomass dryers, biomass collection times, labor costs, and fuel properties. For this chapter, a preliminary run of the spreadsheet program was conducted. Some of the more important inputs for this test run were:

- 500 MW_{elec} plant capacity
- 10,280 kJ_{th}/kWh_{elec} (9750 Btu/kWh) heat rate
- 10 % of overall heat rate supplied by biomass reburn fuel
- NOx levels of 26 g NO_x/GJ (0.06 lb/MMBtu) were assumed to be achieved by the CB reburn system
- 217 km (135 mile) transport distance for CB reburn fuel
- $ 3.86/metric ton CO_2 ($ 3.50/ton) carbon penalty or avoided cost of carbon sequestration

However, all of these inputs can be changed to model different power facilities or circumstances. This preliminary run is simply a demonstration of the CB reburn economic model's capabilities.

Some of the results for this preliminary run are presented here. The capital investment components for a CB reburn system are presented in Figure 4.32a. The cost of installing the reburner and ancillary system components on the power plant was found to constitute 54 % of the total investment cost, while constructing drying and compost facilities comprised 36 % and purchasing transport vehicles comprised the final 10 %. Yet, as can be seen in Figure 4.32b, dryer O&M and CB transportation costs dominated the annual overall O&M of the reburn system.

The juxtaposition of results for different NO_x control technologies is presented in Table 4.5. Primary controls include low-NO_x burners and/or air staging. Secondary technologies such as reburning or SCR typically do not replace primary technologies; rather they supplement initial attempts to reduce NO_x. Therefore, primary controls are assumed to run even after a secondary system is retrofitted. The overall O&M costs of CB reburning and SCR are similar (about $ 80 million per year); however, SCR systems require a greater investment. SNCR systems require the cheapest investment, but these systems typically reduce NO_x in large boilers by only 35 % and hence do not earn as much revenue from NO_x credits.

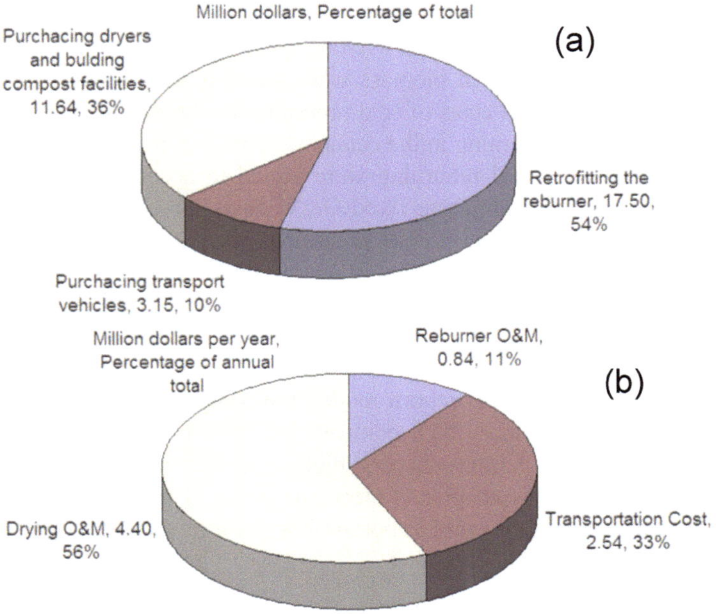

Figure 4.32 Breakdown of computed (**a**) capital costs and (**b**) operation and maintenance costs for a CB reburn system installed on a 500 MW coal-fired power plant, 10 % heat supplied by CB reburn fuel, 217 km (135 mile) CB transportation distance (note: US dollars)

The first column in Table 4.5 may be thought of as the cost breakdown of the plant's NO_x control systems before any secondary retrofits are installed. If the annualized cost for this case is compared to that of column two for the CB reburn retrofit, a difference of overall costs can be computed:

$$\begin{aligned} \%\,Annulized\ \cos t\ difference &= \frac{Cost_{primary,only} - Cost_{primary+reburn}}{Cost_{primary,only}} \times 100 \\ &= \frac{84,220,00 - 86,460,00}{84,220,000} \times 100 = -2.66\% \end{aligned} \tag{4.18}$$

This difference can be monitored as various inputs are incrementally changed to gage the CB reburn system's sensitivity to certain parameters. For example, in Figure 4.33, annualized cost difference is plotted against the dollar amount of non-renewable CO_2 released from the plant. Since CB would replace part of the coal required to fuel the plant, non-renewable CO_2 emissions will drop, and the plant would save on any avoided carbon penalties. From this sensitivity analysis, a minimum profitable dollar amount of CO_2 can be determined; allowing plant owners to decide if installing a CB reburn system is indeed in their interests. A similar plot can be made for the variation of nearly all 90 input parameters of the reburn economic model in order to complete a full sensitivity analysis of the proposed CB reburn system retrofit.

Table 4.5 Economic spreadsheet results and comparison of four different NOx control technology configurations on a 500 MW coal-fired power plant, 10 % heat supplied by CB reburn fuel, 217 km (135 mile) CB transportation distance, SCR and CB reburning achieve similar NOx levels of 26 g/GJ, while primary controls alone achieve 86 g/GJ NOx levels, SNCR is assumed to achieve a 65 g/GJ NOx level (note: all numbers in US dollars per year, unless otherwise stated)

	Primary control only	Primary control + reburn	Primary control + SCR (equal annual no_x reduction as reburn)	Primary control + SNCR (continuous operation)
Fixed O&M cost	87,407	925,347	391,816	156,234
Variable O&M cost*	94,662	4,492,812	2,264,172	3,530,542
Biomass delivery cost	-	2,536,042	-	-
Coal delivery cost	73,172,488	65,855,239	73,172,488	73,172,488
NO_x credits	-	(5,454,205)	(5,459,666)	(1,844,761)
CO_2 penalty	13,008,942	11,718,393	13,008,942	13,008,942
Ash revenue	(3,074,291)	(3,035,376)	(3,074,291)	(3,074,291)
Ash disposal cost**	-	3,375,760	-	-
Annualized capital cost	737,613	4,775,457	6,472,915	1,315,998
Total cost (w/o capital)	83,289,207	80,414,010	80,303,461	84,949,153
Annualized cost ($/yr)***	84,223,417	86,462,270	88,501,601	86,615,905
Specific NO_x reduction ($/ton)	15,878	11,339	11,603	14,224
Cost of NO_x reduction (¢/kWh)	2.40	2.47	2.52	2.47

*For biomass reburing, variable O&M includes the cost of drying the biomass.
**For this preliminary run, coal plant operators were assumed to sell 100 % of ash when not reburning, but any additional ash produced when reburning was considered a disposal cost to the power plant, as a worst case scenario.
***Includes annualized capital and straight-line depreciation of capital.

Another important factor to the success of a CB reburn retrofit is the distance between the feedlot or dairy and the power plant. Ideally, the CB combustion facilities and the animal feeding operations should be within close proximity of each other in order for coal and emission credit savings to be realized. Figure 4.34 is a plot of the overall cost of reducing NO_x versus the biomass transportation distance. With such a plot, certain existing coal-fired power plants located near feedlots or dairies can be further investigated to determine the applicability of a CB combustion retrofit on current operations.

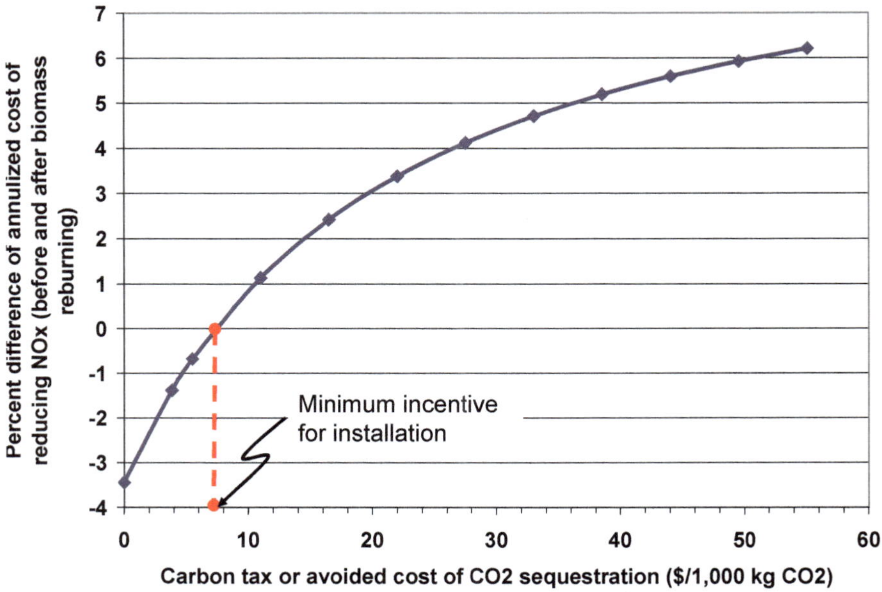

Figure 4.33 Percentage difference of annualized cost of reducing NOx before and after a CB reburn system is retrofitted on an existing 500 MW coal-fired power plant versus dollar values of CO_2

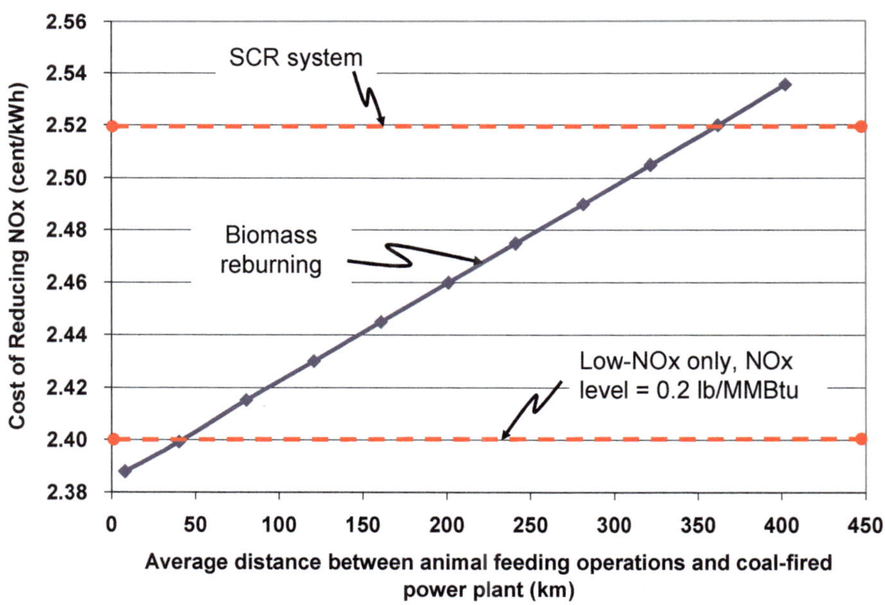

Figure 4.34 Effect of CB transport distance on the overall cost of reducing NOx (note: US cent)

4.3.5 Small-scale, On-the-Farm Combustion of Cattle Biomass

More recently, animal biomass has also been considered a possible feedstock for smaller, on-the-farm combustion systems designed to properly dispose of animal manure solids and wastewater. Using commercially available equipment like solid separators, augers, and dryers, DB can be prepared for smaller combustion processes. If these systems are constructed on or near a confined animal feeding operation, the benefits discussed in the previous sections of this chapter can be realized without much of the transportation and processing costs required to burn cattle biomass in large electric utility boilers. Typically, the manure is split into a liquid wastewater stream and a solid waste stream of mechanically separated solids and/or scraped solids from open lots.

The objective of this research is to design a combustion system that can be integrated into current waste processing procedures illustrated in Figure 4.3. The combustion system would need to accept the manure solids, the wastewater stream, and combustion air to produce products of combustion, dry ash, and recyclable water (or steam) that may be used as a thermal commodity or for further manure flushing.

Thermodynamically, a black box method can be utilized to determine the greatest amount of waste that can be converted into the desired end products. This method is shown in Figure 4.35, with the inputs and outputs to the system crossing through the control volume (CV) fixed around the combustion system. A complete mass and energy balance of the system can now be conducted. The ash and moisture percentages may be treated as variables in order to determine their required values to convert all material to combustion gases, water vapor, dry ash, and to maintain a desired system temperature (in this case, 373 K).

Figure 4.36 displays the results of the black box methodology. According to the figure, if the flushed manure emanating from dairy or feedlot has a moisture percentage of more than 85 %, then no amount of combustible material in the solids can produce enough heat during combustion to fully vaporize all of the moisture portion (wastewater) of the manure. However, ash also plays a limiting role in the effectiveness of independent manure combustion systems. Depending on the ma-

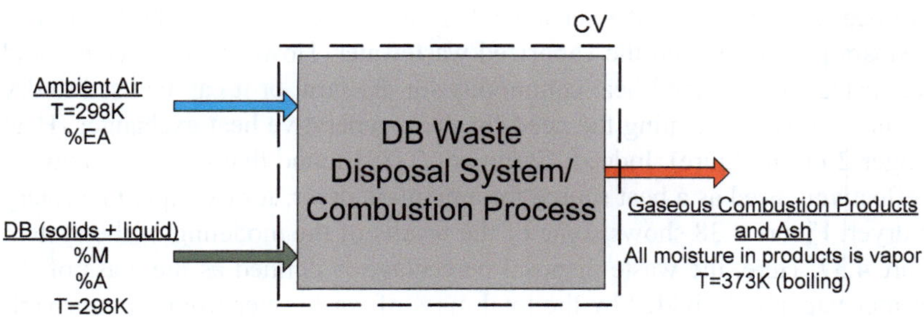

Figure 4.35 Black box method of determining the maximum effectiveness of a CB combustion system (adopted from Carlin [21])

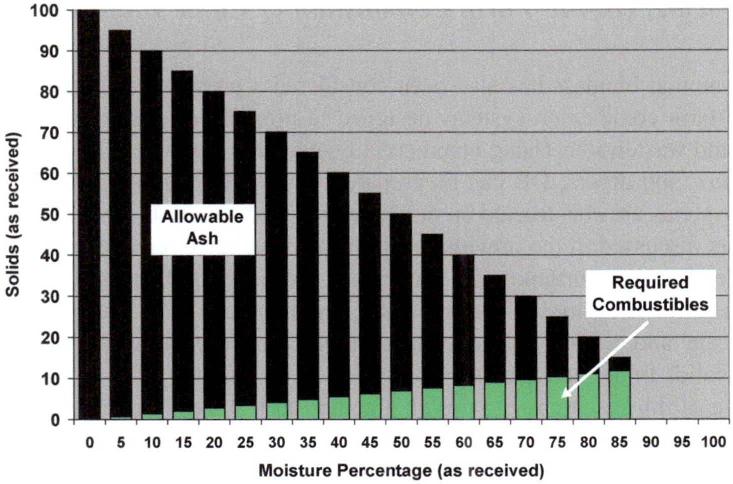

Figure 4.36 Required solids composition to vaporize all moisture in CB fuel (10% excess air) (adopted from Carlin [21])

nure collection process, the bedding used in the dairy freestalls, or the pavement surfacing of the feed yards and open lots, the ash content of the solid manure material can make direct combustion impossible due to fouling and inadequate fuel heating value.

The system shown in Figure 4.37 has the potential to combust most of the manure solids and vaporize at least a portion of the wastewater stream. Just as in Figure 4.3, the flushed manure is mechanically separated into solid and liquid streams. The solids are injected into a combustor, furnace or perhaps a gasifier with a subsequent product gas burner. The combustion air is preheated by some of the heat of the hot products of combustion. Meanwhile, some of the remaining wastewater is sent to a fire-tube boiler where it is sprayed onto heat pipes containing the combustion gases. The remaining solids from the wastewater can be removed periodically from the boiler (similar to blow down in conventional fire-tube boilers) and either sent back to the combustor or used as fertilizer.

This system was modeled by Carlin [21, 82]. Carlin *et al.* [83] added a recuperative heat exchanger (Heat Exchanger 1 in the figure) to utilize the latent heat of the steam produced from the vaporized wastewater. However, the steam could be used instead as a general heat commodity for the farm or it can be used to dry the manure solids, eliminating the need for the regenerative heat exchanger (Heat Exchanger 2 in the figure). Indeed, Carlin *et al.* [83] found that the steam may be more effectively used as a heat source for a biomass dryer, for example in a rotary steam dryer. Figure 4.38 shows some of the results of the modeling of the system in Figure 4.37. Here, the waste disposal percentage is defined as the mass of the wastewater vaporized divided by the total mass of wastewater from the mechanical solid separator.

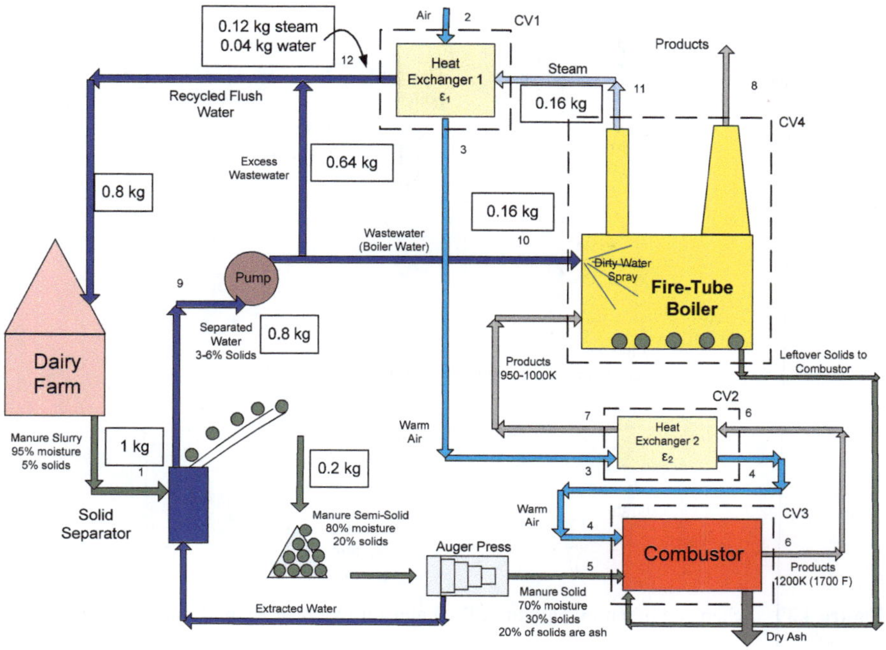

Figure 4.37 Diagram of proposed waste management plan for flushed cattle biomass (adopted from Carlin *et al.* [82, 83])

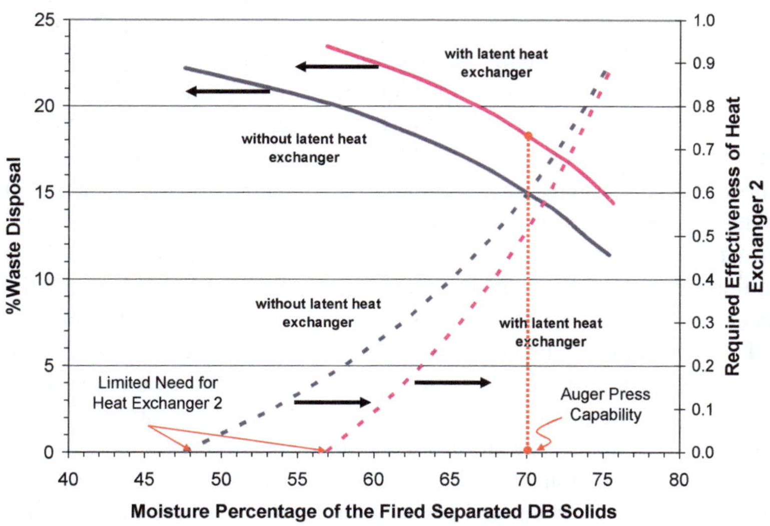

Figure 4.38 Percentage waste disposal and required effectiveness of heat exchanger 2 versus moisture percentage of separated solids (with and without heat exchanger 2) (adopted from Carlin *et al.* [83])

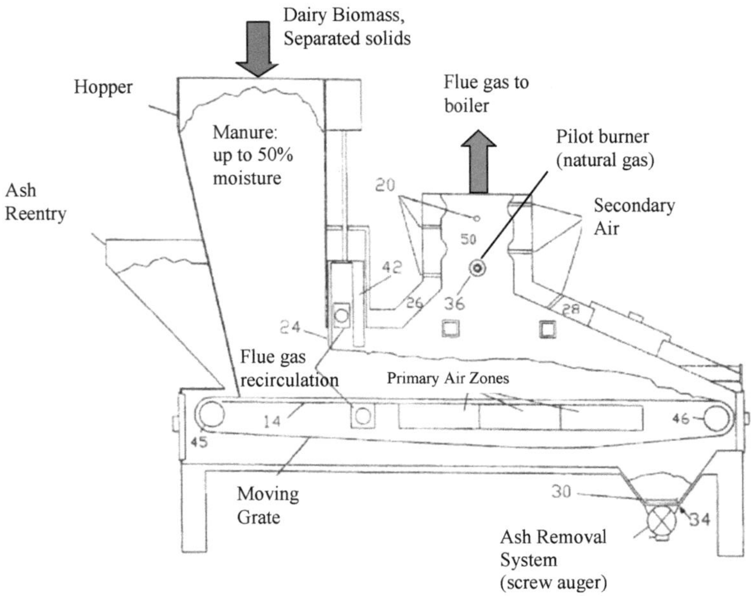

Figure 4.39 Schematic of a moving grate CB combustor (adopted from Mooney *et al.* [85])

There have been other studies on small-scale, onsite combustion systems, such as the gasification system discussed by Young *et al.* [49] and the prototype system installed by Skill Associates [50]. There are also some patents claiming effective ways to dispose of animal biomass such as the on-the-farm system designed by Kolber [84]. Most of these systems assume that high temperature gasification would be the most appropriate means by which the manure solids would be combusted. However, there are some claims to directly firing manure solids such as a patent for a moving grate combustor by Mooney *et al.* [85] (see Figure 4.39). However, most of these systems are essentially two-stage gasification systems in which the released volatile gases are immediately fired, in this case, by a natural gas pilot burner. In this sense, these systems become co-firing furnaces like those discussed in the previous section, only now the manure is the primary fuel and the fossil fuel is an igniter. Figure 4.40 shows how the amount of wastewater that can be vaporized in the proposed disposal system, in Figure 4.37, can increase when using additional fuel. Here, disposal efficiency is defined as the heat released by the biomass solids and the additional fuel divided by the total amount of heat required to vaporize all of the wastewater from the mechanical separator.

On-the-farm CB gasification systems might also solve many of the economic and practical issues with reburning and co-firing on larger coal-fired power plants discussed above. For example, synthesis gas from CB gasification may be a viable and effective reburn fuel itself. Synthesis gas can be piped to the power plant from CAFOs or centralized CB gasification facilites, instead of hauled by truck. Plus, no additional ash loading would be incurred by the coal plant. Moreover, reburn-

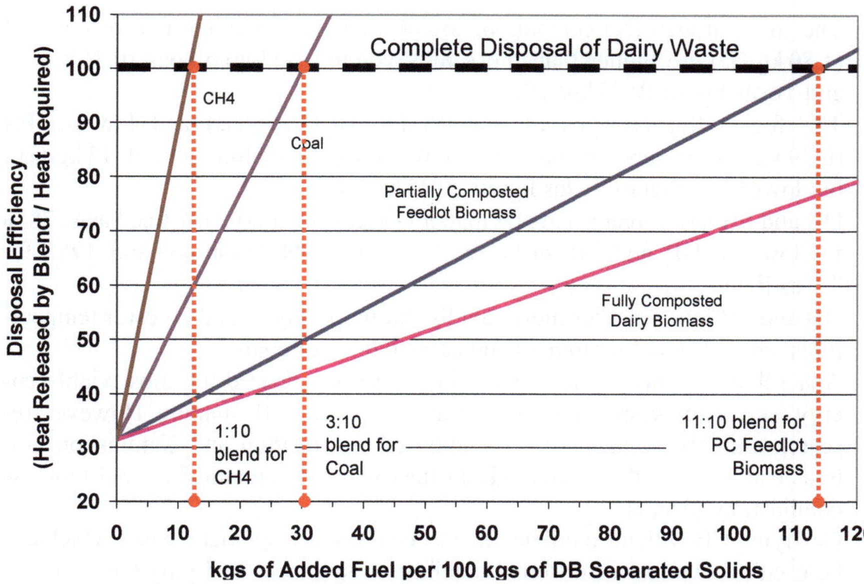

Figure 4.40 Manure disposal improvements with blending fossil fuels with separated CB solids (adopted from Carlin [21])

ing with gas requires significantly less capital costs to solid fuel reburning systems; although, the capital cost of constructing enough gasifiers to supply a suitable amount of synthetic gas to the coal plant must be taken into account. Studies by Rudiger *et al.* [86, 87] investigated the fuel nitrogen content in pyrolysis gases from both coal and wood and grass-based biomass that could possibly be used as reburn fuel. Future investigation into the nitrogen content of pyrolysis gases from CB should also be undertaken.

4.4 Summary

1. The utilization of cattle biomass (manure, CB) in combustion facilities can help ease the impacts large CAFOs, including dairies, have on the environment.
2. Large power plant facilities can benefit from CB co-combustion due to displacement of current fossil fuels such as coal and emissions reduction.
3. Dairy biomass (DB) and feedlot biomass (FB) are not high quality fuels compared to coal because of their high ash and moisture contents. Plus DB and FB have poorer heating values: 12,800 kJ/kg and 13,195 kJ/kg, respectively (both low-ash and as received basis), compared to Wyoming subbituminous which has a higher heating value of 18,190 kJ/kg.

4. The fuel nitrogen (N) contents of low-ash DB (1.51 kg/GJ) and low-ash FB (1.80 kg/GJ) are higher than those of Wyoming subbituminous (0.36 kg/GJ) and Texas lignite (0.48 kg/GJ).
5. The fuel sulfur contents of low-ash DB (0.336 kg/GJ) and low-ash FB (0.29 kg/GJ) are greater than that of Wyoming sub-bituminous (0.15 kg/GJ), but lower than that of Texas lignite (0.43 kg/GJ).
6. DB and FB have greater volatile matter contents on a dry, ash-free basis; 78 % for low ash DB and FB, 46 % for Wyoming subbituminous, and 49 % for Texas lignite.
7. DB and FB lose volatiles more rapidly during pyrolysis and at lower temperatures; about 500 K for biomass and about 620 K for coals.
8. Anaerobic digestion systems have long been considered the most viable energy conversion system for high ash and moisture cattle manure. However, research and development of non-biological gasification and combustion systems can improve the likelihood of effective and economical animal biomass combustion systems.
9. Gasifying CB with air oxidizing agents can produce synthetic gases which can be used in a variety of different combustion systems. Gasifying CB with air-steam mixture oxidizing agents can produce synthetic gases richer in hydrogen (H_2). Based on an adiabatic chemical equilibrium model, H_2 formation from CB gasification can be optimized at equivalence ratios (ERs) of 1.5–3.0 and air-steam ratios (ASRs) of 0.2–0.4. Cattle biomass gasification experiments with air-steam mixtures are currently being conducted.
10. Co-firing coal with DB or FB in boiler burners was found to produce lower or similar levels of NO (about 150 g/GJ) compared to burning pure coal, despite the fact that CB has a higher N content. This may be explained by a faster release of fuel N from CB in the form of NH_3 and biomass's high volatile content, which create local areas of rich combustion, even when the overall combustion process is slightly lean. Moreover, it was found that at most combustion regimes, the fuel N conversion efficiency to NO were lower for CB than coal.
11. Lower elemental mercury (Hg^0) emissions were measured during co-firing experiments when greater amounts of DB were blended with coal. This suggests that co-firing with CB can allow more oxidation of mercury (H^{2+}) in primary and secondary burn regions, which would in turn increase the ability of flue gas desulphurization-type devices to capture mercury.
12. Reburning coal with CB can be just as effective as, and possibly more economical than, reburning with conventional fuels like natural gas and micro-ionized coal. NO_x reductions while utilizing pure DB or FB as reburn fuel can be as high as 80 % from a base NO_x level of 600 ppm (approximately 300 g/GJ or 0.70 lb/MMBtu), mostly due to DB's and FB's high volatile matter contents and higher ammonia (NH_3)-type fuel nitrogen contents. However, newer, more effective low-NO_x burners and air-staging systems might make reburn application impractical as they reduce NO emissions from primary burn zones below 150 ppm (approximately 100 g/GJ or 0.20 lb/MMBtu).

13. Smaller-scale, on-the-farm combustion systems can be used to dispose of both solid and liquid manure waste streams, without the high costs of preparing and transporting biomass feedstock to a large coal plant. Thermal energy from on-site combustion (or gasification and subsequent combustion of volatile gases) of CB solids can then be used to vaporize between 15 and 25 % of wastewater generated by the farm, reducing required capacities of storage and treatment structures (*e.g.*, lagoons) and reducing the cost of hauling raw solid manure waste from animal farms.
14. A deterministic spreadsheet computer model for a single coal-fired unit utilizing CB as a reburn fuel was developed to gage the economic viability of retrofitting CB co-combustion systems in existing coal facilities. The relative cost of coal and the distance between the combustion facility and the animal feeding operations can greatly impact the economic viability of a biomass combustion retrofit project on an existing coal plant. However, possible future CO_2 taxes or cap and trade legislation can greatly improve the economics of biomass co-combustion.

4.5 Notation

%A	Ash percentage
AMU	Atomic mass unit
ASR	Air/steam ratio
Btu	British thermal unit
c	carbon content of fuel blend (kmoles)
CAFO	Concentrated animal feeding operation
CB	Cattle biomass
CEM	Continuous emission monitors
CH_4	Methane
CO	Carbon monoxide
CO_2	Carbon dioxide
CV	Control volume
DAF	Dry, ash free
DB	Dairy biomass
DOE	Department of Energy
DSC	Differential scanning calorimeter
%EA	Excess air percentage
EPA	Environmental Protection Agency
ER	Equivalence ratio
ESP	Electrostatic precipitator
FB	Feedlot biomass
FGD	Flue gas desulphurization
FiC	Finished/fully composted
GJ	Giga Joule, 10^9 Joules (Newton-meter)

H_2	Hydrogen
HA	High ash
HCN	Hydrogen cyanide
Hg	Mercury
Hg^0	Elemental mercury
Hg^{2+}	Oxidized mercury
Hg_p	Particulate mercury
HHV	Higher heating value
ΔH_R	Enthalpy of reaction (kJ/kg)
H_2S	Hydrogen sulfide
IC	Internal combustion
IPM	Integrated Planning Model
LA	Low ash
LB	Litter biomass
LNB	Low NO_x burner
LNBO	Low NO_x burner with over fire air
LNC1	Low NO_x burner with closed coupled over fire air
LNC2	Low NO_x burner with separated over fire air
LNC3	Low NO_x burner with both closed coupled and separated over fire air
%M	Moisture percentage
MJ	Mega Joule, 10^6 Joules (Newton-meter)
MSW	Municipal solid waste
n	Nitrogen content of fuel blend (kmoles)
N	Fuel nitrogen
N'_{CH4}	Variable used to balance CH_4 production in anaerobic digestion reaction equation
N'_{CO2}	Variable used to balance CO_2 production in anaerobic digestion reaction equation
N_{conv}	Fuel nitrogen converted to NO_x
NETL	National Energy Technology Laboratory
NG	Natural gas
N'_{H2O}	Variable used to balance reactive H2O in anaerobic digestion reaction equation
NH_3	Ammonia
NO_x	Oxides of nitrogen, may include NO, NO_2, N_2O, *etc.*
OFA	Over fire air
O&M	Operation and maintenance
PC	Partially composted
PM	Particulate matter
SCFM	Standard cubic feet per minute
SCR	Selective catalytic reduction
SNCR	Selective non-catalytic reduction
SO_x	Sulfur oxides
SRI	Southern Research Institute

TAMU	Texas A&M University
TGA	Thermal gravimetric analyzer
Ton	2,000 lb (907.18 kg)
TSP	Total suspended particles
TXL	Texas lignite
VG ProLAB	Mass spectrometer
VM 3000	Vapor monitoring instrument
WYPRB	Wyoming Powder River Basin
X	kmole fraction
$	US dollars

Acknowledgements

The present work has been supported with grants from the DOE-National Renewable Energy Laboratory, Grant #DE-FG36-05GO85003 and Texas Commission on Environmental Quality (TCEQ), Grant #582-5-65591 0015.

References

[1] McFarland A, Millican J. Assessment of preexisting and post-implementation effects for the North Bosque River Watershed: Clean Water Act section 319 report. Texas Institute for Applied Environmental Research (TiAER), Tarleton State University, Stephenville, Texas. Available online at: http://www.tiaer.tarleton.edu. 2006.

[2] TNRCC. Two total maximum daily loads for phosphorus in the North Bosque River: for segments 1226 and 1255. Prepared by the Strategic Assessment Division of the Texas Natural Resource Conservation Commission, 2001.

[3] Centner TJ. Empty pastures: confined animals and the transformation of the rural landscape. Urbana and Chicago, IL: University of Illinois Press, 2004.

[4] Keplinger K, Tanter A, Hauck L. Manure transportation and application: model development and a general application. Texas Institute for Applied Environmental Research (TiAER), Tarleton State University, Stephenville, Texas. Available online at: http://www.tiaer.tarleton.edu. 2004.

[5] NASS. U.S. dairy herd structure: large operations increase share of milk production. National Agricultural Statistics Service. Available online at: http://usda.mannlib.cornell.edu/reports/nassr/livestock/dairy-herd/specda02.pdf. 2002.

[6] Probstein RF, Hicks RE (2006b). Synthetic fuels, Chapter 8: Biomass conversion. Mineola, New York: Dover Publications, Inc., 2006.

[7] Probstein RF, Hicks RE. Synthetic fuels, Chapter 4: Gas from coal. Mineola, New York: Dover Publications, Inc., 2006.

[8] NASS. Cattle on feed: US cattle on feed down 5 percent, August 2007. National Agricultural Statistics Service of the United States Department of Agriculture. Available online at: http://usda.mannlib.cornell.edu/usda/current/CattOnFe/CattOnFe-08-17-2007.pdf. 2007.

[9] Harman WL. Pen cleaning costs for dust control, Southern Great Plains feedlots. Texas Agricultural Experiment Station, Texas A&M University, BRC Report No. 04-01, 2004.

[10] DPI&F. Feedlot waste management series: manure production data. Department of Primary Industries and Fisheries. Government of Queensland, Australia. Available online at: http://www.dpi.qld.gov.au/environment/5166.html. 2003.
[11] CDFA. California dairy facts. California Dairy Research Foundation of the California Department of Food and Agriculture. Available online at: http://www.cdrf.org. 2006.
[12] Schmidt GH, Van Vleck LD, Hutjens MF. Principles of dairy science, second edition, Chapter 27: Dairy cattle housing. Englewood Cliffs, NJ: Prentice Hall, 1988.
[13] USDA. Economics, statistics, and market information system. United States Department of Agriculture. Available online at: http://usda.mannlib.cornell.edu/. 2007.
[14] Osei E, Gassman PW, Jones RD, Pratt SJ, Hauck LM, Beran WD, *et al.* Economic and environmental impacts of alternative practices on dairy farms in an agricultural watershed. J Soil Water Conservation 2000;55(4):466–472.
[15] Sweeten JM. Texas Agricultural Extension Service Publication L-1094. Texas Agricultural Extension Service, Texas A&M University, College Station, Texas, 1979.
[16] Mukhtar S. Proper lagoon management to reduce odor and excessive sludge accumulation. Texas A&M University, Texas Agricultural Extension Service. Available online at: http://tammi.tamu.edu/pdf%20pubs/lagoonmanagement.pdf. 1999.
[17] Sweeten JM, Annamalai K, Thien B, McDonald L. Co-firing of coal and cattle feedlot biomass (FB) fuels, Part I: feedlot biomass (cattle manure) fuel quality and characteristics. Fuel 2003;82(10):1167–1182.
[18] Rodriguez P, Annamalai K, Sweeten JM. Effects of drying on heating values of biomass. Soc Agric Trans 1998;41:1083–1087.
[19] Annamalai K, Ishwar P. Combustion science and engineering. Orlando, FL: Taylor and Francis, 2006.
[20] Annamalai K, Sweeten JM, Ramalingam SC. Estimation of the gross heating values of biomass fuels. Trans Soc Agric Eng 1987;30:1205–1208.
[21] Carlin NT. Thermo-chemical conversion of dairy waste based biomass through direct firing. MS Thesis, Texas A&M University, College Station, Texas, 2005.
[22] Martin B, Annamalai K, Sweeten JM, Heflin K. Pyrolysis, ignition, and fuel characteristics of coal, feedlot biomass, and coal:feedlot biomass blends. Submitted to the Texas Commission on Environmental Quality, 2006.
[23] Raman P, Walawender WP, Fan LT, Howell JA. Thermogravimetric analysis of biomass. Devolatilization studies on feedlot manure. Ind Eng Chem Process Des Dev 1981; 20:630–636.
[24] Global Energy Decisions. Coal view interactive: short-term forecast for 16 coals. Available online at: http://www.globalenergy.com. 2006.
[25] TAMU Fuel Data Bank. Texas A&M University Coal and Biomass Energy Laboratory. Available online at:
http://www1.mengr.tamu.edu/REL/TAMU%20FDB.htm. 2006.
[26] Arcot Vijayasarathy U. Mercury emission control for coal fired power plants using coal and biomass. MS Thesis, Texas A&M University, College Station, Texas, 2007.
[27] DOE. Clean coal technology: reburning technologies for the control of nitrogen oxides emissions from coal-fired boilers. United States Department of Energy Topical Report Number 14. 1999. Available online at:
http://www.fossil.energy.gov/programs/powersystems/publications.
[28] Srivastava RK, Hall RE, Khan S, Culligan K, Lani BW. Nitrogen oxides emissions control options for coal-fired electric utility boilers. J Air Waste Manage Assoc 2005; 55:1367–1388.
[29] Pacala S, Socolow R. Stabilization wedges: solving the climate problem for the next 50 years with current technologies. Sci Mag 2004;305:968–972.
[30] Di Nola G. Biomass fuel characterization for NO_x emissions in co-firing applications. Doctoral Thesis, Delft University of Technology, Delft, the Netherlands, 2007.
[31] EIA. US carbon dioxide emissions from energy sources 2006 flash estimate. Energy Information Administration. Available online at: http://www.eia.doe.gov. 2007.

[32] Annamalai K, Priyadarsan S, Arumugam S, Sweeten JM. Energy conversion: principles for coal, animal waste, and biomass fuels. Encyclopedia Energy Eng Technol 2007; 1(1):476–497.
[33] Wen Z, Frear C, Chen S. Anaerobic digestion of liquid dairy manure using a sequential continuous-stirred tank reactor system. J Chem Technol Biotechnol 2007;82:758–766.
[34] Kohl A, Riesenfeld F. Gas Purification, 2nd edn. Houston, TX: Gulf Publishing Co., 1974.
[35] Krich K, Augenstein D, Batmale JP, Benemann J, Rutledge B, Salour D. Biomethane from dairy waste: a sourcebook for the production and use of renewable natural gas in California. Prepared for The Western United Dairymen, Funded in part through USDA Rural Development, 2005.
[36] Monnet F. An introduction to anaerobic digestion of organic wastes: final report, 2003. Remade Scotland. Available online at: http://www.remade.org.uk/Organics/organics_documents/IntroAnaerobicDigestion.pdf.
[37] Ghaly AE. A comparative study of anaerobic digestion of acid cheese whey and dairy manure in a two-stage reactor. Bioresource Technol 1996;58:61–72.
[38] Chen YR, Varel VH, Hashimoto AG. Chemicals from cellulosic materials. Ind Eng Chem Prod Res Dev 1980;19:471–477.
[39] USEPA. Guide to operational systems: U.S. operating digesters by state. The AgSTAR program of the United States Environmental Protection Agency. Last update November 2. Available online at: http://epa.gov/agstar/operation/bystate.html. 2007.
[40] Associated Press. Texas plant uses dairy waste to make natural gas. By Brown AK, Associated Press Writer. The Associated Press State and Local Wire, Business News Section, November 5, 2007.
[41] Goodrich PR, Schmidt D, Haubenschild D. Anaerobic digestion for energy and pollution control. Agric Eng Int: the CIGR Ejournal. Manuscript EE 03 001, Volume VII, 2005.
[42] Meyer D. Digesters on the farm: making electricity from manure. BioCycle: J Composting Organics Recycling 2003;44(1):52–55.
[43] Chang FH. Energy and sustainability comparisons of anaerobic digestion and thermal technologies for processing animal waste. 2004 ASAE/CSAE Annual International Meeting, Fairmont Chateau Laurier, The Westin, Government Centre, Ottawa, Ontario, Canada, 2004.
[44] Simons G, Zhang Z. California continues push for dairy power. BioCycle: J Composting Organics Recycling 2003;44(7):63–64.
[45] Hotchkiss R. Coal gasification technologies. Proc Inst Mech Eng Part A, J Power Energy 2003;217(A1):27–33.
[46] Howard-Smith I. Coal conversion technology. Noyes Data Corporation/Noyes Publications, Park Ridge, New Jersey, 1976.
[47] Nowacki P. Coal gasification processes. Noyes Data Corporation, Park Ridge, New Jersey, 1981.
[48] Quaak P, Knoef H, Stassen H. Energy from biomass: a review of combustion and gasification technologies. World Bank Technical Paper No. 422, The World Bank, Washington, DC, 1999.
[49] Young L, Pian CCP. High-temperature, air-blown gasification of dairy-farm wastes for energy production. Energy 2003;28:655–672.
[50] Skill Associates. Elimanure™. Available online at: http://www.burnmanure.com/management/elimanure.html. 2005.
[51] Sami M, Annamalai K, Wooldridge M. Co-firing of coal and biomass fuel blends. Prog Energy Combust Sci 2001;27:171–214.
[52] Priyadarsan S, Annamalai K, Sweeten JM, Holtzapple M. Co-gasification of blended coal with feedlot and chicken litter biomass. Proc Combust Inst 2005;30:2973–2980.
[53] Sweeten JM, Korenberg J, LePori WA, Annamalai K, Parnell CB. Combustion of cattle feedlot manure for energy production. Energy in Agric 1986;5:55–72.
[54] Annamalai K, Ibrahim MY, and Sweeten JM. Experimental studies on combustion of cattle manure in a fluidized bed combustor. J Energy Res Technol 1987;109:49–57.

[55] Ràman P, Walawender WP, Fan LT, Chang CC. Mathematical model for the fluid-bed gasification of biomass materials. Application to feedlot manure. Ind Eng Chem Process Des Dev 1981;20:686–692.
[56] Panda Ethanol Inc. Manure gasification: using renewable fuel to create renewable fuel. Available online at: http://pandaethanol.com/gas/index.html. 2007.
[57] Megel AJ, DeOtte RE, Robinson CA. Investigation of economically viable coproducts developed from ash from the combustion of manure. 2007 ASABE Annual International Meeting, Minneapolis Convention Center, Minneapolis, Minnesota, 2007.
[58] Priyadarsan S, Annamalai K, Thien B, Sweeten JM, Mukhtar S. Animal waste as a source of renewable energy. In Outlaw JL, Collins KJ, Duffield JA (eds), Chapter 16 of the USDA-Source Book on Animal Waste as a Source of Renewable Energy, 2005.
[59] NASA. Chemical equilibrium with applications (CEA), chemical equilibrium computer program. Curator: Chao MM. Last updated on January 17, 2007. Available online at: http://www.grc.nasa.gov/WWW/CEAWeb/. 2007.
[60] Gordillo G, Annamalai K, Arcot VU. Gasification of coal and animal waste using an air-steam mixture as oxidizing agent. 19th National and 8th ISHMT-ASME, Heat and Mass Transfer Conference, JNTU Hyderabad, India, 2008.
[61] Priyadarsan S. Fixed bed gasification studies on coal-feedlot biomass and coal-chicken litter biomass under batch mode operation. MS Thesis, Texas A&M University, College Station, Texas, 2002.
[62] Priyadarsan S, Annamalai K, Sweeten JM, Mukhtar S, Holtzapple MT. Fixed bed gasification of feedlot and poultry litter biomass. Trans ASAE 2005;47(5):1689–1696.
[63] Frazzitta S, Annamalai K, Sweeten JM. Performance of a burner with coal and coal:feedlot manure blends. J Propul Power 1999;15:181–186.
[64] Arumugam S, Annamalai K, Thien B, Sweeten JM. Feedlot biomass co-firing: a renewable energy alternative for coal-fired utilities. Int J Green Energy 2005;2:409–419.
[65] Annamalai K, Sweeten JM, Mukhtar S, Thien B, Wei G, Priyadarsan S. Co-firing coal: feedlot and litter biomass (CFB and CLB) fuels in pulverized fuel and fixed bed burners. Final Department of Energy Report. DOE-Pittsburgh Contract #40810, 2003.
[66] Annamalai K, Thien B, Sweeten JM. Co-firing of coal and cattle feedlot biomass (FB) fuels, Part II: performance results from 100,000 Btu/hr laboratory scale boiler burner. Fuel 2003;82(10):1183–1193.
[67] Thomas N, Lucht A, Lucht RP, Priyadarsan S, Annamalai K, Caton JA. *In situ* measurements of nitric oxide in coal-combustion exhaust using a sensor based on a widely-tunable external-cavity GaN diode laser. J Appl Optics 2006;46(19):3946–3957.
[68] Annamalai K, Sweeten JM, Freeman M, Mathur M, O'Dowd W, Walbert G, *et al.* Co-firing of coal and cattle feedlot biomass (FB) fuels, Part III: fouling results from a 500,000 Btu/hr pilot plant scale boiler burner. Fuel 2003;82(10):1195–1200.
[69] Annamalai K, Sweeten JM. Reburn system with feedlot biomass. US Patent #6,973,883 B1, 2005.
[70] Lawrence BD. Co-firing of coal and dairy biomass in a 100,000 Btu/hr furnace. MS Thesis, Texas A&M University, College Station, Texas, 2007.
[71] USDA, Cornell University, the University of Vermont. Nitrogen management on dairy farms: types of N in manure. Available online at: http://www.dairyn.cornell.edu/pages/20cropsoil/250credits/251manTypes.shtml. 2007.
[72] Robinson AL, Rhodes JS, Keith DW. Assessment of potential carbon dioxide reductions due to biomass-coal co-firing in the United States. Environ Sci Technol 2003; 37(22):5081–5089.
[73] Lissianski VV, Zamansky VM, Maly PM. Effect of metal-containing additives on NO_x reduction in combustion and reburning. Combust Flame 2001;125(3):1118–1127.
[74] Yang YB, Naja TA, Gibbs BM, Hampartsoumian E. Optimisation of operating parameters for NO reduction by coal reburning in a 0.2 MWt furnace. J Inst Energy 1997;70:9–16.
[75] Annamalai K, Thien B. Feedlot manure as reburn fuel for NOx reduction in coal fired plants. National Combustion Conference. Oakland, California, 2001.

[76] Oh H, Annamalai K, Sweeten JM. Investigations of ash fouling with cattle wastes as reburn fuel in a small boiler burner under transient conditions. Submitted to the J. Air Waste Manage Assoc on March 14, 2007.
[77] Arumugam S. Nitrogen oxides emission control through reburning with biomass in coal-fired power plants. MS Thesis, Texas A&M University, College Station, Texas, 2004.
[78] Colmegna G, Annamalai K, Udayasarathy AV. Modeling of Hg reduction during reburn process with the use of cattle biomass as a reburn fuel. Third International Green Energy Conference (IGEC-III), Västerås, Sweden, 2007.
[79] USEPA. Documentation for EPA base case 2006 (V.3.0) using the integrated planning model. Clean Air Markets Division of the United States Environmental Protection Agency, EPA Contract No. 430-R-05-011. Available online at: http://www.epa.gov/airmarkets/progsregs/epa-ipm/. 2006.
[80] USEPA. Analyzing electric power generation under the CAAA, Appendix No. 5. A technical report by the United States Environmental Protection Agency. 1998.
[81] Zamansky VM, Sheldon MS, Lissianski VV, Maly PM, Moyeda DK, Marquez A, *et al.* Advanced biomass reburning for high efficiency NO_x control and biomass reburning – modeling/engineering studies. USDA Phase II SBIR, Contract #97-33610-4470 and DOE-NETL, Contract #DE-FC26-97FT-97270. Available online at: http://www.osti.gov/bridge/servlets/purl/786516-Dddr9I/native/786516.pdf. 2000.
[82] Carlin N, Annamalai K, Sweeten JM, Mukhtar S. Thermo-chemical conversion analysis on dairy manure-based biomass through direct combustion. Int J Green Energy 2007;4:1–27.
[83] Carlin N, Annamalai K, Sweeten JM, Mukhtar S. Utilization of latent heat derived from vaporized wastewater in high moisture dairy manure combustion schemes. International Symposium on Air Quality and Waste Management for Agriculture. Broomfield, Colorado, 2007.
[84] Kolber SN. Treatment of waste produced by farm animals raised under confined conditions. US Patent #6,190,566, 2001.
[85] Mooney R, Mooney D, Latulippe D. Method and apparatus for the gasification and combustion of animal waste, human waste, and/or biomass using a moving grate over a stationary perforated plate in a configured chamber. US Patent #6,948,436 B2, 2005.
[86] Rudiger H, Kicherer A, Greul U, Spliethoff H, Hein KRG. Investigations in combined combustion of biomass and coal in power plant technology. Energy Fuels 1996;10:789–796.
[87] Rudiger H, Greul U, Spliethoff H, Hein KRG. Distribution of fuel nitrogen in pyrolysis products used for reburning. Fuel 1997;76(3):201–205.
[88] Sweeten JM, Heflin K. Preliminary interpretation of data from proximate, ultimate and ash analysis, results of June 7, 2006, samples taken from feedlot and dairy biomass biofuel feedstocks at TAES/USDA-ARS, Bushland, TX. Amarillo/Bushland/Etter, Texas, Texas A&M Agricultural Research & Extension Center, 2006.
[89] Maly PM, Zamansky VM, Ho L, Payne R. Alternative fuel reburning. Fuel 1999; 78:327–334.
[90] Sweterlitsch JJ, Brown RC (2002) Fuel lean biomass reburning in coal-fired boilers. Final technical report to the DOE, Award number DE-FG26-00NT40811. Submitted by the Center for Sustainable Environmental Technologies, Iowa State University.

Chapter 5
Polymer Electrolyte Fuel Cell Modeling – a Pore-scale Perspective

Partha P. Mukherjee and Chao-Yang Wang

5.1 Introduction

Fuel cells, being highly energy efficient, environmentally benign, and minimally noisy, are widely considered as the 21st century energy-conversion devices for mobile, stationary, and portable power. Unlike the conventional Carnot cycle based energy conversion devices with intermediate heat and mechanical energy generation, fuel cells convert the chemical energy of a fuel directly into electricity. Among the several types of fuel cells, the polymer electrolyte fuel cell (PEFC) has emerged as the most promising power source for a wide range of applications.

A typical PEFC, shown schematically in Figure 5.1, consists of seven subregions: the anode gas channel, anode gas diffusion layer (GDL), anode catalyst layer (CL), ionomeric membrane, cathode CL, cathode GDL, and cathode gas channel. The proton-exchange membrane electrolyte is a distinctive feature of the PEFC. Usually, the two thin CLs are coated on both sides of the membrane, forming a membrane–electrode assembly (MEA). The anode feed, generally, consists of hydrogen, water vapor, and nitrogen or hydrogen/water binary gas, whereas humidified air is fed into the cathode. Hydrogen and oxygen combine electrochemically within the active CL to produce electricity, water and waste heat. The GDL allows the transport of reactants to and products out from the reaction sites and also conducts electrons. Gottesfeld and Zawodzinski [1] provided a comprehensive overview of the PEFC function and operation.

Partha P. Mukherjee
Los Alamos National Laboratory, New Mexico, USA
e-mail: partham@lanl.gov

Chao-Yang Wang
Director, Electrochemical Engine Center, Department of Mechanical Engineering and Materials Science and Engineering, The Pennsylvania State University, 338 A Reber Building, University Park, PA 16802, USA
e-mail: cxw31@psu.edu

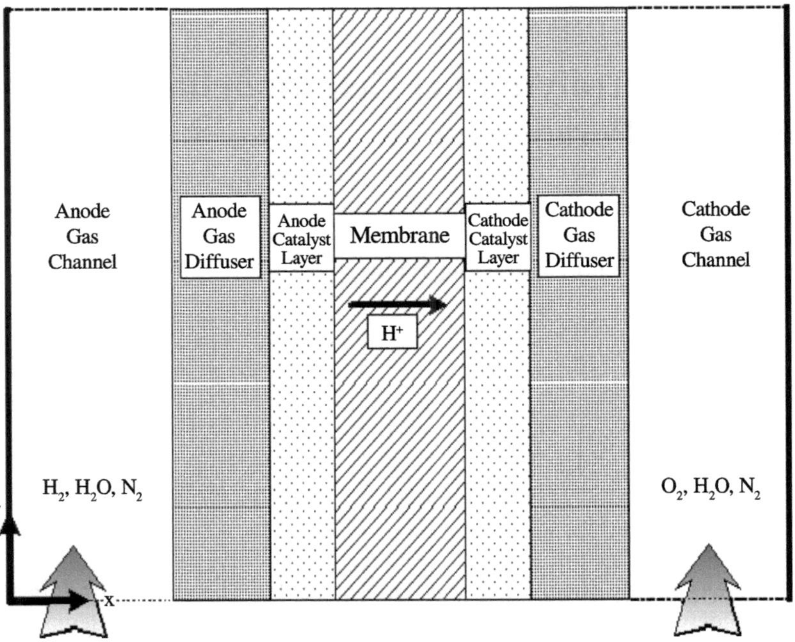

Figure 5.1 Schematic diagram of a polymer electrolyte fuel cell

The hydrogen oxidation reaction (HOR) occurs at the anode side CL and generates protons and electrons according to the following reaction:

$$H_2 \rightarrow 2H^+ + 2e^- \tag{5.1}$$

The electrons traverse through the external circuit, while the protons migrate through the ionomeric membrane and combine with oxygen at the cathode CL to produce water according to the following oxygen reduction reaction (ORR).

$$O_2 + 4H^+ + 4e^- \rightarrow 2H_2O \tag{5.2}$$

Thus, the overall cell reaction is:

$$2H_2 + O_2 \rightarrow 2H_2O \tag{5.3}$$

HOR has orders of magnitude higher reaction rate than ORR, which leaves ORR as a potential source of large voltage loss in PEFCs and hence the cathode CL is the electrode of primary importance in a PEFC. Due to the acid nature of the polymer membrane and low-temperature operation, Pt or Pt alloys are the best-known catalysts for PEFCs.

The performance of a PEFC is characterized by the *polarization curve* giving the relation between cell voltage and current density. Figure 5.2 shows a typical polarization curve for a PEFC with three distinct voltage loss regimes. At low current density operation, the voltages loss is primarily due to the sluggish ORR at

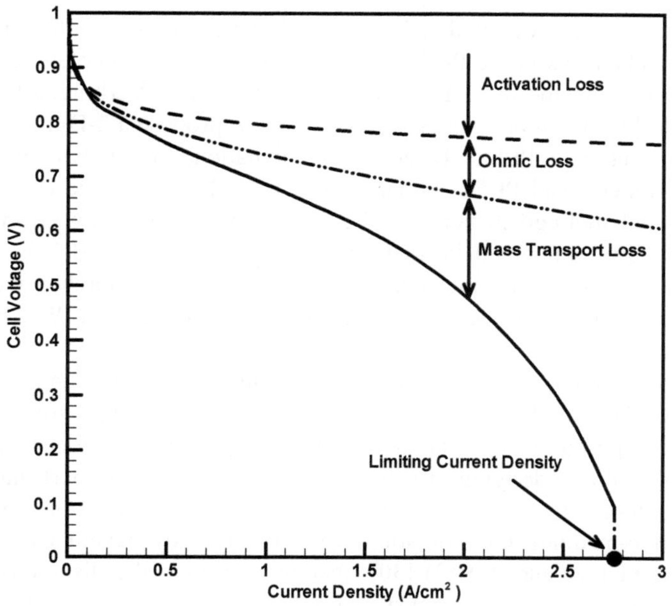

Figure 5.2 Typical polarization curve of a PEFC

the cathode CL and is referred to as the "kinetic loss" or "activation loss". At intermediate current densities, the voltage loss characterized by resistance to ion transport in the polymer electrolyte membrane and the CLs dominates and is known as "ohmic loss". At high current density operation, which is of particular interest to vehicular applications to obtain high power density, "mass transport limitations" come into play due to the excessive liquid water build up mainly in the cathode side. Liquid water blocks the porous pathways in the CL and GDL thus causing hindered oxygen transport to the reaction sites as well as covers the electrochemically active sites in the CL thereby increasing surface overpotential. This phenomenon is known as "flooding" and is perceived as the chief mechanism leading to the limiting current behavior in the cell performance.

In addition to the voltage loss incurred due to the slow kinetics of the ORR, it is apparent from the polarization curve that the cell performance is primarily limited by voltage losses stemming from the widely disparate underlying transport mechanisms in the CL and GDL, namely: (1) oxygen transport resistance through the gas phase as well as the electrolyte phase; (2) resistance to proton and electron transport; and (3) transport loss in the presence of liquid water and flooding. It is worth mentioning that liquid water is needed to maintain sufficient humidification level in the ionomeric membrane to reduce protonic resistance, while excessive liquid water causes flooding in the CL and GDL. Water management, aimed at maintaining a delicate balance between reactant transport from the gas channels and water removal from the electrochemically active sites, has received wide attention in the PEFC research.

Different approaches have been undertaken in the literature to model the underlying transport mechanisms in the PEFC with different levels of complexities. Notable earlier attempts include the 1-D models by Bernardi and Verbrugge [2, 3] and Springer *et al.* [4, 5]. Gurau *et al.* [6] was the first to propose a 2-D model of transport phenomena in a PEFC. Later, the computational fluid dynamics (CFD) based multi-dimensional PEFC model by Um *et al.* [7] led to the emergence of computational fuel cell dynamics (CFCD) as a major modeling effort. Several studies have been attempted in recent years to specifically model two-phase behavior and flooding phenomena in the PEFC [8–27]. Comprehensive overviews of the various PEFC models were furnished by Wang [28] and Weber and Newman [29].

In most of the macroscopic models reported in the literature, the CL and the GDL, responsible for the complex transport mechanisms in a PEFC, were treated as macrohomogeneous porous layers. Due to the macroscopic nature, the current models fail to resolve the underlying structural influence on the transport and performance. Furthermore, these models employ constitutive closure relations based on the effective-medium approximation (EMA), originally developed by Bruggeman (1935) and Landauer (1952) [30], thereby replacing the disordered porous medium with an equivalent uniform system with certain effective transport properties, which mimic the actual medium. The situation is even worse in the two-phase regime where liquid water covers the active reaction sites in the CL rendering reduced catalytic activity as well as blocks the porous pathways in the CL and the GDL which causes hindered oxygen transport to the active reaction sites and results in CL and GDL flooding and severe performance degradation. This leads to an extra level of complexity in macroscopic two-phase fuel cell models in terms of employing appropriate two-phase closure relations, namely capillary pressure and relative permeability as functions of liquid water saturation. It is important to note that liquid water saturation refers to the volume fraction of the total void space of the porous medium occupied by liquid water. Unfortunately, no such correlations for the CL are available in the literature, primarily because a viable experimental approach in obtaining such constitutive relations in the 10 μm thick structure is exceedingly difficult and probably impossible in the foreseeable future. Similarly, there is a serious scarcity of reliable two-phase correlations tailored specifically for the non-woven and woven GDLs. Current two-phase fuel cell models often deploy a capillary pressure–saturation relation for modeling liquid water transport in hydrophobic gas diffusion media adapted by Pasaogullari and Wang [9] and Nam and Kaviany [21] from Udell's work [31] in the form of the Leverett-J function [32]. Very recently a few attempts to experimentally evaluate the capillary pressure for the PEFC GDL have emerged in the literature [33–36], while the CL still remains experimentally intractable. Furthermore, no experimental efforts to measure relative permeability for both GDL and CL, more important for two-phase transport than capillary pressure, have been made. Despite substantial research, both theoretical and experimental, there is serious paucity of fundamental understanding about the overall structure–transport–performance interactions as well as the underlying two-phase dynamics in the CL and the GDL. This leads to the following outstanding issues pertaining to transport modeling in the PEFC CL and GDL:

- What is a reasonable estimate of the resistance to oxygen diffusion in the CL and GDL as well as ion transport in the CL owing to the underlying complex morphology of the respective structures? Can we reasonably predict Bruggeman factors [30] for effective transport property estimation which could be used as valuable inputs to the macrohomogeneous models?
- What is a viable solution toward predicting reliable two-phase correlations, namely capillary pressure and relative permeability as functions of liquid water saturation in the CL and GDL, which can be employed as valuable closure relations in two-phase computational fuel cell dynamics (CFCD) models?
- What could be a viable solution toward quantitative estimation of the catalytic site coverage effect in the CL and the pore blockage effect in the CL and GDL due to liquid water, which can be further used as inputs into macroscopic two-phase models?
- What is the influence of the structural and wetting characteristics on liquid water transport in the CL and GDL?

Pore-scale modeling approaches offer excellent versatility in the fundamental investigation of transport phenomena owing to the detailed information available at the microscopic scale. Specifically, pore-scale modeling shows tremendous prospect in the following aspects.

- Pore-scale modeling provides unique opportunities to understand the underlying processes, *e.g.*, multiphase dynamics occurring within the complex porous structure, which are still unknown due to the scarcity of the viable experimental methods as well as to predict the inherent structure–performance dependence.
- Pore-level simulation tools offer excellent flexibility in designing numerical experiments conforming to the experimental operating conditions in order to evaluate constitutive closure relations, *e.g.*, two-phase correlations in terms of capillary pressure and relative permeability as functions of saturation, which are extremely difficult and probably impossible to be obtained by laboratory experiments.

Recent advances in microstructure generation techniques, allowing accurate representation of the underlying pore-morphology, and the quantum jump in computational power through the availability of relatively inexpensive high-performance computers over the last decade are the major reasons for the explosion of interest in predictive pore-scale modeling.

This chapter presents a systematic account of recent efforts in pore-scale modeling to gain fundamental insight into two-phase transport along with the evaluation of the salient transport properties pertaining to the CL and the GDL of a PEFC and demonstrates a vertical approach encompassing a pore-scale to macro-scale modeling strategy.

5.2 Pore-scale Modeling

Pore-scale models for solving flow, specifically multiphase flow, and transport through porous media can be broadly classified into rule-based and first-principle-based models [37].

Rule-based models try to capture physical processes pertaining to certain flow regimes, namely capillary fingering, stable displacement and viscous fingering, which can be perceived as limiting cases where certain assumptions can be specifically made. Percolation-based models, diffusion-limited aggregation (DLA) model and anti-DLA model belong to this category. Such models, based on diluted physical description, nonetheless provide useful predictions of the transport properties of a given medium or network. Simulations based on these models are generally faster since there is no need to solve large systems of equations. The most prominent among these rule-based models is the percolation-based pore-network modeling approach. Such network models are commonly used as investigative tools to study a variety of processes in porous media [37]. Dullien [38], Sahimi [39], and Blunt [40] provided extensive literature reviews of pore-network models. The pore-network model, being computationally less demanding compared to the first-principle-based models, could be used to simulate flow at a laboratory scale, or larger. However, the use of pore-network modeling is primarily limited to simplified idealized porous media owing to its difficulty in adequately describing the complex pore morphology in real porous media. Additionally, the deployment of overly simplified invasion rules in describing the underlying displacement processes deters such network models from being applicable to generalized flow processes, namely a transition zone where viscous as well as capillary effects might be important.

Unlike the rule-based approaches, the first-principle-based approaches resolve the underlying transport processes by solving the governing partial differential equations (PDE), namely the Navier–Stokes (NS) equations. The PDEs can be solved by employing either the fine-scale conventional computational fluid dynamics (CFD) methods, the so-called "top-down approach" or by the coarse-grain approach *i.e.*, the "bottom-up" approach, shown schematically in Figure 5.3 [41]. The molecular dynamics (MD), lattice gas (LG) and, lattice Boltzmann (LB) methods fall under the "bottom-up" approach category. With a given set of suitable boundary conditions, the governing differential equations can be properly discretized on a computational grid using standard CFD techniques, namely finite difference, finite volume or finite element methods. However, the lack of versatility of implementing the boundary conditions for arbitrary grain shapes precludes the application of CFD methods in a porous medium involving two-phase flow. Even though several CFD-based two-phase models, such as multi-field, interface-tracking, and volume of fluid approaches, have been developed, their applicability is limited to idealized solid wall structures (channels, plates, stair-step, *etc.*) [37]. Hardly any such attempt is known yet which utilizes fine-scale CFD to simulate two-phase flow in real porous media.

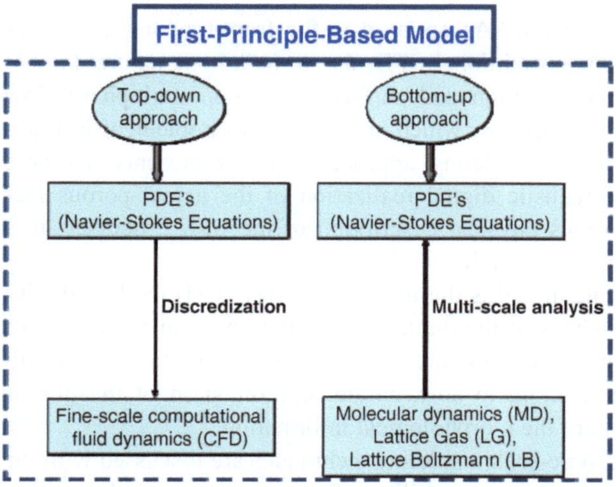

Figure 5.3 Top-down and bottom-up numerical approaches [41]

On the other hand, the MD approach [42, 43] takes into account the movements and collisions of all individual molecules constituting the fluid with detail description of the inter-molecular interactions and thereby providing realistic equations of state characterizing the real fluid. However, the complexity of interactions as well as the number of molecules representative of the actual fluid make the molecular dynamics models computationally prohibitive for application to macroscopic flows in porous media.

The LB method and its predecessor the LGlattice gas method, instead of tracking all the individual molecules, consider the behavior of a collection of particles, comprised of large number of molecules, moving on a regular lattice, thereby reducing the degrees of freedom of the system and make the pore-scale simulation computationally tractable. Owing to its excellent numerical stability and constitutive versatility, the lattice Boltzmann method has developed into a powerful technique for simulating fluid flows in recent years and is particularly successful in fluid flow applications involving interfacial dynamics and complex geometries [44, 45]. Wolf-Gladrow [41] and Succi [46] provided a formal description of the evolution of the LB method from its predecessor, the LG method [47, 48]. Unlike the conventional Navier–Stokes solvers based on the discretization of the macroscopic continuum equations, LB methods consider flows to be composed of a collection of pseudo-particles residing on the nodes of an underlying lattice structure which interact according to a velocity distribution function. The LB method is an ideal scale-bridging numerical scheme which incorporates simplified kinetic models to capture microscopic or mesoscopic flow physics and yet the macroscopic averaged quantities satisfy the desired macroscopic equations [44]. Due to its underlying kinetic nature, the LB method has been found to be particularly useful in applications involving interfacial dynamics and complex boundaries, *e.g.*, multiphase or multicomponent flows, and flow in porous medium. As opposed to the front-tracking and front-capturing multiphase models in traditional CFD, due to its kinetic nature, the LB model incorporates phase segregation and surface tension in multiphase flow through interparticle force/interactions, which are difficult to implement in traditional methods. While the LB modeling approach better represents the pore morphology in terms of a realistic digital realization of the actual porous medium and incorporates rigorous physical description of the flow processes, it is however computationally very demanding.

Another type of pore-scale model is the full morphology (FM) model [49, 50] based on morphological analysis of the digital representation of an actual porous medium. The FM model aims to link the macroscopic static properties, *e.g.*, capillary pressure – saturation relation, to an accurate representation of the porous medium through direct input of the morphological information.

In this chapter, different pore-scale modeling techniques are discussed with the objective to understand the structure-wettability influence on the underlying two-phase dynamics and evaluate the associated constitutive transport parameters in the CL and the GDL of a PEFC.

5.3 Microstructure Reconstruction

Detailed description of a porous microstructure, which is an essential prerequisite to pore-scale modeling, can be obtained in the form of 3-D volume data either by experimental imaging or by stochastic simulation method.

Several experimental techniques can be deployed to image the pore structure in three dimensions. Earlier attempts include employing destructive serial sectioning of pore casts to reconstruct the complex pore space. Recently, non-invasive techniques such as X-ray and magnetic resonance micro-tomography and scanning laser confocal microscopy are the preferred choices over the earlier destructive methods. Additionally, 3-D porous structure can be generated using stochastic simulation technique, which creates 3-D replicas of the random microstructure based on specified statistical information obtained from high-resolution 2-D micrographs of a porous sample. The low cost and high speed of data generation make the stochastic generation methods the preferred choice over the experimental imaging techniques.

5.3.1 CL Structure Generation

For the electrochemical reaction to occur, the state-of-the-art CL of a PEFC is a three-phase composite comprising: (1) ionomer, *i.e.*, the ionic phase which is typically Nafion® to provide a passage for protons to be transported in or out; (2) Pt catalysts supported on carbon, *i.e.*, the electronic phase for electron conduction; and (3) pores for the reactant to be transferred in and product water out. Gottesfeld and Zawodzinski [1] provided a good overview of the CL structure and functions. In the present study, the CL is delineated as a two-phase (pore/solid) structure consisting of the gas phase (*i.e.*, the void space) and a mixed electrolyte/electronic phase (*i.e.*, the solid matrix). The assumption of the mixed electrolyte/electronic phase is well justified from the perspective of ion transport in the electrolyte phase as the limiting mechansim as compared to the electron conduction via the electronic (C/Pt) phase within the CL and henceforth is referred to as the "electrolyte" phase [51, 52].

The stochastic reconstruction method is based on the idea that an arbitrarily complex porous structure can be described by a binary phase function, which assumes the value 0 in the pore space and 1 in the solid matrix [53]. The intrinsic randomness of the phase function can be adequately qualified by the low order statistical moments, namely porosity and two-point autocorrelation function [53]. Adler and Thovert [54] and Torquato [55] provided comprehensive overviews of the statistical description of porous microstructures.

Details of the CL microstructure reconstruction along with the underlying assumptions are elaborated in recent work [51], which is based on the slightly modified version by Bentz *et al.* [56] of the stochastic generation method originally proposed by Quiblier [57] and was later applied to the reconstruction of Fontainbleau sandstone by Adler *et al.* [58]. In brief, the stochastic reconstruction technique starts with a Gaussian distribution which is filtered with the two-point autocorrelation function and finally thresholded with the porosity, which creates the 3-D realization of the CL structure [51, 56]. The autocorrelation function is computed from a 2-D TEM (transmission electron microscope) image of an actual CL based on the image processing technique proposed by Berryman [59]. The porosity can be calculated by converting the mass loading data of the constituent components available from the CL fabrication process [60]. The pore/solid phase is further distinguished as "transport" and "dead" phase. The basic idea is that a pore phase unit cell surrounded by solid phase-only cells does not take part in species transport and hence in the electrochemical reaction and can, therefore, be treated as a dead pore and similarly for the electrolyte phase. The interface between the transport pore and the transport electrolyte phases is referred to as the electrochemically active area (ECA) and the ratio of ECA and the nominal CL cross-sectional area provides the "ECA-ratio". Figure 5.4 shows the reconstructed microstructure of a typical catalyst coated membrane (CCM) CL with nominal porosity of 60 % along with the input TEM image and the evaluated pore and electrolyte phase volume fraction distributions across the CL thickness.

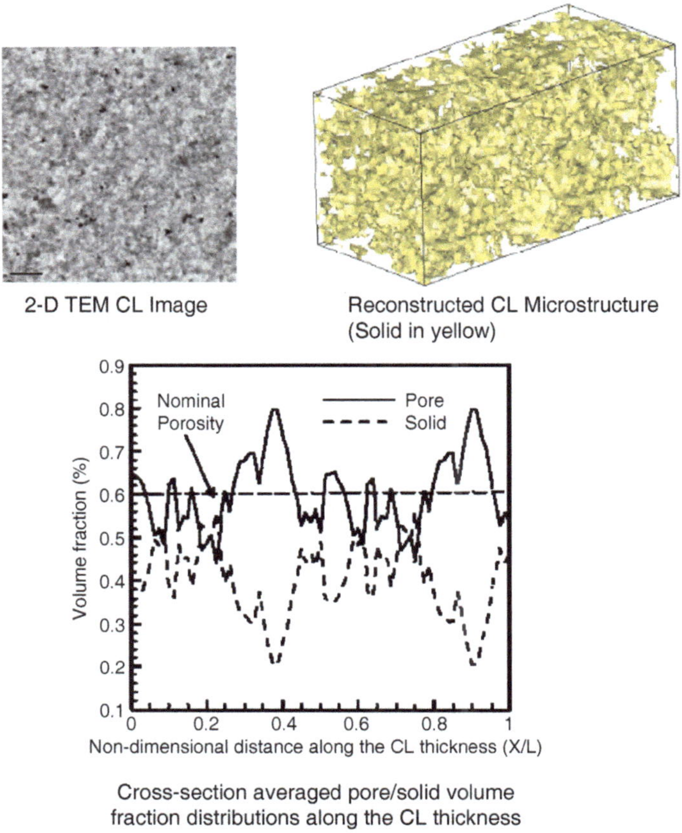

Figure 5.4 Reconstructed catalyst layer microstructure along with pore and electrolyte phase volume fractions distribution [51, 52]

5.3.2 GDL Structure Generation

The multi-faceted functionality of a GDL includes reactant distribution, liquid water transport, electron transport, heat conduction and mechanical support to the membrane–electrode–assembly. Carbon-fiber based porous materials, namely non-woven carbon paper and woven carbon cloth, have received wide acceptance as materials of choice for the PEFC GDL owing to their high porosity (~ 70 % or higher) and good electrical/thermal conductivity. Mathias *et al.* [61] provided a comprehensive overview of the GDL structure and functions. In this chapter, the non-woven carbon paper GDL is considered, which consists of well-defined carbon fibers with a fixed diameter and the fibers are randomly oriented leading to anisotropy in material properties along the through-plane and in-plane directions.

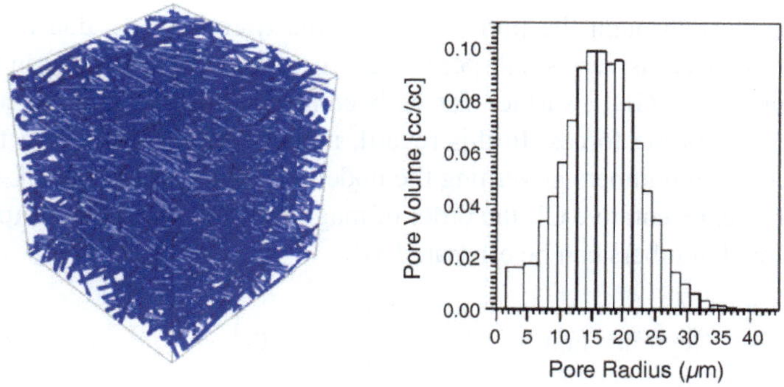

Reconstructed Non-woven Carbon Paper GDL Microstructure

Pore Size Distribution

Permeability	In-plane	Through-plane	In-plane/Through-plane
Reconstructed GDL	30.9 darcy	21.1 darcy	1.46
Measured data	33 darcy	18 darcy	1.83

Figure 5.5 Reconstructed non-woven carbon paper GDL microstructure along with the evaluated structural properties [62]

The stochastic simulation technique creates 3-D realization of the non-woven carbon paper GDL based on structural inputs, namely fiber diameter, fiber orientation and porosity, which can be obtained either directly from the manufacturer's specifications or indirectly from the SEM micrographs or by experimental techniques. Details about the carbon paper GDL microstructure reconstruction along with the underlying assumptions are elaborated in recent work [62], which is based on the non-woven structure generation technique originally proposed by Schladitz *et al.* [63]. Briefly, the stochastic technique is based on a Poisson line process with one-parametric directional distribution where the fibers are realized as circular cylinders with a given diameter [62, 63]. Figure 5.5 shows the reconstructed digital structure of a typical carbon paper GDL with porosity of 72 % along with the structural parameters in terms of the evaluated pore size distribution (PSD) and the anisotropy in the in-plane versus through-plane permeability values.

5.4 Two-phase Transport in the PEFC CL and GDL

Prior to developing the specific numerical experiments for evaluating the two-phase closure relations, namely the capillary pressure and relative permeability as functions of liquid water saturation, as well as the estimation of the detrimental consequence of liquid water in terms of the pore blockage and catalytic site coverage effects, it is imperative to briefly discuss the primary mechanisms governing the two-phase transport in the PEFC CL and GDL [52, 64].

For two-phase flow through the porous CL and the fibrous GDL, due to the complex structure as well as very small pore size, *e.g.*, around 0.05–0.1 μm in the CL and 20–30 μm in the GDL, surface forces become dominant compared with gravity, viscous, and inertia forces. In this regard, representative values of a few salient non-dimensional numbers governing the underlying transport in the CL are listed below. For a representative CL the order of magnitude of some of the important non-dimensional numbers can be estimated as:

$$\text{Reynolds number: } \mathrm{Re} = \frac{\rho_2 U_2 D}{\mu_2} \sim 10^{-4}$$

$$\text{Capillary number: } \mathrm{Ca} = \frac{\mu_2 U_2}{\sigma} \sim 10^{-6}$$

$$\text{Bond number: } \mathrm{Bo} = \frac{g'(\rho_2 - \rho_1) D^2}{\sigma} \sim 10^{-10}$$

$$\text{Weber number: } \mathrm{W}e = \frac{\rho_2 U_2^2 D}{\sigma} = \mathrm{Re.Ca} \sim 10^{-10}$$

Here U_2 and μ_2 are the non-wetting phase velocity and dynamic viscosity, respectively; σ is the surface tension and g' is the gravitational acceleration. Similarly, the afore-mentioned non-dimensional parameters also exhibit very low values in the GDL within the comparable order of magnitude variations. It should be noted that for the hydrophobic CL and GDL representative of a PEFC system, water is the non-wetting phase (NWP) and air the wetting phase (WP). From the Bond number, defined as the ratio of gravitational force to the surface tension force, it is evident that the effect of gravity is negligible with respect to the surface force, thus indicating strong capillary force dominance. Also within PEFC electrodes, velocity, U_2 is very small. Now, from the Reynolds number, representing the ratio of inertia force to viscous force, it is obvious that the inertial effect is negligible in the CL, compared with the viscous force. Combining the implications of low Reynolds and Bond numbers, it can be inferred that the density difference, which is ~ 1000 for air–water two-phase flow in the PEFC operation, should have only a small influence on the overall transport in the CL and GDL. Now, from the Capillary number, Ca, which represents the ratio of viscous force to the surface tension force, it can be observed that the effect of viscous force is also negligible as compared to the surface tension force. Apart from these non-dimensional numbers, the Weber number, We, defining the ratio of inertial force to the surface tension force, which is also a product of Re and Ca, emphasizes the fact that the effects of inertia and viscous forces are truly insignificant compared with the surface tension force representative of two-phase transport in the PEFC CL and GDL. It should be noted that the viscosity ratio for the non-wetting and wetting phases in a fuel cell operating at 80 °C is estimated to be, $\mathrm{M} = \mu_2 / \mu_1 \sim 18$. Based on the calculated representa-

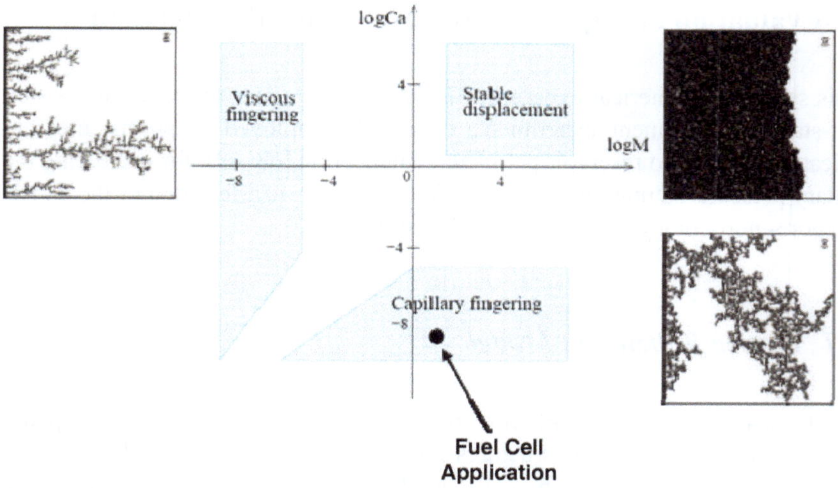

Figure 5.6 Phase diagram along with fluid displacement patterns [65, 66]

tive viscosity ratio and capillary number values, the typical operating two-phase regime for the CL and GDL belongs to the capillary fingering zone on the "phase diagram" proposed by Lenormand *et al.* [65], and is shown in Figure 5.6. The notion of the phase diagram, proposed by Lenormand *et al.* [65], is based on their experiment, involving immiscible displacement of a wetting phase by a non-wetting phase, in a flat and horizontal porous medium where gravity forces were neglected.

This phase diagram further bolsters the validity of our assumption that for air–water two-phase flow in the CL of a PEFC, the viscous forces are truly negligible and the principal force is due to the action of capillarity, which might consequently lead to capillary fingering type displacement pattern. Viscous fingering pattern is observed at low M, where the principal force is the viscous force of the resident fluid and the capillary force and pressure drop in the displacing fluid can be neglected. In the stable displacement flow regime, representative of high M and high Ca, the principal force is due to the viscosity of the displacing fluid and capillary effects in the resident fluid are negligible. Typical fluid displacement patterns pertaining to the three flow regimes are also shown in Figure 5.6 along with the phase diagram and are adapted from the work by Ewing and Berkowitz [66], who extended the 2-D phase diagram by Lenormand *et al.* [65] to a 3-D phase diagram by adding the effect of gravity force through Bond number (Bo) in the third direction. However, in general, the effect of gravity force in the overall fuel cell system operation, let alone in the porous CL and GDL, has been shown to be truly insignificant [67]. Using this analysis, it can safely be adjudged that for modeling air–water two-phase transport in the PEFC CL and GDL the effects of high density ratio (~ 1000) and viscosity ratio (~ 18) variation can be assumed to be negligible represented by the significantly low capillary number.

5.5 Evaluation of Capillary Pressure–Saturation Relation

In this section, a numerical experiment is devised analogous to an actual, ex-situ, quasi-static displacement experiment, commonly employed in porous medium applications related to reservoir/petroleum engineering [68, 69], for evaluating the capillary pressure–saturation relation, which is a direct manifestation of the underlying pore-morphology of the CL and GDL structures.

5.5.1 Lattice Boltzmann Model

The LB model offers great promise in the investigation of detailed two-phase dynamics in complex porous sructures representative of the CL and GDL in a PEFC. In this regard, a two-phase LB model has been developed based on the interaction potential based model originally proposed by Shan and Chen [70–73] and henceforth will be referred to as the S-C model.

In brief, the S-C model [70–73] introduces k distribution functions for a fluid mixture comprising of k components. Each distribution function represents a fluid component and satisfies the evolution equation. The non-local interaction between particles at neighboring lattice sites is included in the kinetics through a set of potentials. The evolution equation for the kth component can be written as:

$$f_i^k(\mathbf{x}+\mathbf{e}_i\delta_t, t+\delta_t) - f_i^k(\mathbf{x},t) = -\frac{f_i^k(\mathbf{x},t) - f_i^{k(eq)}(\mathbf{x},t)}{\tau_k} \tag{5.4}$$

where $f_i^k(\mathbf{x},t)$ is the number density distribution function for the kth component in the ith velocity direction at position $\mathbf{x}$ and time t, and δ_t is the time increment. In the term on the right-hand side, τ_k is the relaxation time of the kth component in lattice unit, and $f_i^{k(eq)}(\mathbf{x},t)$ is the corresponding equilibrium distribution function. The right-hand-side of Equation 5.4 represents the collision term based on the BGK (Bhatnagar–Gross–Krook), or the single-time relaxation approximation [74]. The current LB model is designed for a 3-D 19-speed lattice (D3Q19, where D is the dimension and Q is the number of velocity directions) [72]. The fluid/fluid interaction via the surface tension force and the fluid/solid interaction through the wall adhesion force are taken into account through the modified equilibrium distribution function corresponding to the interparticle interactions and hence the collision term. The macroscopic fluid phase properties are obtained through appropriate averaging of the particle distribution function.

In the present LB model, comparable density and viscosity values of the NWP and WP are assumed primarily due to numerical instability arising from spurious currents at the phase interface. However, as it has been amply discussed earlier, the assumption of negligible effects of the high density ratio (~ 1000) and viscos-

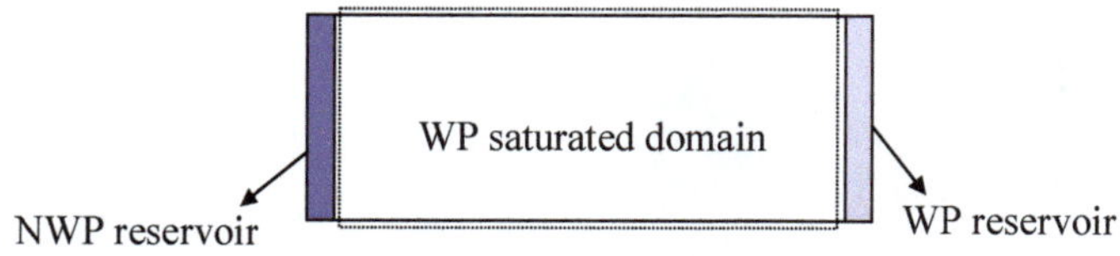

Figure 5.7 Schematic diagram of the primary drainage simulation setup

ity ratio (~ 18) of air–water two-phase flow through the PEFC CL and GDL could well be valid, because the system is characterized by very low values of Ca and Re, and consequently the surface forces indeed dominate over the inertial and viscous forces. In addition, in the current LB model we do not consider the effect of electrochemical reaction since the primary purpose is to simulate two-phase transport in an ex-situ porous CL, which is not a part of an operational fuel cell. The idea is to mimic an actual ex-situ experimental set-up, commonly employed in geologic porous medium applications, *e.g.*, involved in reservoir/petroleum engineering [68, 69], for evaluating the two-phase correlations as functions of saturation.

Based on the afore-mentioned two-phase LB model a quasi-static capillary pressure experiment is designed for simulating two-phase transport through the reconstructed CL and GDL structures. Figure 5.7 shows schematically the numerical experiment, based on the work by Pan *et al.* [75], where immiscible displacement for oil–water two-phase flow through a simulated sphere-packed porous structure was conducted for capillary pressure–saturation relation estimation.

A NWP reservoir is added to the porous structure at the front end and a WP reservoir is added at the back end. It should be noted that for the primary drainage (PD) simulation in the hydrophobic CL and GDL, liquid water is the NWP and air is the WP. The primary drainage process was simulated starting with zero capillary pressure, by fixing the NWP and WP reservoir pressures to be equal. Then the capillary pressure was increased incrementally by decreasing the WP reservoir pressure while maintaining the NWP reservoir pressure at the fixed initial value. The pressure gradient drives liquid water into the initially air-saturated CL and GDL by displacing it. More details of the current LB model, the simulation set up, the model input parameters and the boundary conditions can be found in [52, 76, 77].

Figure 5.8 displays the steady state saturation fronts of the invading non-wetting phase, *i.e.*, liquid water, corresponding to increasing capillary pressures from the drainage simulation in the reconstructed CCM CL characterized by hydrophobic wetting characteristics with a static contact angle of 110° [52, 76, 77]. At lower capillary pressures, the liquid water saturation front exhibits finger like pattern, similar to the displacement pattern observed typically in the capillary fingering regime. The displacing liquid water phase penetrates into the body of the resident wetting phase (*i.e.*, air) in the shape of fingers owing to the surface tension driven capillary force. However, at high saturation levels, the invading non-wetting phase tends to exhibit a somewhat flat advancing front. This observation

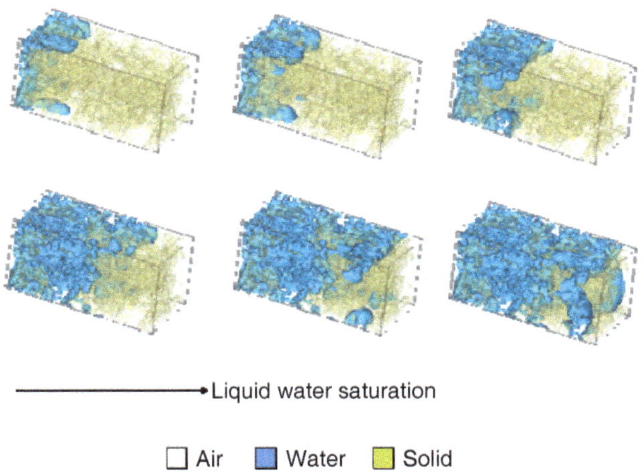

Figure 5.8 Advancing liquid water saturation front with increasing capillary pressure through the reconstructed CL microstructure [52]

indicates that with increasing capillary pressure, even at very low Capillary number (Ca), several penetrating saturation fronts tend to merge and form a stable front and the invasion pattern transitions from the capillary fingering regime to the stable displacement regime and might lie in the transition zone in between.

In an operating fuel cell, the resulting liquid water displacement pattern pertaining to the underlying CL structure and wetting characteristics would play a vital role in the transport of the liquid water and hence the overall flooding behavior.

Figure 5.9 shows the liquid water distribution as well as the invasion pattern with increasing capillary pressure in the initially air-saturated reconstructed carbon paper GDL characterized by hydrophobic wetting characteristics with a static contact angle of 140 ° [64]. At the initially very low capillary pressure, the invading front overcomes the barrier pressure only at some preferential locations depending upon the pore size along with the emergence of a droplet owing to strong hydrophobicity. As the capillary pressure increases, several liquid water fronts start to penetrate into the air occupied domain. Further increase in capillary pressure exhibits growth of droplets at two invasion fronts, followed by the coalescence of the drops and collapsing into a single front. This newly formed front then invades in the less tortuous in-plane direction. Additionally, emergence of tiny droplets and subsequent growth can be observed in the constricted pores in the vicinity of the inlet region primarily due to strong wall adhesion forces from interactions with highly hydrophobic fibers with the increasing capillary pressure. One of the several invading fronts reaches the air reservoir at one location corresponding to the capillary pressure.

The 2-D liquid water saturation maps at different cross-sections along the GDL through-plane direction corresponding to a representative liquid water saturation level are shown in Figure 5.10 which demonstrates the porous pathways actually available for oxygen transport from the channel to the CL reaction sites [64]. It is

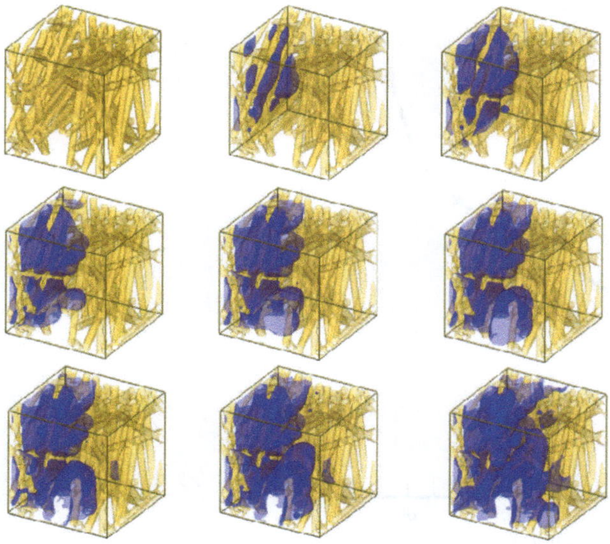

Figure 5.9 Liquid water distribution in the reconstructed non-woven GDL microstructure from the drainage simulation using the two-phase LB model [64]

worth mentioning that the LB simulation is indeed able to capture the intricate liquid water dynamics including droplet interactions through the hydrophobic fibrous GDL structure.

Finally, it is important to note that the liquid water transport and flooding dyanmics through a woven carbon cloth GDL would lead to a very different scenario owing to liquid water motion along individual fibers as well as between fiber bundles, as opposed to that in the non-woven carbon cloth GDL. The mesoscopic LB simulations should provide deatiled understanding of the pore-scale liquid water transport through different GDL structures and would probably enable novel GDL design with better water removal and mitigate flooding.

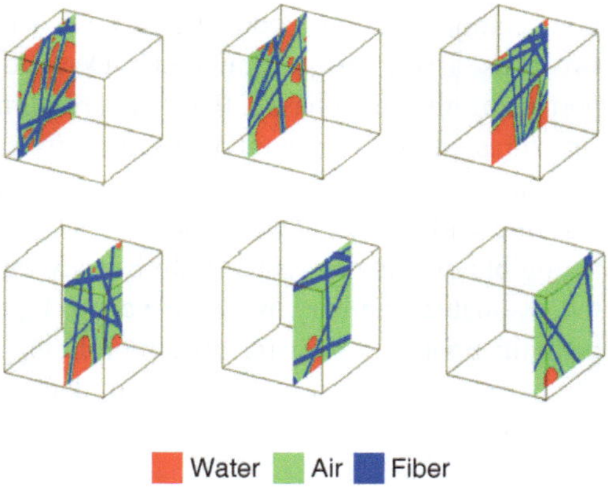

Figure 5.10 2-D phase distribution maps on several cross-sections from the two-phase LB drainage simulation [64]

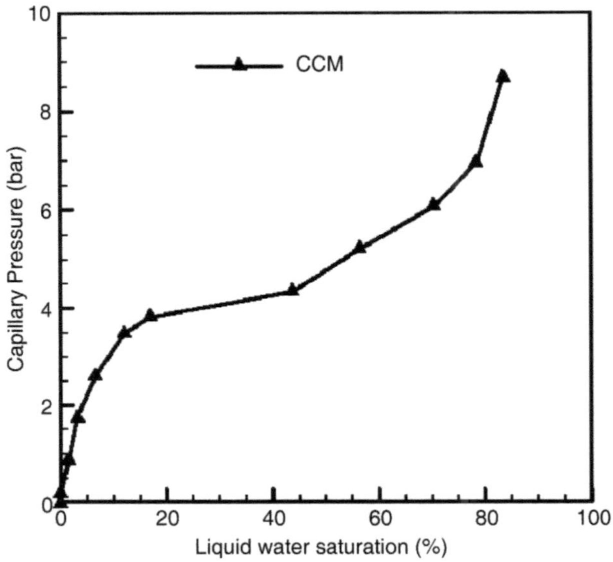

Figure 5.11 Capillary pressure versus liquid water saturation relation for the reconstructed CL microstructure [52]

The liquid water saturation versus capillary pressure response can be evaluated from the two-phase LB drainage simulation and the corresponding transport relation can be devised as shown in Figure 5.11 for the reconstructed CCM CL microstructure [52].

The overall shape of the capillary pressure curve agrees well with those reported in the literature for synthetic porous medium [75]. The capillary pressure–saturation curve exhibits a non-zero entry pressure for the NWP to initiate invasion into the WP-saturated hydrophobic CL structure. Another point to be noted is that for the CCM CL, the NWP saturation increases gradually with increase in capillary pressure until around 20 %, after which NWP saturation jumps to around 45 % with negligible increase in capillary pressure, and finally exhibits a residual WP saturation of around 18 %. Similarly, the capillary pressure–saturation relations for the reconstructed GDL structures can be constructed from the two-phase LB drainage simulations. Detailed investigations of liquid water transport using the two-phase LB model are currently underway with different GDL structures, along with numerical sensitivity study with respect to the sample size, boundary conditions, *etc.* A high density ratio and viscosity ratio LB model is also presently under development which will ultimately test the assumption of negligible effects of density and viscosity ratios in fuel cell applications. Finally, the evaluated capillary pressure versus liquid water saturation correlations depending upon the realistic pore morphology could be employed in the macroscopic two-phase fuel cell models in place of the arbitrary closure relations used otherwise.

5.5.2 Full Morphology Model

With the reconstructed GDL microstructure, the stationary distribution of air and water for diferent capillary pressure variations can also be evaluated using the full morphology (FM) model via a simulated drainage process through an initially air-saturated GDL [50, 62].

The FM model relies on morphological decomposition of the 3-D digital image of the reconstructed GDL to determine the pore space accessible to the non-wetting phase using the pore radius as the ordering parameter corresponding to a specified pressure during drainage [50, 62]. The key steps in the simulated quasi-static drainage process with the FM model include [62]:

- The entire pore space is initially filled with the wetting phase (WP) and the capillary pressure is zero. At one end, the porous medium is connected to a NWP reservoir while at the opposite end it is connected to the WP reservoir.
- The pore space is eroded by a spherical structuring element with radius, r, corresponding to the capillary pressure, p_c, according to the Young–Laplace equation:

$$p_c = \frac{2\sigma \cos\theta}{r} \tag{5.5}$$

σ is the surface tension between NWP and WP, and θ is the contact angle.

- At a given capillary pressure, only those pores within the eroded pore space having a continuous connection to the NWP reservoir are filled with the NWP. The rest of the unconnected pores are removed from the eroded space.
- The phase saturations related to the capillary pressure are subsequently determined by dilating the eroded pore space and evaluating the corresponding occupied volume fractions of the pore space [40, 44].
- The erosion–dilation process is repeated with the next larger spherical element corresponding to varying capillary pressures.

Figure 5.12 shows the liquid water distributions in the reconstructed carbon paper GDL microstructure at various capillary pressure levels as predicted by the FM model [62]. At bubble point, the invading liquid water front reaches the air reservoir via a connected pathway through the GDL structure and the corresponding capillary pressure is referred to as the bubble point pressure (p_{cb}). From the liquid saturation map, it can be observed that with increasing capillary pressure, several liquid water fronts start penetrating into the air-saturated medium based on the pore size distribution and the resulting capillary force in the form of fractal fingering.

Figure 5.13 shows the capillary pressure versus liquid water saturation relation for the GDL structure evaluated from the primary drainage simulation as well as the simulation with arbitrarily mobile fluids using the FM model [62]. The arbitrarily mobile NWP and WP distributions correspond to the physical scenario of liquid water formation due to condensation of water vapor at random locations

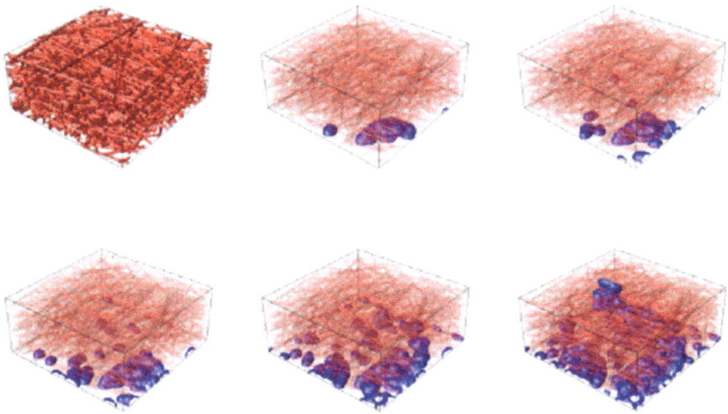

Figure 5.12 Liquid water distribution pattern from the drainage simulation using the FM model [62]

inside the GDL, which subsequently transports owing to the action of capillarity. A typical contact angle of 120 ° is used in the present calculations. The GDL microstructure characterized by hydrophobic wetting characteristics exhibits a finite entry pressure for the initiation of the invasion of the non-wetting phase into the wetting phase saturated domain in the primary drainage experiment. The capillary pressure versus liquid water saturation relation is based on the direct manifestation of the underlying pore morphology and such correlation could prove to be valuable input for macroscopic two-phase fuel cell models.

It is to be noted that the quasi-static two-phase distribution in the FM model is based on purely morphological consideration of overlapping spherical elements and therefore cannot accurately capture the effect of the interfacial shape on the invasion process [62]. However, the FM model is a fast, direct simulation method

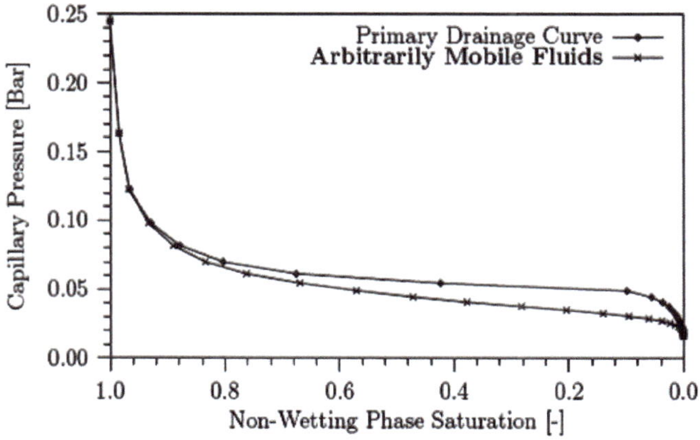

Figure 5.13 Capillary pressure versus liquid water saturation relation for the GDL structure using the FM model [62]

based on the actual pore morphology and the two-phase distributions obtained thereof can be useful in finding saturation-dependent effective transport properties, *e.g.*, effective gas diffusivity and will be explained later. Details about the FM simulations along with the description of the reconstructed structure are provided in recent work [62].

5.5.2.1 Effect of GDL Compression

The importance of cell clamping pressure on fuel cell performance has been studied by several researchers. Notable works include Mathias *et al.* [61], Wilde *et al.* [78], and Ihonen *et al.* [79]. Mathias *et al.* [61] reported compression and flexural behavior of carbon paper and carbon cloth GDLs and indicated the effect of compressive characteristics on the channel flow-field pressure drop. Wilde *et al.* [78] studied the impact of compression force on the GDL properties, namely electrical resistivity, pore size, and permeability for different materials and briefly described the resulting influence on PEFC performance. Ihonen and co-workers [79], on the other hand, tried to assess experimentally the influence of clamping pressure on the flooding behavior of the GDL. However, none of these studies focused on the dependency of capillary pressure–saturation relation on clamping pressure through its dependecne on the underlying GDL structure. In the present study, we briefly report the variation of two-phase correlations in terms of the capillary pressure–saturation relation for the GDL with various levels of cell compression.

The detailed modeling of a porous material under compression is a challenging task of applied structural mechanics. The reduced compression model employed in the current study is based on the unidirectional morphological displacement of solid voxels in the GDL structure under load and with the assumption of negligible transverse strain. The details of the reduced compression model are provided in recent work [62]. However, with our reduced model, it is difficult to find a relation between the compression ratio and the external load. Nevertheless, our approach leads to reliable 3-D morphology of the non-woven GDL structures under compression. Figure 5.14 shows the compressed structures with compression ratio of 0.8 and 0.6 along with the uncompressed structure for the reconstructed non-woven carbon paper GDL along with 2-D cross-sections [62].

The primary drainage simulation was performed using the FM model with the uncompressed and compressed GDL microstructures. The variation of liquid water (NWP) saturation with capillary pressure is shown in Figure 5.15 for the GDL in the uncompressed state as well as with compression ratio of 0.7 and 0.5 from the primary drainage simulation [62]. It is quite evident that increased compression leads to more tortuous pore structure which in turn requires increasing capillary pressure for the invasion of the non-wetting phase into the wetting phase saturated GDL in order to achieve the same level of saturation. The increase in the bubble point pressure (p_{cb}) with increased compression level further testifies the extra level of resistance offered by the compressed GDL owing to the underlying mor-

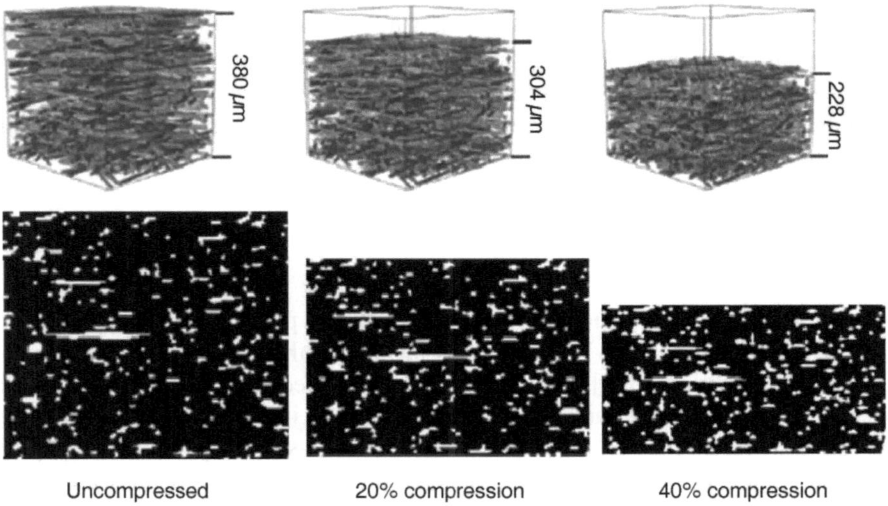

Figure 5.14 Uncompressed and compressed GDL structures [62]

phology as opposed to the uncompressed sample. The capillary pressure (p_c) versus liquid water saturation (S_r) curves are further fitted according to the van Genuchten correlation [62]:

$$p_c(S_r) = p_{cb}\left(S_r^{-n/(n-1)} - 1\right)^{1/n} \tag{5.6}$$

The compression effect can be adequately realized with the same parameter, n=7.38 with, however, better match of the correlation with the simulated data toward the lower saturation limits.

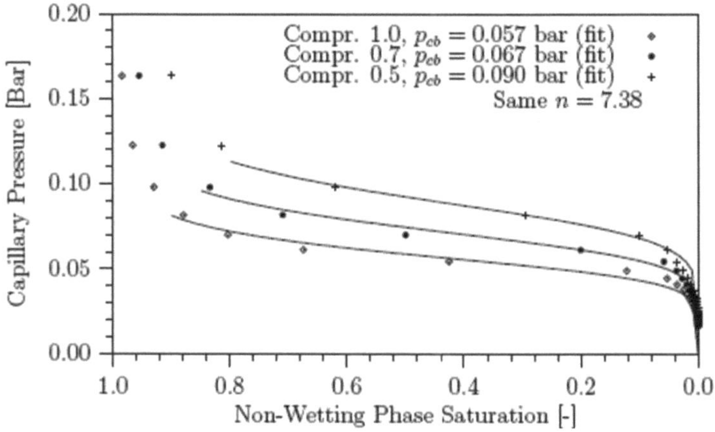

Figure 5.15 Capillary pressure versus liquid water saturation relations for different compression ratios of the GDL structure [62]

5.6 Evaluation of Relative Permeability–Saturation Relation

For simultaneous flow of two immiscible fluid phases in a porous medium, each phase satisfies Darcy's law, written as:

$$v_i = -\frac{\kappa\kappa_{ri}}{\mu_i}(\nabla p_i - \rho_i g') \tag{5.7}$$

where v_i is the Darcy velocity for the wetting phase and non-wetting phase, p_i is the fluid phase pressure, $\rho_i g'$ is the body force, μ_i is the dynamic viscosity of the fluid, κ is the intrinsic permeability determined by the pore structure of the porous medium alone, and κ_{ri} is the relative permeability of each phase that depends upon fluid saturation and the underlying two-phase dynamics. It should be noted that in the case of the PEFC CL and GDL body force can be safely neglected based on the justifications provided earlier. Excellent discussion about the influence of underlying two-phase dynamics on the relative phase mobility is provided by Ehrlich [80] and Avraam and Payatakes [81, 82].

In this section, a numerical experiment is designed based on the steady-state flow experiment, typically devised in the petroleum/reservoir engineering applications and detailed elsewhere in the literature [68, 69, 83] in order to evaluate the phasic relative permeability relations.

The numerical experiment is devised according to the procedure outlined by Li *et al.* [83] within the two-phase LB modeling framework for the reconstructed CCM CL, generated using the stochastic reconstruction technique described earlier. Briefly, in the steady-state flow experiment the two immiscible fluids are allowed to flow simultaneously until equilibrium is attained and the corresponding saturations, fluid flow rates and pressure gradients can be directly measured and correlated using Darcy's law. The term "steady-state", however needs to be duly qualified in the sense that the process is intrinsically a dynamical equilibrium of two moving fluids, although macroscopically stable [82].

The numerical experiment for the immiscible two-phase flow is described below [52, 77, 83]:

- Initially, both the NWP and WP are randomly distributed throughout the CL porous structure such that the desired NWP saturation is achieved. The initial random distribution of the liquid water phase (*i.e.*, NWP) in the otherwise air (*i.e.*, WP) occupied CL closely represents the physically perceived scenario of liquid water generation due to the electrochemical reaction at different catalytically active sites within the CL structure and subsequent transport by the action of capillarity. In the present numerical experiment, the CL domain is assumed to be bounded by walls in the span-wise directions and periodic in the thickness direction.
- Co-current flow is simulated by applying a body force for both the phases along the flow direction which mimics the redistribution of the phases under the capillary force corresponding to the typical capillary number (Ca ~ 10^{-6}).

- Once steady state is achieved, the flux of each phase is calculated. Steady state is considered attained when the saturation and flow rates of both phases do not change any more. The corresponding absolute flux is calculated by modifying the two-phase LB model, where a body force is applied to one phase and the density of the other phase is rendered zero at all locations. Finally, the ratio of flux of each phase from the two-phase calculation to the one obtained from the single-phase calculation gives the relative permeability related to the saturation level.

Figure 5.16 exhibits the 3-D liquid water distributions corresponding to several low saturation levels (below 15 %) for the CCM CL once the steady state is reached [52, 77]. It can be observed that below 10 % saturation level there is hardly any connected pathway for the liquid water phase to transport through the CL structure and hence the relative mobility of the liquid water phase with respect to the incumbent air phase is negligible. As the saturation level increases the liquid water phase finds a connected pathway for transport through the CL structure. The underlying structure and wetting characteristics of the CL would have profound impact on liquid water distribution and dynamics which subsequently influence the relative phase mobility and transport.

Figure 5.17 shows the NWP and WP relative permeability as functions of liquid water (NWP) saturation for the CCM CL [52, 77]. The general trend of the relative permeability curves agrees well with those reported elsewhere in the literature for geological flows [83].

It can be observed that the liquid water relative permeability values below 10 % saturation is negligibly small due to the non-existence of a connected pathway for transport apropos of the 3-D liquid water distributions in Figure 5.16. In general it is observed that at this low Ca the relative permeability curves exhibit non-linear relationships for both phases between the flow rate and the driving pressure gradient and this observation is in agreement with the results reported by Li *et al.* [83]. This observation further emphasizes that for the capillary number of

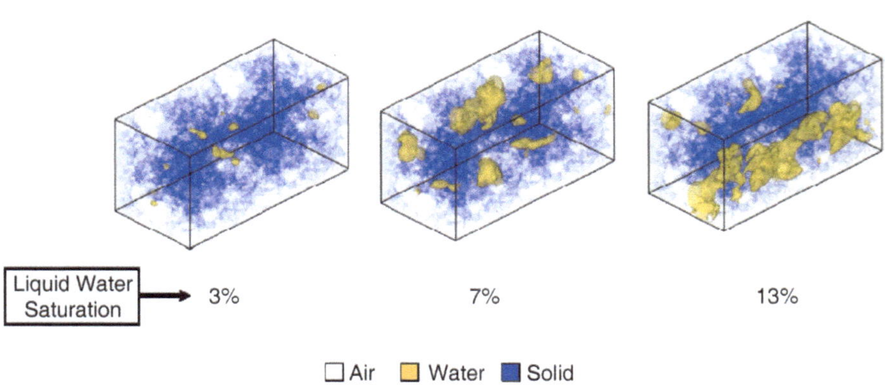

Figure 5.16 3-D liquid water distributions in the reconstructed CL microstructure from two-phase LB simulations [52]

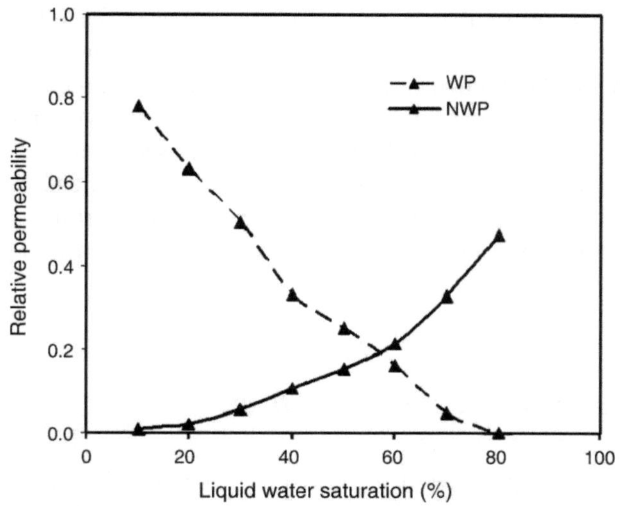

Figure 5.17 Relative permeability versus liquid water saturation relations for the reconstructed CL microstructure [52]

interest for fuel cell operation, *i.e.*, Ca ~ 10^{-6}, the relative permeability relations for both phases will exhibit strong non-linearity. Despite the linear relation between the pressure gradient and the flow-rate demonstrated by the Darcy's law, the current prediction from the numerical experiment further underscores the capability of the LB model in capturing the underlying interfacial dynamics inherent to the low capillary number flow regime of interest. This relation of the relative permeability as function of the liquid water saturation can be further deployed in the two-phase computational fuel cell dynamics models for reliable transport predictions.

The above-mentioned approach for estimating the relative permeability – saturation relation applies exactly for the reconstructed GDL microstructure as well and work is currently under progress which will be reported in a forthcoming article. It is important to note that for the evaluation of the relative permeability – saturation relation, detailed description of the underlying interfacial dynamics is of paramount importance which cannot be discerned otherwise by the morphological image analysis based full morphology model.

5.7 Effect of Liquid Water on CL and GDL Performance

The detrimental consequence of liquid water on the electrochemical performance manifests in terms of coverage of the electrochemically active area in the CL leading to reduced catalytic activity and blockage of the porous pathways in the CL and GDL, rendering hindered oxygen transport to the active reaction sites. In the macroscopic fuel cell models, where the CL and GDL are treated as macrohomogeneous porous layers, the site coverage and pore blockage effects owing to liquid water are taken into account through an electrochemical area reduction relation and the Bruggeman type correction for the effective oxygen diffusivity, respectively. These two empirical correlations, however, cannot be separately discerned through experimental techniques. A combination of the predictions of liquid water distributions from two-phase simulations and species transport using a direct numerical simulation (DNS) model can be deployed to quantify the site coverage and pore blockage effects.

5.7.1 CL Site Coverage and Pore Blockage Effects

The computational approach couples the two-phase LB model for the liquid water transport and the DNS model for the species and charge transport for the CL [51, 52, 76, 77]. The two-phase simulation using the LB model is designed based on the ex-situ, steady-state flow experiment for porous media, detailed earlier in Section 5.6, in order to obtain the liquid water distributions within the CL microstructure for different saturation levels resulting from the dynamic interactions between the two phases and the underlying pore morphology. Briefly, with an initial random distribution of the liquid water phase corresponding to a saturation level in the otherwise air occupied CL structure, the imposed driving force representative of the typical Capillary number (*i.e.*, Ca $\sim 10^{-6}$) allows the two immiscible phases to transport and redistribute owing to capillarity characterized by the surface tension force and the wall adhesion force depending upon the pore wall wetting characteristics. Once steady state is attained the corresponding saturations, fluid flow rates and pressure gradients can be directly measured. As mentioned earlier, the initial random distribution of the liquid water phase conforms very closely to the physical scenario of active reaction sites dispersed randomly within the electrochemically reactive CL microstructure where liquid water is generated and subsequently transported by the action of capillarity. The details of the simulation setup are provided in [52, 76]. Once steady state is achieved, 3-D liquid water distributions can be obtained within the CL. Figure 5.16 shows the 3-D liquid water distribution for a representative saturation level in the CCM CL microstructure. From the liquid water distributions within the CL structure, the information about the catalytic site coverage effect can be extracted directly.

The DNS model can be deployed subsequently on the liquid water blocked CL structure pertaining to a saturation level for the evaluation of the hindered oxygen transport. In brief, the DNS model is a "top-down" numerical approach based on a fine-scale CFD framework which solves point-wise accurate conservation equations for species and charge transport in the CL with appropriate source terms due to the oxygen reduction reaction (ORR) directly on the CL microstructures [51,

52, 84]. The conservation equations for proton, oxygen and water vapor transport, respectively, are given by [51, 52, 84]:

$$\nabla \cdot (\kappa_e \nabla \phi_e) + S_{\phi_e} = 0 \tag{5.8}$$

$$\nabla \cdot (D^g_{O_2} \nabla c_{O_2}) + S_{O_2} = 0 \tag{5.9}$$

$$\nabla \cdot (D^g_{H_2O} \nabla c_{H_2O}) + S_{H_2O} = 0 \tag{5.10}$$

S_{ϕ_e}, S_{O_2} and S_{H_2O} refer to the respective source terms owing to the ORR, ϕ_e is the electrolyte phase potential, c_{O_2} is the oxygen concentration and c_{H_2O} is the water vapor concentration, κ_e is the proton conductivity duly modified w.r.t. to the actual electrolyte volume fraction, $D^g_{O_2}$ is the oxygen diffusivity and $D^g_{H_2O}$ is the vapor diffusivity. The details about the DNS model for pore-scale description of species and charge transport in the CL microstructure along with its capability of discerning the compositional influence on the CL performance as well as local overpotential and reaction current distributions are furnished in [51, 52, 84, 85].

5.7.1.1 Estimation of Catalytic Site Coverage Effect

With the liquid water distribution available from the two-phase LB simulation corresponding to a saturation level, the reduction in electrochemically active interfacial area (ECA) owing to liquid water coverage can be estimated from the 2-D saturation maps and subsequently a correlation between the effective ECA and the liquid water saturation can be established as the following [52, 76, 77]:

$$ECA^{eff} = ECA(1 - S_r)^c \tag{5.11}$$

where S_r is the liquid water saturation and c the site coverage factor.

The 2-D liquid water saturation maps corresponding to 20 % saturation level on several cross-sections along the thickness of the reconstructed CCM CL are shown in Figure 5.18.

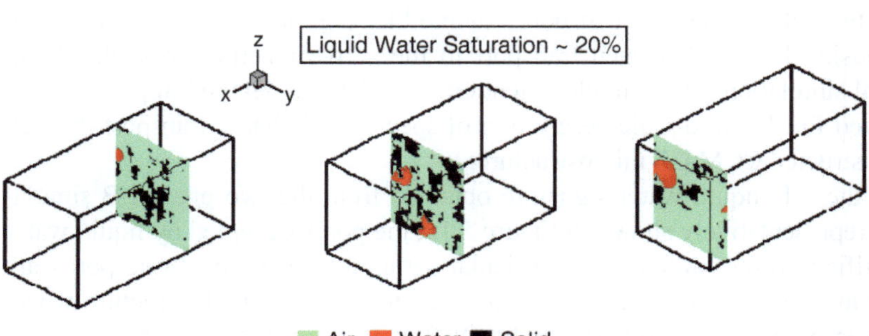

Figure 5.18 2-D liquid water distribution maps in the CL structure from two-phase LB simulations [52, 76]

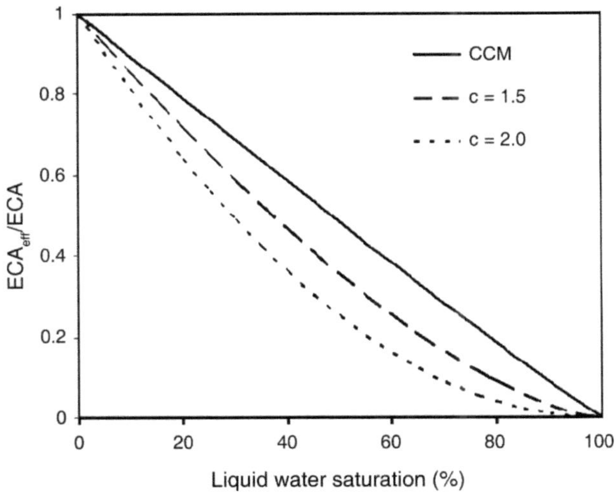

Figure 5.19 Catalytic site coverage relation as function of liquid water saturation for the CL structure [52, 76]

From the 2-D saturation maps, the effective ECA can be evaluated and correlated according to Equation 5.11. Based on several liquid water saturation levels, the catalytic surface coverage factor for the CCM CL microstructure is estimated and the following correlation can be constructed, which can be used as valuable input to macroscopic two-phase fuel cell models [52, 76].

$$ECA^{eff} = ECA(1 - S_r)^{1.05} \tag{5.12}$$

Figure 5.19 shows the variation of the effective ECA with liquid water saturation from the evaluated correlation given in Equation 5.12 along with the typical correlations otherwise used arbitrarily in the macroscopic fuel cell modeling literature.

5.7.1.2 Estimation of Pore Blockage Effect

In order to evaluate the effect of pore volume blockage in the presence of liquid water causing hindered oxygen transport to the active reaction sites, the direct numerical simulation (DNS) model, mentioned earlier and detailed in [51, 52, 84], is deployed for the pore-scale description of species and charge transport through the reconstructed CCM CL microstructure.

From the 3-D liquid water signature obtained from the two-phase LB simulation and representatively shown in Figure 5.16, the pores occupied by liquid water are identified corresponding to a particular saturation level and these pores are rendered as "dead" pores. The idea is to generate a modified CL structure where the blocked pores do not take part in the ORR as well as produce extra resistance by reducing the effective porosity of the structure. With this virtual morphology of the liquid water blocked CL, the point-wise accurate species and charge conserva-

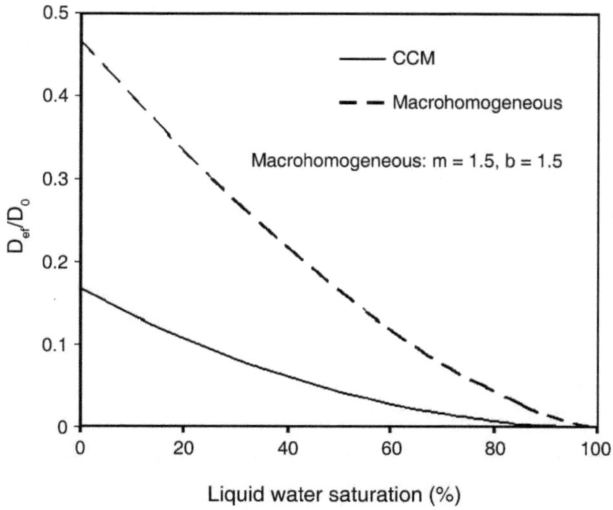

Figure 5.20 Pore blockage relation as function of liquid water saturation for the CL structure [52, 76]

tion equations are solved within the DNS modeling framework. It should be noted that unlike in the GDL, the reactive transport in the CL requires the solution of the electrochemistry coupled oxygen transport for the true oxygen concentration field to evolve. The pore blockage effect is finally evaluated from the oxygen concentration field and can subsequently be correlated in terms of the effective oxygen diffusion coefficient based on the oxygen flux as the following [52, 76]:

$$D_{O_2}^{eff} = D_{O_2,0} f\left(\varepsilon_{CL}\right) g(S_r) = D_{O_2,0}(\varepsilon_{CL})^m (1-S_r)^b \tag{5.13}$$

where ε_{CL} is the CL porosity, S_r the liquid water saturation, m the Bruggeman factor for the oxygen transport through the unblocked CL microstructure and b is the volume blockage factor representing the extra resistance to oxygen transport in the presence of liquid water in the CL. The effect of the reistance due to the tortuous pore pathways to oxygen transport in the absence of liquid water is evaluated using the DNS model in terms of a Bruggeman factor, m, and is detailed in earlier work [51, 84]. Similarly, the resistance to ion transport due to the tortuous electrolyte phase network in terms of Bruggeman correction factor can also be evaluated using the DNS model and is elaborated in [51, 52]. With $\varepsilon_{CL}=0.6$ and the Bruggeman factor, $m=3.5$, as estimated in [51, 84], the pore blockage factor for the CCM CL is evaluated and the following correlation is constructed [52, 76].

$$D_{O_2}^{eff} = D_{O_2,0}(\varepsilon_{CL})^{3.5}(1-S_r)^{1.97} \tag{5.14}$$

This estimate could prove to be valuable input for more accurate representation of the pore blockage effect in the macroscopic two-phase fuel cell models. Figure 5.20 shows the variation of the effective oxygen diffusivity with liquid water saturation from the correlation in Equation 5.14 along with the typical macrohomogeneous correlation with $m = 1.5$ and $b = 1.5$ otherwise used arbitrarily in the macroscopic fuel cell modeling literature.

5.7.2 GDL Pore Blockage Effect

The methodology for estimating the pore blockage effect due to liquid water leading to oxygen transport resistance in the GDL is identical to that outlined for the CL in Section 5.7.1.

Stationary liquid water distribution can be obtained from the full morphology (FM) model for arbitrary capillary pressures [62, 86]. The FM model can be used either to determine the two-phase distribution from a drainage process or for the scenario of arbitrarily mobile phase distribution. In the present work, we use the 3-D liquid water signature corresponding to arbitrarily mobile phase distribution since we are interested in the determination of an average phase distribution. This phase distribution can be perceived as the physical scenario of pore space being occupied by liquid water condensed from water vapor at random locations and redistributed according to the capillary pressure and the underlying pore size distribution in the GDL. Figure 5.21 shows a typical configuration of liquid water distribution in an uncompressed GDL microstructure pertaining to liquid water saturation of around 18 % [86].

On the contrary, the prediction from two-phase LB model not only captures the pore occupancy by liquid water due to pore size distribution but also phase redistribution owing to the underlying interfacial dynamics and wetting characteristics

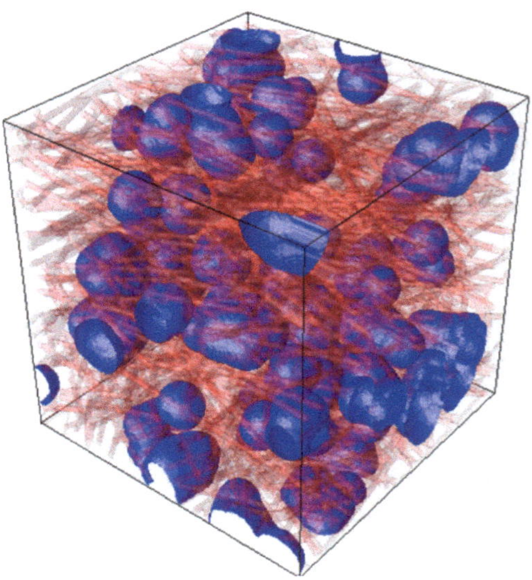

Figure 5.21 Liquid water distribution in the GDL structure pertaining to arbitrarily mobile phase configuration using the FM model [86]

[64]. Figure 5.22 shows the equilibrium liquid water distribution after the randomly dispersed initial liquid water redistributes by the action of capillarity for a purely hydrophobic GDL with contact angle of around 140 ° and a mixed wettability GDL with hydrophilic and hydrophobic contact angles of 80 ° and 140 ° respectively [64]. In the mixed wettability GDL, a hydrophilic pore fraction of 50 % is considered and the hydrophilic pores are assumed to be randomly distributed through the GDL structure.

It can be observed that at the same saturation level, the liquid water distribution is quite different for the two GDLs, underscoring the influence of the wetting characteristics and interfacial dynamics on transport processes. The far reaching implication of this study is to understand the liquid water behavior and flooding dynamics in a beginning-of-life GDL which is predominantly hydrophobic as opposed to that in a mxed-wettability GDL after prolonged operation which leads to partially hydrophilic fibers thereby causing enhanced flooding and severe performance degradation. Additionally, Figure 5.23 exhibits the 2-D phase distributions on several cross-sections for 20 % saturation level. The liquid water saturation distributions from such study could be further used to quantify the averaged saturation-dependent effective transport properties (*e.g.*, effective species diffusivity).

With the 3-D liquid water distribution available in the GDL structure from either the LB model or the FM model, the DNS model [51, 52] can be applied to solve the Laplace equation for oxygen transport given below [86].

$$\nabla \cdot (D^g_{O_2} \nabla c_{O_2}) = 0 \tag{5.15}$$

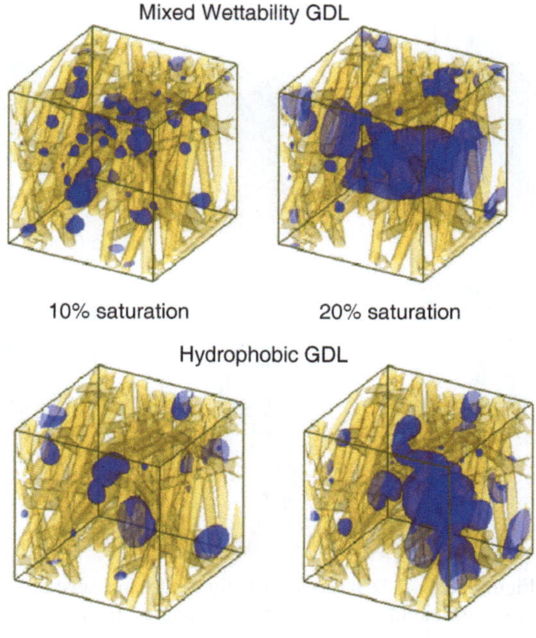

Figure 5.22 Liquid water distributions for a hydrophobic and a mixed wettability GDL using the two-phase LB model [64]

Similar to the approach outlined for the CL, the liquid water blocked pores are considered as "dead pores" which act as solid obstacles leading to extra resistance to oxygen diffusion. The pore blockage effect is evaluated from the oxygen concentration field pertaining to different liquid water satuartion levels and can be subsequently correlated in terms of the effective oxygen diffusion coefficient based on the oxygen flux [86].

$$D_{O_2}^{eff} = D_{O_2,0} f\left(\varepsilon_{GDL}\right) g(S_r) = D_{O_2,0} (\varepsilon_{GDL})^m (1-S_r)^b \qquad (5.16)$$

ε_{GDL} is the GDL porosity, S_r the liquid water saturation, m the Bruggeman factor for the oxygen transport through the unblocked GDL microstructure and b is the volume blockage factor representing the extra resistance to oxygen transport in the presence of liquid water in the GDL. The effect of the resistance due to the tortuous pore pathways to oxygen transport in the absence of liquid water can be evaluated using the DNS model in terms of a Bruggeman factor, m. It should be noted that the anisotropy resulting from the fiber orientations in the GDL structure manifests in terms of disparate resistance behavior to oxygen transport in the through-plane and in-plane directions.

Figure 5.24 shows the oxygen concentration contours on several 2-D cross-sections in an uncompressed GDL structure for different saturation levels with the liquid water distributions obtained from the FM simulations. With GDL porosity of 0.72, the pore blockage effect in terms of the effective oxygen diffusivity is

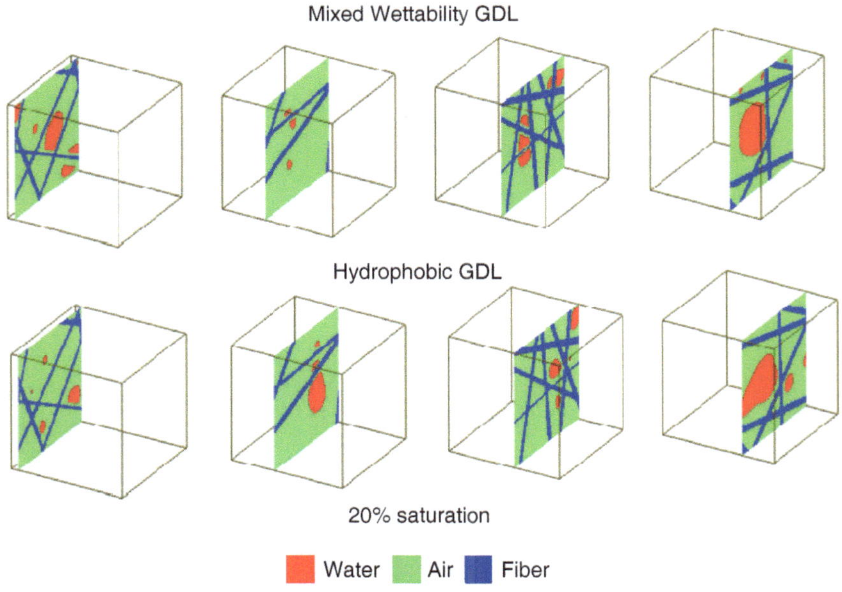

Figure 5.23 2-D phase distribution maps on several cross-sections for a hydrophobic and a mixed wettability GDL from the two-phase LB simulations [64]

evaluated and the following correlations are constructed for the through-plane and in-plane variations, respectively [86]:

$$D^{eff}_{O_2,th} = D_{O_2,0}(\varepsilon_{GDL})^{5.2}(1-S_r)^{2.3} \tag{5.17}$$

$$D^{eff}_{O_2,in} = D_{O_2,0}(\varepsilon_{GDL})^{2.3}(1-S_r)^{2.3} \tag{5.18}$$

Figure 5.25 displays the anisotropic effective oxygen diffusivity variations with liquid water saturation based on the evaluated pore blockage correlations.

The effect of GDL compression on the pore blockage effect is also studied using the combination of the predictions from the FM model for liquid water distribution and the DNS model for oxygen concentration distribution [86]. Figure 5.26 shows the variation of the effective oxygen diffusivity with varying liquid water saturation and GDL compression ratios [86]. The compression of the gas diffusion medium has significant impact on the oxygen diffusion coefficient toward the lower liquid water saturation limits. Interestingly, increasing amount of liquid water saturation has little influence on the effective diffusion coefficient with enhanced compression ratio.

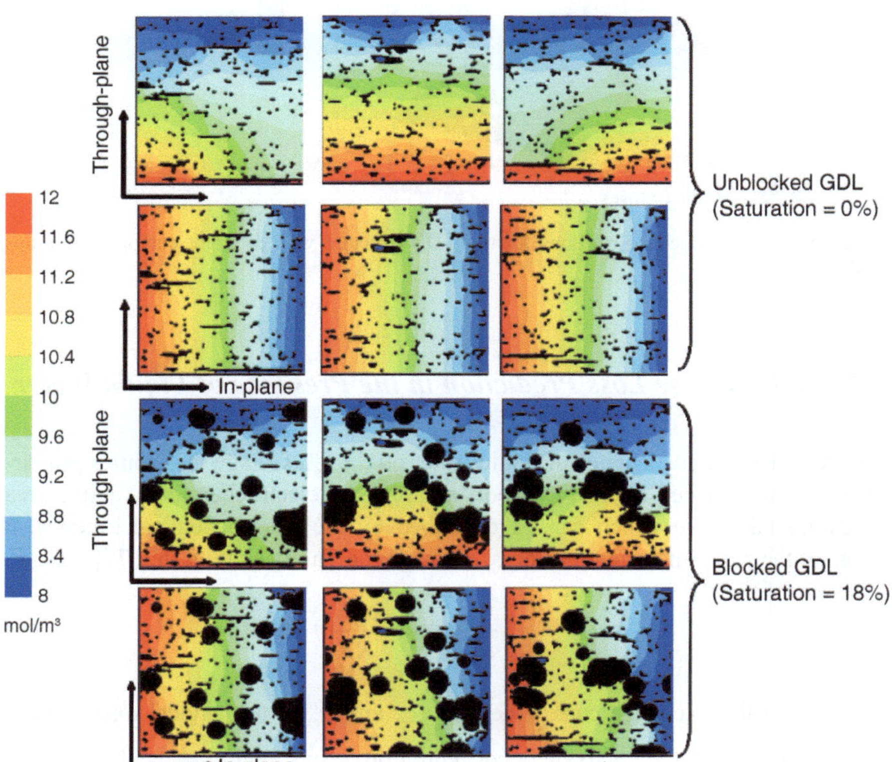

Figure 5.24 Oxygen concentration contours in the unblocked and liquid water blocked GDL structures from the DNS calculations (fiber and water in black) [86]

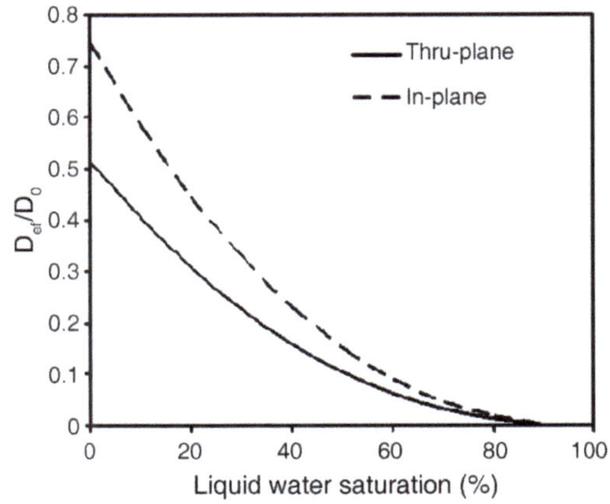

Figure 5.25 Pore blockage correlations for the uncompressed GDL structure [86]

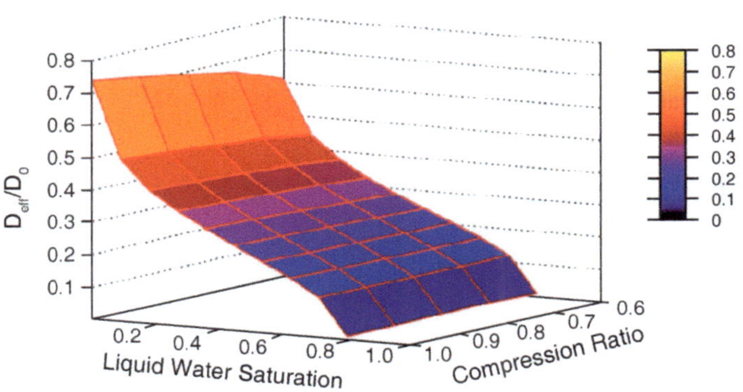

Figure 5.26 Pore blockage relations for the GDL structure with different compression ratios [86]

5.7.3 CL Voltage Loss Prediction in the Presence of Liquid Water

The detrimental consequence of liquid water on the CL voltage loss comes primarily from the impeded oxygen transport and reduced electrochemically active area as explained above which can be described by the electrochemical kinetics in terms of the reaction current density, j, through the Tafel equation [52, 77]:

$$j = i_0 a^{eff} \frac{c_{O_2}}{c_{O_2}^{ref}} \exp\left(-\frac{\alpha_c F}{RT} \eta \right) \tag{5.19}$$

where i_0 is the exchange current density, c_{O_2} and $c_{O_2}^{ref}$ refer to local oxygen concentration and reference oxygen concentration respectively, α_c is the cathode transfer co-efficient for ORR, F is the Faraday's constant, R is the universal gas constant, and T is the cell operating temperature. In the above expression, η is the cathode overpotential and a^{eff} represents the effective ECA due to the catalytic site coverage effect. Details about the above equation along with the pertinent assumptions are furnished in [51, 52].

The pore blockage effect comes into play through the oxygen concentration, c_{O_2} distribution given by the following equation:

$$\nabla \cdot \left(D_{O_2}^{eff} \nabla c_{O_2} \right) = \frac{j}{4F} \tag{5.20}$$

With the evaluated site coverage and pore blockage correlations for the effective ECA and oxygen diffusivity, respectively, and the intrinsic active area available from the reconstructed CL microstructure, the electrokinetics coupled species and charge transport equations can be solved with different liquid water saturation levels within the 1-D macrohomogeneous modeling framework, as detailed in [51, 52, 77], and the cathode overpotential, η can be estimated.

Figure 5.27 exhibits the polarization curves in terms of the cathode overpotential variation with current density for the CCM CL [52, 77] obtained from the 3-D, single-phase DNS model prediction [51], the experimental observation [51], and the liquid water transport corrected 1-D macrohomogeneous model [52]. The "polarization curve" refers to the cathode overpotential versus current density curve in the present study and hence differs from the standard performance curve in terms of the cell voltage versus current density variation shown in Figure 5.2 otherwise used popularly in the literature. It is worth mentioning that the polarization curve in Figure 5.2 is purely for the purpose of illustrating the different voltage loss regimes observed in a typical PEFC operation and does not correspond to any of the calculations reported in this chapter. In Figure 5.27, "DNS" refers to the polarization curve predicted by the single-phase, electrochemistry coupled DNS

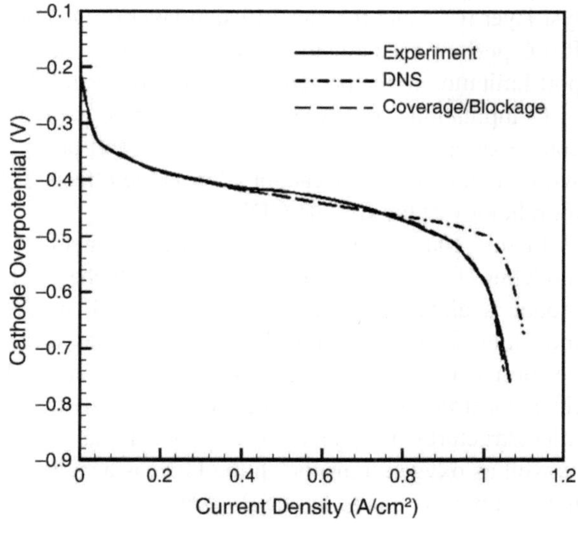

Figure 5.27 Polarization curves for the reconstructed CL structure [52]

model for the reconstructed CL microstructure [51], whereas "Coverage/Blockage" refers to the performance curve predicted by the 1-D macrohomogeneous model with correction for liquid water transport taken into account via the correlations for pore blockage and surface coverage effects evaluated from the two-phase simulations as detailed in [52]. It should be noted that the disagreement between the DNS calculations and experimental data in the transport-control regime is due to the fact that the resistance from liquid water transport was not considered in the DNS model. The details about the DNS calculations, the corresponding operating and boundary conditions along with the experimental data are furnished in [51, 52]. It can be observed that the estimated catalytic site coverage and pore blockage correlations for the CCM CL from the combined two-phase LB and the DNS modeling framework can indeed capture the transport limiting regime and agrees well with the experimental data.

Finally, a key highlight of this investigation is that the systematic estimation of the effective transport parameters for the porous CL and GDL from pore-scale modeling can predict quantitatively the fuel cell performance from the macroscopic fuel cell models. This further indicates that a vertical approach comprising pore-scale to macroscopic modeling could possibly transition the current macroscopic PEFC models beyond empiricism.

5.8 Summary and Outlook

In the present scenario of a global initative toward sustainable energy future, fuel cells are perceived to play a key role and the polymer electrolyte fuel cell (PEFC) is arguably the front runner in the fuel cell race. Despite tremendous progess in recent years, a pivotal performance limitation in the PEFC comes from the underlying competeing transport mechanisms in the constituent components. The catalyst layer (CL) and the gas diffusion layer (GDL) play a crucial role in the overall PEFC performance due to the sluggish oxygen reduction reaction as well as transport limitation in the presence of liquid water and flooding phenomena.

Computational modeling has been employed extensively at different levels of complexities to study fuel cell transport and performance. However, the macroscopic fuel cell models cannot address the effects of the underlying complex pore morphology of the CL and GDL.

In this chapter, we discuss the development of a comprehensive pore-scale modeling framework comprising of a stochastic microstructure reconstruction model, a direct numerical simulation (DNS) model and mesoscopic two-phase modeling formalism in order to reveal the underlying structure–wettability–performance relationship and to predict two-phase closure relations. The stochastic reconstruction model generates 3-D, statistically meaningful CL and GDL microstructures. Pore-level description of charge and species transport in the CL as well as oxygen transport in the GDL is achieved through the DNS model. The mesoscopic two-phase lattice Boltzmann (LB) model simulates liquid water transport through the CL and GDL microstructures in order to gain insight into the influence of structure and wettability on the pore-scale two-phase dynamics as well as to evaluate the two-phase constitutive relations in terms of capillary pressure and relative permeability as functions of liquid water saturation. Additionally, the full morphology (FM) model is discussed which is a fast direct simulation tool and simulates stationary fluid displacement experiment to provide capillary pressure – saturation response based on morphological analysis of the digital structure. A quantitative estimate of the detrimental consequence of liquid water transport in the CL and GDL on the cell performance in terms of the pore blockage and catalytic site coverage effects is predicted by combining LB and DNS models. In the dearth of two-phase correlations for the PEFC CL and GDL, the two-phase transport parameters evaluated from such pore-scale study could be adapted into two-phase computational fuel cell dynamics (CFCD) models for more reliable performance predictions.

Finally, the primary objective of this study is to gain insight into underlying two-phase dynamics and understand the intricate structure–transport–performance interplay in the PEFC CL and GDL. The pore-scale modeling also underscores its capability to transition the current macroscopic fuel cell models beyond empiricism through systematic estimation of the otherwise empirically constructed transport parameters. Extensive future research effort is warranted to develop a "bottom-up", vertically integrated modeling paradigm in order to study the abiding performance limiting mechanisms encompassing multiple length scales and enable virtual material design in the PEFC.

Acknowledgements

PPM would like to thank Drs. V. P. Schulz, A. Wiegmann and J. Becker from Fraunhofer ITWM, Germany for collaboration with GDL structure generation and FM modeling. Financial support from NSF Grant no. 0609727 and ECEC industrial sponsors is gratefully acknowledged.

References

[1] Gottesfeld S, Zawodzinski T. Polymer electrolyte fuel cells. In C. Tobias *et al.* (eds), Advances in Electrochemical Sciences and Engineering. New York: John Wiley and Sons, 1997; 195–301.

[2] Bernardi DM, Verbrugge MW. Mathematical model of a gas diffusion electrode bonded to a polymer electrolyte. AIChE J 1991;37:1151–1163.

[3] Bernardi DM, Verbrugge MW. A mathematical model of the solid-polymer-electrolyte fuel cell. J Electrochem Soc 1992;139:2477–2491.

[4] Springer TE, Zawodzinski TA, Gottesfeld S. Polymer electrolyte fuel cell model. J Electrochem Soc 1991;138:2334–2342.

[5] Springer TE, Wilson MS, Gottesfeld S. Modeling and experimental diagnostics in polymer electrolyte fuel cells. J Electrochem Soc 1993;140:3513-3526.

[6] Gurau V, Liu H Kakac S. Two-dimensional model for proton exchange membrane fuel cells. AIChE J 1998;44:2410–2422.
[7] Um S, Wang CY, Chen KS. Computational fluid dynamics modeling of proton exchange membrane fuel cells. J Electrochem Soc 2000;147:4485–4493.
[8] Wang ZH, Wang CY, Chen KS. Two-phase flow and transport in the air cathode of proton exchange membrane fuel cells. J Power Sources 2001;94:40–50.
[9] Pasaogullari U, Wang CY. Liquid water transport in gas diffusion layer of polymer electrolyte fuel cells. J Electrochem Soc 2004;151:399–406.
[10] Pasaogullari U, Wang CY. Two-phase modeling and flooding prediction of polymer electrolyte fuel cells. J Electrochem Soc 2005;152:A380–A390.
[11] Pasaogullari U, Mukherjee PP, Wang CY, Chen KS. Anisotropic heat and water transport in a PEFC cathode gas diffusion layer. J Electrochem Soc 2007;154:B823–B834.
[12] Wang Y, Wang CY. A non-isothermal, two-phase model for polymer electrolyte fuel cells. J Electrochem Soc 2006;153:A1193–A1200.
[13] Wang Y, Wang CY. Two-phase transients of polymer electrolyte fuel cells. J Electrochem Soc 2007;154:B636–B643.
[14] Weber AZ, Darling RM, Newman J. Modeling two-phase behavior in PEFCs. J Electrochem Soc 2004;151:A1715–A1727.
[15] Weber AZ, Newman J. Effects of microporous layers in polymer electrolyte fuel cells. J Electrochem Soc 2005;152:A677–A688.
[16] Berning T, Djilali N. A 3D, multiphase, multicomponent model of the cathode and anode of a PEM fuel cell. J Electrochem Soc 2003;150:A1589–A1598.
[17] Noponen M, Birgersson E, Ihonen J, Vynnycky M, Lundblad A, Lindbergh G. A two-phase non-isothermal PEFC model: theory and validation. Fuel Cells 2004;4:365–377.
[18] Dutta S, Shimpalee S, Van Zee JW. Three-dimensional numerical prediction of straight channel PEM fuel cells. J Appl Electrochem 2000;30:135–146.
[19] Dutta S, Shimpalee S, Van Zee JW. Numerical prediction of mass-exchange between cathode and anode channels in a PEM fuel cell. Int J Heat Mass Transfer 2001;44:2029–2042.
[20] Mazumder S, Cole JV. Rigorous 3-D modeling of PEM fuel cells II: model predictions with liquid water transport. J Electrochem Soc 2003;150:A1510–A1517.
[21] Nam J, Kaviany M. Effective diffusivity and water-saturation distribution in single- and two-layer PEMFC diffusion medium. Int J Heat Mass Transfer 2003;46:4595–4611.
[22] Lin G, He W, Nguyen TV. Modeling liquid water effects in the gas diffusion and catalyst layers of the cathode of a PEM fuel cell. J Electrochem Soc 2004;151:A1999–A2006.
[23] Natarajan D, Nguyen JV. Three-dimensional effects of liquid water flooding in the cathode of a PEM fuel cell. J Power Sources 2003;115:66–80.
[24] Natarajan D, Nguyen TV. A two-dimensional, two-phase, multicomponent, transient model for the cathode of a proton exchange membrane fuel cell using conventional gas distributors. J Electrochem Soc 2001;148:A1324–A1335.
[25] He W, Yi JS, Nguyen TV. Two-phase flow model of the cathode of PEM fuel cells using interdigitated flow fields. AIChE J 2000;46:2053–2064.
[26] You L, Liu H. A two-phase flow and transport model for the cathode of PEM fuel cells. Int. J. Heat Mass Transfer 2002;45:2277–2287.
[27] Baschuk JJ, Li X. Modeling of polymer electrolyte membrane fuel cells with variable degrees of water flooding. J Power Sources 2000;86:181–196.
[28] Wang CY. Fundamental models for fuel cell engineering. Chem Rev 2004;104:4727–4766.
[29] Weber AZ, Newman J. Modeling transport in polymer electrolyte fuel cells. Chem Rev 2004;104:4679–4726.
[30] Sahimi M. Application of Percolation Theory. Bristol: Taylor & Francis, 1994.
[31] Udell KS. Heat transfer in porous media considering phase change and capillarity - the heat pipe effect. Int J Heat Mass Transfer 1985;28:485–495.
[32] Leverett MC. Capillary behavior in porous solids. AIME Trans 1941:142:152–168.

[33] Gostick JT, Fowler MW, Ioannidis MA, Pritzker MD, Volfkovich YM, Sakars A. Capillary pressure and hydrophilic porosity in gas diffusion layers for polymer electrolyte fuel cells. J Power Sources 2006;156:375–387.
[34] Fairweather JD, Cheung P, St-Pierre J, Schwartz DT. A microfluidic approach for measuring capillary pressure in PEMFC gas diffusion layers. Electrochem Commun 2007;9:2340–2345.
[35] Koido T, Furusawa T, Moriyama K, Keiichiro T. Two-phase transport properties and transport simulation of the gas diffusion layer of a PEFC. ECS Trans 2006;3:425–434.
[36] Nguyen TV, Lin G, Ohn H, Hussey D, Jacobson D, Arif M. Measurements of two-phase flow properties of the porous media used in PEM fuel cells. ECS Trans 2006;3:415–423.
[37] Pan C. Use of pore-scale modeling to understand flow and transport in porous media. PhD Dissertation: University of North Carolina at Chapel Hill, 2003.
[38] Dullien FAL. Porous Media: Fluid Transport and Pore Structure. San Diego: Academic Press, 1992.
[39] Sahimi M. Fractal and superdiffusive transport and hydrodynamic dispersion in heterogeneous porous media. Transport Porous Media 1995;21:189–194.
[40] Blunt MJ. Flow in porous media - pore-network models and multiphase flow. Curr Op Colloid Interface Sci. 2001;6:197–207.
[41] Wolf-Gladrow DA. Lattice Gas Cellular Automata and Lattice Boltzmann Models: An Introduction. Heidelberg: Springer, 2000.
[42] Bird GA. Molecular Gas Dynamics and the Direct Simulation of Gas Flows. Oxford: Clarendon Press, 1994.
[43] Rapaport DC. The Art of Molecular Dynamics Simulation. Cambridge: Cambridge University Press, 1995.
[44] Chen SY, Doolen GD. Lattice Boltzmann method for fluid flows. Annu Rev Fluid Mech. 1998:30:329–364.
[45] Raabe D. Overview of the lattice Boltzmann method for nano- and microscale fluid dynamics in materials science and engineering. Model Simul Mater Sci Eng. 2004:12:13–46.
[46] Succi S. The Lattice Boltzmann Equation for Fluid Dynamics and Beyond. Oxford: Oxford University Press, 2001.
[47] Rothman DH, Zaleski S. Lattice Gas Cellular Automata. Cambridge: Cambridge University Press, 1997.
[48] Rivet JP, Boon JP. Lattice Gas Hydrodynamics, Cambridge: Cambridge University Press, 2001.
[49] Hazlet RD. Simulation of capillary-dominated displacements in microtomographic images of reservoir rocks. Transport Porous Media 1995;20:21–35.
[50] Hilpert M, Miller CT. Pore-morphology-based simulation of drainage in totally wetting porous media. Adv Water Resources 2001;24:243–255.
[51] Mukherjee PP, Wang CY. Stochastic microstructure reconstruction and direct numerical simulation of the PEFC catalyst layer. J Electrochem Soc 2006;153:A840–A849.
[52] Mukherjee PP. Pore-scale modeling and analysis of the polymer electrolyte fuel cell catalyst layer. PhD Dissertation: The Pennsylvania State University, 2007.
[53] Adler PM. Porous Media: Geometry and Transports. Stoneham: Butterworth-Heinemann, 1992.
[54] Adler PM, Thovert JF. Real porous media: local geometry and macroscopic properties. Appl Mech Rev 1998:51:537–585.
[55] Torquato S. Statistical description of microstructures. Annu Rev Mater Res 2002:32:77–111.
[56] Bentz DP, Martys NS. Hydraulic radius and transport in reconstructed model three-dimensional porous media. Transport Porous Media 1994;17:221–238.
[57] Quiblier JA. New three-dimensional modeling technique for studying porous media. J Colloid Interf Sci 1984:98:84–102.

[58] Adler PM, Jacquin CJ, Quiblier JA. Flow in simulated porous media. Int J Multiphase Flow 1990:16:691–712.
[59] Berryman JG. Measurement of spatial correlation functions using image processing techniques. J Appl Phys 1985:57:2374–2384.
[60] Gasteiger HA, Gu W, Makharia R, Mathias MF, Sompalli B. Beginning-of-life MEA performance – efficiency loss contributions. In Vielstich W, Lamm A, Gasteiger HA (eds). Handbook of Fuel Cells—Fundamentals, Technology and Applications, Vol. 3. Chichester: Wiley, 2003; 593–610.
[61] Mathias M, Roth J, Fleming J, Lehnert W. Diffusion media materials and characterization. In Vielstich W, Lamm A, Gasteiger HA (eds). Handbook of Fuel Cells—Fundamentals, Technology and Applications, Vol. 3. Chichester: Wiley, 2003; 517–537.
[62] Schulz VP, Becker J, Wiegmann A, Mukherjee PP, Wang CY. Modeling of two-phase behavior in the gas diffusion medium of PEFCs via full morphology approach. J Electrochem Soc 2007;154:B419–B426.
[63] Schladitz K, Peters S, Reinel-Bitzer D, Wiegmann A, Ohser J. Design of acoustic trim based on geometric modeling and flow simulation for non-woven. Comput Mater Sci 2006;38:56–66.
[64] Sinha PK, Mukherjee PP, Wang CY. Impact of GDL structure and wettability on water management in polymer electrolyte fuel cells. J Mater Chem 2007;17:3089–3103.
[65] Lenormand R, Touboul E, Zarcone C. Numerical models and experiments on immiscible displacements in porous media. J Fluid Mech 1988;189:165–187.
[66] Ewing RP, Berkowitz B. Stochastic pore-scale growth models of DNAPL migration in porous media. Adv Water Resour 2001;24:309–323.
[67] Yang XG, Zhang FY, Lubawy AL, Wang CY. Visualization of liquid water transport in a PEFC. Elecrochem Solid-State Lett 2004;7:A408–A411.
[68] Bear J. Dynamics of Fluids in Porous Media. New York: Dover, 1972.
[69] Tiab D, Donaldson E. Petrophysics: Theory and Practice of Measuring Reservoir Rock and Transport Properties. Houston: Gulf Publishing Company, 1996.
[70] Shan X, Chen H. Lattice Boltzmann model for simulating flows with multiple phases and components. Phys Rev E 1993;47:1815–1819.
[71] Shan X, Chen H. Simulation of nonideal gases and liquid-gas phase transitions by the lattice Boltzmann equation. Phys Rev E 1994;49:2941–2948.
[72] Shan X, Doolen GD. Multicomponent lattice-Boltzmann model with interparticle interaction. J Stat Phys 1995;81:379–393.
[73] Shan X, Doolen GD. Diffusion in a multicomponent lattice Boltzmann equation model. Phys Rev E 1996;54:3614–3620.
[74] Bhatnagar PL, Gross E, Krook M. A model for collision processes in gases. I. Small amplitude processes in charged and neutral one-component systems. Phys Rev 1954;94:511–525.
[75] Pan C, Hilpert M, Miller CT. Lattice-Boltzmann simulation of two-phase flow in porous media. Water Resources Res 2004;40:W1501–W1514.
[76] Mukherjee PP, Wang CY. Modeling of catalyst layer surface coverage and volume blockage owing to liquid water in a PEFC. ECS Trans 2006;3:1085–1094.
[77] Mukherjee PP, Wang CY, Kang Q. Mesoscopic modeling of two-phase behavior and flooding phenomena in polymer electrolyte fuel cells. Electrochimica Acta 2009; 54:6861
[78] Wilde PM, Mandel M, Murata M, Berg N. Structural and physical properties of GDL and GDL/BPP combinations and their influence on PEMFC performance. Fuel Cells 2004;4:180–184.
[79] Ihonen J, Mikkola M, Lindbergh G. Flooding of gas diffusion backing in PEFCs: Physical and electrochemical characterization. J Electrochem Soc 2004;151:A1152–A1161.
[80] Ehrlich R. Viscous coupling in two-phase flow in porous media and its effect on relative permeabilities. Transport Porous Media 1993;11:201–218.
[81] Avraam DG, Payatakes AC. Flow regimes and relative permeabilities during steady-state two-phase flow in porous media. J Fluid Mech 1995;293:207–236.

[82] Avraam DG, Payatakes AC. Generalized relative permeability coefficients during steady-state two-phase flow in porous media, and correlation with the flow mechanisms. Transport Porous Media 1995;20:135–168.
[83] Li H, Pan C, Miller CT. Pore-scale investigation of viscous coupling effects for two-phase flow in porous media. Phys Rev E 2005;72:026705–026714.
[84] Mukherjee PP, Wang G, Wang CY. Direct numerical simulation modeling of polymer electrolyte fuel cell catalyst layers. In White RE, Vayenas CG, Gamboa-Aldeco ME (eds). Modern Aspects of Electrochemistry, Vol. 40. New York: Springer, 2007; 285–341.
[85] Mukherjee PP, Wang CY. Direct numerical simulation modeling of bilayer cathode catalyst layers in polymer electrolyte fuel cells. J Electrochem Soc 2007;154:B1121–B1131.
[86] Schulz VP, Mukherjee PP, Becker J, Wiegmann A, Wang CY. Numerical evaluation of effective gas diffusivity – saturation dependence of uncompressed and compressed gas diffusion media in PEFCs. ECS Trans 2006;3:1069–1075.

Chapter 6
Nanostructured Hydrides for Solid State Hydrogen Storage for Vehicular Applications

Robert A. Varin, Tomasz Czujko and Zbigniew S. Wronski

6.1 Introduction

The energy supply to mankind in the last two centuries was solely based on fossil fuels such as coal in the 19th century and crude oil and natural gas in the 20th century. Unfortunately, this fossil fuel-based economy has led to a number of new challenges facing all of mankind in the 21st century, such as global warming and the following climate changes due to the release of growing amounts of greenhouse gas CO_2, poor urban air quality, and reduction in the world crude oil supply, which could reach the so-called Hubbert's Peak around 2011–2020. It is also noted that no major oil field has been discovered since 1970 [1]. Since the mid-1970s the concept of an ecologically clean "Hydrogen Economy" has been gaining momentum as essentially the only viable remedy for the growing world energy problems. The Hydrogen Economy offers a potential solution to satisfying the global energy requirements while reducing (and eventually eliminating) carbon dioxide and other greenhouse gas emissions and improving energy security. Hydrogen is a very attractive alternative fuel or more precisely energy vector. It is ubiquitous, clean, efficient, and can be produced directly from sunlight and water by biological organisms and using semiconductor-based systems similar to photo-

Robert A. Varin
Department of Mechanical and Mechatronics Engineering, University of Waterloo, Waterloo, Ontario N2L 3G1, Canada
e-mail: ravarin@mecheng1.uwaterloo.ca

Tomasz Czujko
CanEnergy Technology Centre, Hydrogen Fuel Cells and Transportation Energy, Natural Resources Canada, 1 Haanel Drive, Ottawa, Ontario K1A 1M1, Canada

Zbigniew S. Wronski
CanEnergy Technology Centre, Hydrogen Fuel Cells and Transportation Energy, Natural Resources Canada, 1 Haanel Drive, Ottawa, Ontario K1A 1M1, Canada

voltaics. Hydrogen can also be produced indirectly via thermal processing of biomass or fossil fuels where the development of advanced technological processes combined with CO_2 sequestration [2] have the potential to produce essentially unlimited quantities of hydrogen in a sustainable manner. For example, electricity produced by wind turbines or nuclear power plants during off-peak periods [3] can be used for the electrolysis of water into hydrogen [4] and the latter stored for future distribution to places of use [1]. When hydrogen burns, it releases energy as heat and produces water: $2H_2 + O_2 \rightarrow 2H_2O$. Since no carbon is involved then using hydrogen produced from renewable or nuclear energy as an energy resource would eliminate carbon monoxide and CO_2 emissions and reduce greenhouse warming. However, if air is used for flame combustion of hydrogen, small amounts of NO_x may be produced. In a fuel cell, hydrogen is converted directly to electricity in a similar reaction to the one above for burning hydrogen and in essence just electricity and water are produced. Pure hydrogen enters the anode channel in a fuel cell, diffuses through a porous anode towards the catalyst (Pt) where the hydrogen molecules H_2 are stripped of their electrons and become positively charged ions (protons) ($H_2 \rightarrow 2H^+ + 2\bar{e}$). Protons then migrate through the proton-permeable polymeric (Nafion) membrane (Proton Exchange Membrane also called Polymer Electrolyte Membrane – PEM) and the electrons generated during oxidation pass through the external circuit to the cathode, thereby creating electric current. On the cathode side, humidified air enters the cathode channel and diffuses towards the cathode-side catalyst layer. At the catalyst Pt surface, hydrogen protons recombine with electrons and oxygen molecules in air to produce water and heat according to the overall reaction $1/2O_2 + 2H^+ + 2\bar{e} \rightarrow H_2O +$ heat. This waste heat gives a PEM fuel cell an operating temperature of ~60–80 °C [5]. And last but not least is the fact that a hydrogen fuel-cell car can convert hydrogen energy into motion about 2–3 times as efficiently as a normal car converts gasoline energy into motion: depending on how it's designed and run, a good fuel-cell system is about 50–70 % efficient, hydrogen-to-electricity, while a typical car engine's efficiency from gasoline to output shaft averages only about 15–17 % [6].

David Sanborn Scott [7] succintly summarized the inevitability of hydrogen becoming the fuel of the future by two rationales: a depletion-based rationale and a climate-based rationale. Driven by depletion, civilization must move from fossil fuels to sustainable energy sources. Realistically, the only way sustainable sources can be harvested to make chemical fuels is via hydrogen. Otherwise, how else can we get energy from wind, solar or nuclear power to fuel an airplane? On the other hand, atmosphere CO_2 growth is such that the concentration of CO_2 in the atmosphere increased from 280 to 370 ppm over the past 150 years. CO_2 emissions can only be slowed by the extensive use of hydrogen and can only be stopped with the supremacy of sustainable-derived H_2 among chemical fuels.

The realization of the enormous benefits of the Hydrogen Economy has triggered over the last 15 years intense activities into the development of hydrogen related technologies. There are three major technological obstacles to the full implementation of the Hydrogen Economy in the next few decades. The first is the cost of safe, efficient and ecologically friendly production of hydrogen gas. At present, 48 % of hydrogen is produced from methane steam reforming, 30 % from oil/naphtha reforming, 18 % from coal gasification and only 3.9 % from the electrolysis of water [8]. Apparently, the bulk of hydrogen production still relies upon fossil fuels a by-product of which is CO_2. However, the fossil fuel-based processes are much cheaper than the electrolysis of water. Efforts are underway to reduce the price of electrolysis-derived hydrogen to $ 2–3/kg. The second obstacle is further development of the PEM fuel cell (PEMFC) which is the primary cell most suitable for transportation. Most importantly, the research is focused on the extension of the usable service life, water flooding, dynamics, and reliability. The present cost of energy derived from a PEMFC is around $ 200/kW and this must be reduced to around $ 30/kW before PEMFC could be fully commercialized. The third obstacle is hydrogen storage for supplying PEMFC. There are three major competing technologies for hydrogen storage: compressed gas cylinders, liquid hydrogen tanks, and metal hydrides [9, 10]. Their comparison is shown in Table 6.1. A major drawback of compressed hydrogen storage for transportation applications is the small amount of hydrogen that may be stored in a reasonable volume (volumetric capacity/density). As can be seen in Table 6.1 compressed hydrogen gas technologies, even at such enormous pressures as ~ 80 MPa, suffer from low volumetric densities not exceeding ~ 40 kgH_2/m^3. As pointed out by Sandi [10] even at such a high pressure as 70–80 MPa the energy content of compressed hydrogen is significantly less than that for the same volume of gasoline, 4.4 MJ/L (at 70 MPa) for hydrogen compared with 31.6 MJ/L for gasoline. Even though considered to be quite simple and inexpensive, the high pressure of 80 MPa involved in hydrogen gas cylinders raises safety concerns. There is also some cost involved with compression to such high pressures. Another consideration is the large pressure drop during use. In addition, because most of the system parts exposed to hydrogen will be metallic, there is concern regarding the well-known hydrogen embrittlement [10].

The liquid hydrogen tank for its part, offers almost twice as high storage capacity by volume as pressurized hydrogen, however, this is still less than half that required by the Department of Energy (DOE FreedomCAR goal)(the DOE targets

Table 6.1 Comparison of three major competing technologies for hydrogen storage (based on [9, 10])

Storage system	Volumetric hydrogen capacity (kgH_2/m^3)	Drawbacks
Compressed hydrogen gas under 80 MPa pressure	~ 40	Safety problem since enormous pressures are required; cost of pressurization; large pressure drop during use; hydrogen embrittlement of storage tanks
Liquid hydrogen at cryogenic tank at –252 °C (21 K)	~ 71	Large thermal losses (open system); safety; cost of liquefaction
Solid state hydrides	80–160	None of the above

will be discussed later). A major drawback of liquid storage is the high cost of liquefaction, which today can add as much as 50% to the cost of H_2 [6, 9, 10]. There are also safety issues associated with the handling of cryogenic liquids and the problem of evaporative loss.

Solid state hydrides, which include metal/intermetallic and complex hydrides, are characterized by the highest volumetric capacities and they do not suffer drawbacks such as those experienced by compressed and liquid hydrogen. Because of the low pressures involved in metal hydride technologies and the fact that the release of hydrogen takes place via an endothermic process, this method of hydrogen storage is the safest of all. Moreover, the hydrogen released from a metal hydride is of very high purity and therefore, can be used directly to feed a PEMFC.

The US DOE introduced a number of targets for onboard hydrogen storage systems within the framework of its FreedomCAR program for the years 2007, 2010 and 2015 [11, 12] which are listed in Table 6.2.

It is now appropriate to discuss solid state hydrogen storage in hydrides in the context of the targets shown in Table 6.2.

Conventional metal hydrides based on metals such as V, Nb, Pd, Li, Na, *etc.*, have gravimetric capacities too low for any commercial consideration in hydrogen storage with the exception of LiH, which has high capacity accompanied by extremely high desorption temperature [13]. A notable exception of a metal hydride is Mg, which has a relatively high gravimetric capacity and can desorb around 300 °C after nanostructuring treatment (this will be discussed later). In essence, none of metal hydrides can meet the DOE targets [13, 14]. Similarly, hydrides based on intermetallic compounds AB (FeTi, ZrNi), AB_2 ($ZrMn_2$/$TiMn_2$/$TiCr_2$) AB_5 ($LaNi_5$ or $MmNi_5$ where Mm is mischmetal) and A_2B (Mg_2Ni) have relatively low gravimetric storage capacities, as shown in Table 6.3, which are unsuitable for transportation storage although a number of them desorb hydrogen within the

Table 6.2 US DOE FreedomCAR hydrogen storage system targets [11, 12]

Targeted factor	2007	2010	2015
Specific energy (MJ/kg)	–	7.2	10.8
System gravimetric capacity (wt.%)	4.5	6	9
System volumetric capacity (kgH_2/m^3)	36	45	81
Energy density (MJ/L)	–	5.4	9.72
Storage system cost ($/$kgH_2$)	200	133	67
System cost ($/kg/system)	–	6	3
Operating temperature (°C)	–20/50	–30/50	–40/60
Min/max delivery temperature (°C)	–30/85	–40/85	–40/85
Cycle life-time (absorption/desorption cycles)	500	1000	1500
Flow rate (full throttle) (g/s)	3	4	5
Delivery pressure from tank to FC (bar)	2.5	2.5	2.5
Transient response (s) (10–90% and 90–0%)	30	15	15
Refueling rate (kgH_2/min)	0.5	1.5	2.0

Table 6.3 Hydrogen storage properties of intermetallic compounds [12]

Maximum hydrogen capacity				
Type	Intermetallic	Hydride	wt.%	Temperature for 1 atm $P_{desorption}$ (°C)
A_2B	Mg_2Ni	Mg_2NiH_4	3.6	255
AB	FeTi	$FeTiH_{2\,(1.7)}$	1.86	–8
AB	ZrNi	$ZrNiH_3$	1.85	292
AB_2	$ZrMn_2$	$ZrMn_2H_{3.6}$	1.77	167
AB_5/MmB_5	$LaNi_5$	$LaNi_5H_6$	1.49	12
AB_2	$TiV_{0.62}Mn_{1.5}$	$TiV_{0.62}Mn_{1.5}H_{2.5}$	2.15	–6

temperature range targeted by DOE at the desorption pressure ($P_{desorption}$) of 1 atm (0.1 MPa) (which is more or less the operating pressure of a PEMFC).

However, there exist many complex hydrides available having high and very high gravimetric storage capacities, some of which are shown in Table 6.4. Their theoretical capacity is calculated as the ratio of the atomic mass of hydrogen in the hydride formula to the molecular mass of hydride. Some of these hydrides called "complex" hydrides (like borohydrides) decompose in a multi-stage sequence such as, for example, $LiBH_4$, which decomposes in the first stage into $LiH + B + (3/2)H_2$ and in the second stage LiH decomposes into Li and H. However, only the first reaction, which releases about 13.8 wt.%H_2 (3/2 mol of H per 1 mol of $LiBH_4$) is potentially reversible [23]. Therefore, the fourth column in Table 6.4 includes so-

Table 6.4 Hydrogen storage properties of selected high-capacity hydrides [13–22]

Metal–hydrogen system	Hydride	Theoretical maximum gravimetric H_2 capacity (wt.%)	Theoretical reversible gravimetric capacity (wt.%)	Desorption temperature (°C)
Li-B-H	$LiBH_4$	18.4	~13.8	~470
Mg-B-H	$Mg(BH_4)_2$	14.9	~11.2	~300
Fe-B-H	$Fe(BH_4)_3$	12.1	Unknown	Unknown
Ca-B-H	$Ca(BH_4)_2$	11.6	Unknown	~320(?)
Na-B-H	$NaBH_4$	10.6	10.6	400–600
Li-Al-H	$LiAlH_4$	10.6	~7.9	110–260
Al-H	AlH_3	10.0	10.0	~150
Mg-Al-H	$Mg(AlH_4)_2$	9.3	~7.0	110–160
Li-N-H	$LiNH_2(+LiH+TiCl_3)$	8.8	~6.0	150–280
Zn-B-H	$Zn(BH_4)_2$	8.5	8.5(?)	85–140
Ca-Al-H	$Ca(AlH_4)_2$	7.9	~5.9	80–180
Mg-H	MgH_2	7.6	7.6	300–400
Na-Al-H	$NaAlH_4$	7.5	~5.6	229–247
Mg-N-H	$Mg(NH_2)_2(+LiH)$	7.2	~7.0	140–250
Mg-Fe-H	Mg_2FeH_6	5.5	5.5	>300
Na-N-H	$NaNH_2$	5.3	Unknown	<200 (?)

called "theoretical reversible gravimetric capacity", which is the amount of hydrogen that is potentially feasible to be reversibly desorbed/absorbed from such a complex hydride. Unfortunately, as shown in the last column in Table 6.4, the major problem of complex hydrides is that their desorption temperatures are not even close to the operating temperature range required by the DOE targets (Table 6.2). Some of the borohydrides such as $Zn(BH_4)_2$, which start desorbing around 80 °C, close to the operating temperature of PEMFC, release a toxic borane gas B_2H_6 together with hydrogen [24, 25], which can quickly destroy a PEMFC membrane.

Therefore, the major focus of research in the last decade has been on finding the means to substantially reduce the desorption/absorption temperature of high temperature complex hydrides and in addition to improve their absorption/ desorption kinetics where applicable.

Before turning to the characterization of the advances made in the fundamental understanding and technology of high capacity hydrides it will be prudent first to characterize briefly some fundamental thermodynamic properties of hydrides.

6.2 Thermodynamics

The hydriding and dehydriding of metals *M* by dissociative chemisorption of H_2 is deceptively simple:

$$M + \frac{x}{2} H_2 \leftrightarrow MH_x + heat \tag{6.1}$$

The general mechanism of the absorption of hydrogen gas by a metal using a simplified one-dimensional model is shown in Figure 6.1. As the H_2 molecule approaches the surface of a metal, it is first weakly physisorbed on the interface.

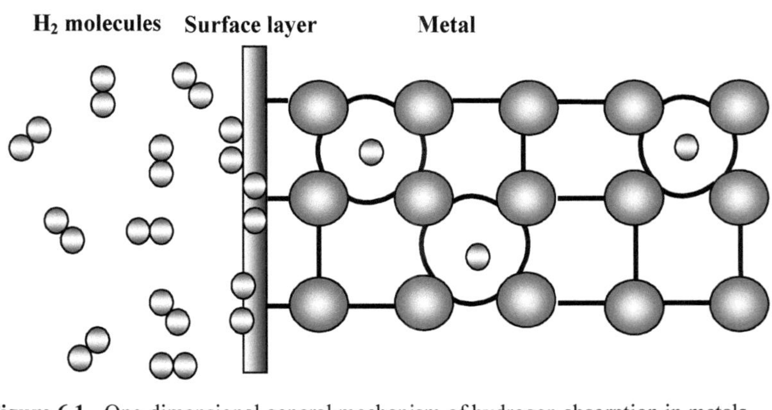

Figure 6.1 One-dimensional general mechanism of hydrogen absorption in metals

After this, the molecule is dissociated and then chemisorbed as strongly bonded, individual H-atoms. H-atoms are light and small so that they can quickly diffuse away from the surface into periodic sites (often interstitial) in the crystal lattice. Once in the crystal lattice, H-atoms can take the form of a random solid solution or an ordered hydride phase if the local hydrogen concentration exceeds a certain limit.

To be able to compare various metal hydrides we must begin by carefully defining the pressure–temperature–composition (PCT) and other properties related to engineering applications, such as plateau pressure value, plateau pressure slope and hysteresis, H capacity, activation, reaction kinetics, thermal conductivity, cycling stability, volume change, decrepitation, gaseous impurity resistance, safety, and cost. In this chapter we will concentrate only on the most important of them.

6.2.1 Pressure–Composition–Temperature Properties

The pressure–composition–temperature (PCT) curve, also called the pressure–composition isotherm (PCI) curve, can be a source of important fundamental information related to the thermodynamic properties of solid hydrides. There are several methods of determining PCT properties ranging from thermogravimetric to precise volumetric measurements obtained by using classical Sieverts-type apparatus. The thermogravimetric methods, unfortunately, are extremely limited in pressure applied during test, to a maximum of 2 MPa, which is usually around equilibrium pressure only for low capacity hydrides. The plateau pressure at the decomposition temperature for high capacity hydrides is much higher and reaches from 2.5 up to 15 MPa.

A typical isotherm of a reversible hydride is shown in Figure 6.2. By measuring the changes in hydrogen pressure and corresponding changes in hydrogen concen-

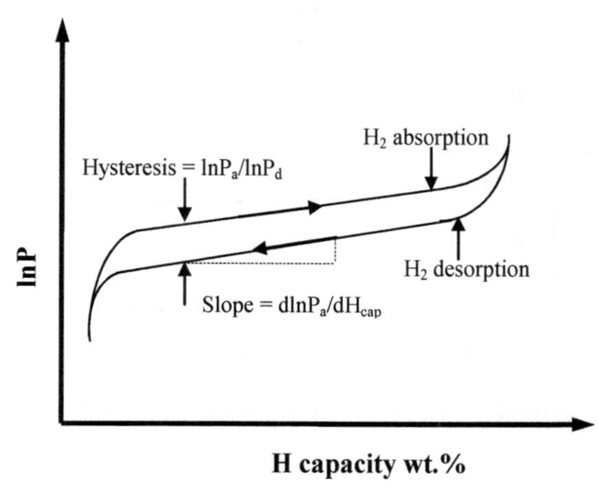

Figure 6.2 Schematic isothermal pressure–composition hysteresis loop

tration in metal at a given temperature, PCT curves can be constructed that are expected to give a flat plateau. Most practical hydrides do not show a perfectly flat plateau or zero hysteresis. Sloping behavior is observed possibly due to different equilibrium pressure, localized defects, and surface inhomogenities [21]. The effect of temperature on the PCT curves is shown in Figure 6.3a. The metal initially dissolves only a small amount of hydrogen (<0.3 wt.%), which creates a solid solution of hydrogen in the metal matrix (α-phase). As the hydrogen pressure together with hydrogen concentration in the metal are increased, interactions between hydrogen and metal atoms become locally important and nucleation and growth of a metal hydride β-phase is observed. In the plateau region there exists a mixture of solid solution α-phase and metal hydride β-phase. The length of plateau determines how much H_2 can be stored reversibly with small pressure variation. It can be seen in Figure 6.3a that increasing the temperature increases plateau pressure, and beyond the critical temperature T_c, the plateau region disappears and the α-phase converts to β-phase continuously.

The relation between mid-plateau pressure P and temperature T is given by the Van't Hoff equation:

$$\ln(P/P_0) = \frac{\Delta H}{RT} = \frac{\Delta S}{R} \tag{6.2}$$

where P_0 is atmospheric pressure, ΔH and ΔS are enthalpy and entropy changes of the hydriding/dehydriding reaction, respectively, T is the absolute temperature and R is the gas constant. For almost all hydrides (but a few exceptions exist) the enthalpy and entropy of the *hydriding* reaction are negative, *i.e.*, the hydriding reaction is exothermic and dehydriding reaction is endothermic. The knowledge of ΔH especially is important to the heat management required for practical engineering devices and is a fundamental measure of the M–H bond strength.

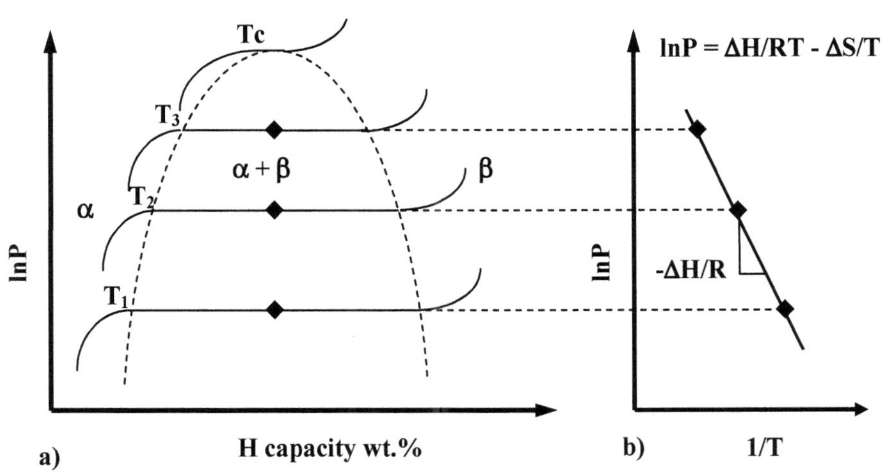

Figure 6.3. **a** Pressure–concentration–temperature (PCT) plot; **b** Van't Hoff plot

The enthalpy of absorption and desorption process ΔH can be determined from the slope ($-\Delta H/R$) using the Van't Hoff plot (logarithm of the mid-plateau pressure against the reciprocal temperature: $\ln P$ versus $1/T$ (or more preferably $1000/T$)), presented in Figure 6.3b. The enthalpy term characterizes the stability of the metal–hydrogen bond and the operating temperature of the metal hydride is fixed by the plateau pressure thermodynamically and by the overall reaction kinetics. The entropy term corresponds mostly to the change from molecular hydrogen gas to dissolved atomic hydrogen and is more or less constant for all hydrides. Substituting $P = 1$ bar (or 1 atm) in Equation 6.2 one can find a simple relationship between the equilibrium temperature (T_{plateau}) required to give a mid-plateau pressure of 1 atm (bar) H_2, ΔH and ΔS in the following form

$$\Delta H = \Delta S T_{\mathrm{plateau}} \tag{6.3}$$

Equation 6.3 is plotted for a number of hydrides in Figure 6.4. As can be seen, all the data points fit very well a simple straight line whose slope is equal to $\Delta S \approx -130$ J/mol K [26]. This clearly shows that the entropy term is, indeed, a constant value for all the solid state hydrogen systems. Figure 6.4 also shows that a low desorption temperature at 1 bar of pressure (more or less the operating pressure of a PEMFC) can only be achieved with hydrides having formation/decomposition enthalpies not larger than 50 kJ/molH_2. For example, hydrides which desorb at room temperature such as $LaNi_5$ and TiFe have $\Delta H \sim 30$ and 33.3 kJ/molH_2, respectively [27]. From this point of view the enthalpy term is one of the most important factors characterizing any hydride.

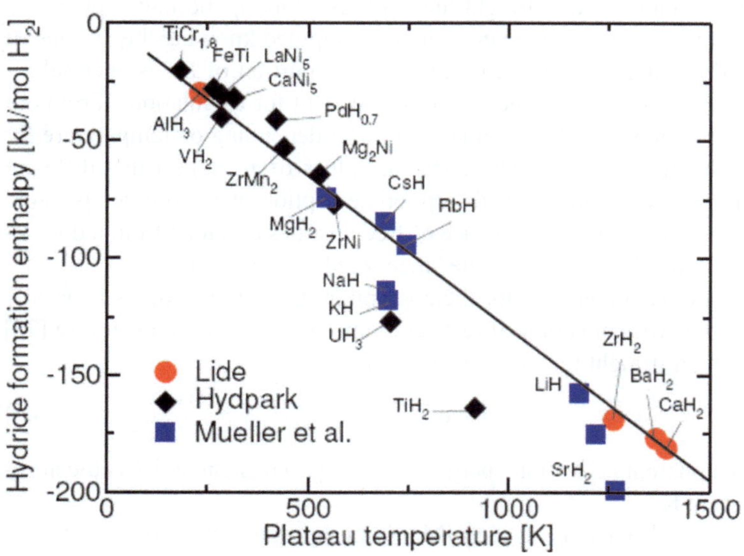

Figure 6.4 Hydride formation enthalpy, ΔH, per mole H_2 as a function of the plateau temperature at 1 bar. The plateau temperature is calculated from reported thermodynamic parameters using the Van't Hoff equation [26]

6.2.2 *Kinetics of Hydrogen Absorption/Desorption*

For hydride powders, the absorption/desorption kinetics are usually analyzed by applying the JMAK (Johnson–Mehl–Avrami–Kolmogorov) theory of phase transformations, which is based on nucleation and growth events [28–30] where α is the fraction transformed at time t or, alternatively, for hydrides the fraction absorbed

$$\alpha = 1 - e^{-(k \cdot t)^{\eta}} \tag{6.4}$$

or desorbed at time t. It must be kept in mind that the JMAK model applies when growth of a new phase begins randomly in the bulk and at the surface (nucleation is spatially random), the sample size is much greater than any individual transformed region, growth proceeds homogeneously throughout the sample, and nucleation rate is constant [28–30]. The parameters describing the reaction kinetics, such as the nucleation and growth rates, are contained within an effective kinetic parameter, k, while the exponent, η, called the Avrami exponent or reaction order, provides some information about the dimensionality of the transformation, *i.e.*, whether it is one, two-or three-dimensional and whether it is interface-limited or diffusion-limited. Equation 6.4 can be rearranged to the following linear equation

$$\ln\left[-\ln(1-\alpha)\right] = (\eta \ln k) + \eta \ln t \tag{6.5}$$

from which the values of the reaction order η and subsequently the rate constant k can be interpolated by plotting $\ln[-\ln(1-\alpha)]$ versus $\ln(t)$. Such a plot for each constant temperature should give a straight line with the slope η and intercept $\eta\ln(k)$. From the latter, the rate constant k can easily be computed knowing the η value. It must be pointed out that only a nearly linear initial portion of the isothermal kinetic curve α versus time (t) is to be taken into account for calculations. From our experience it is common that the η values can differ depending on temperature for which they are being calculated. The different values of η suggest that different mechanisms are rate controlling of absorption/desorption at various temperature ranges. Therefore, we recommend the use of free η values obtained from a double-logarithm fitting procedure as more "true" than fixed η values [31].

The apparent activation energy for the absorption/desorption process is usually evaluated from the Arrhenius plot of rate constant k values with temperature [28] by simply plotting a straight line $\ln k$ versus $1/RT$:

$$k = k_o e^{-E_A/RT} \tag{6.6}$$

where E_A is the apparent activation energy, R is the gas constant and T is the absolute temperature in K.

It must be pointed out that besides JMAK the hydrogen desorption data can also be fitted to other models such as contracting volume and surface reaction model (chemisorption) [32]. We tested these three respective models: JMAK, contracting volume $1-[1-\alpha]^{1/2} = kt$ (two-dimensional growth with constant interface velocity), and surface reaction $\alpha = kt$ for MgH_2 hydride doped with 5 wt.% of micrometric Ni catalyst and obtained activation energy equal to 105, 101 and 105 kJ/mol, respectively. This shows that in practical situation each of these three models gives almost identical activation energy.

6.3 Nanoprocessing of Solid State Hydrides by Ball Milling

Most hydrides are in powder form, having an average particle size in the range of tens of micrometers (μm). These individual particles are internally divided into grains having an average size in the range of a few micrometers or at best, a fraction of a micrometer. The aim of the nanoprocessing of solid state hydrides is to reduce relatively large grains to a size smaller than 100 nm, and additionally to reduce the size of hydride powder particles. In summary, nanoprocessing creates a truly nanocrystalline hydride material or "nanohydride".

In general, nanocrystalline materials are single-phase or multi-phase polycrystals with grain sizes varying from a few nanometers to ~ 100 nm in at least one dimension [33–35]. In powder form, nanostructured/nanocomposite means that each phase present in the individual powder particle is in the form of grains with nanometer size. One particle is one "nano-polycrystal".

In a nanocrystalline/nanostructured material two types of atoms can be distinguished: crystallites and boundary regions/intercrystalline regions [34]. The atomic structure of all crystallites is identical. The only difference between them is their crystallographic orientation. In the grain boundary regions, the average atomic density, interatomic spacing and the coordination between nearest neighbor atoms deviates from those in the crystallites and differs from region to region. The presence of these two structural constituents (crystallites and boundaries) of comparable volume fractions and with typical crystal sizes of a few nanometers is crucial for the properties of nanocrystalline materials [33–35]. As a consequence, many of the physical and mechanical properties of nanocrystalline solids, such as thermal expansion, elastic constants, fracture stress, and ductility, are widely different from those of the same material with conventional grain sizes. This is a direct consequence of the significant fraction of atoms belonging to intercrystalline positions. Consequently, interface structures in these materials are bound to play a major role in material properties. Once the crystal size and boundary dimensions become comparable with certain length scales new physical effects are to be expected.

A primary method of nanostructuring hydrides is processing by mechanical (ball) milling. Processes of manufacturing of nanocrystalline/nanostructured hydrides by ball milling are shown in Figure 6.5. There are three major processes, all of which start from raw metallic and other required elements. They are discussed in detail in [36].

In the first process, most suitable for intermetallic hydrides, the metallic elements are mixed in the required proportion and arc melted into an intermetallic

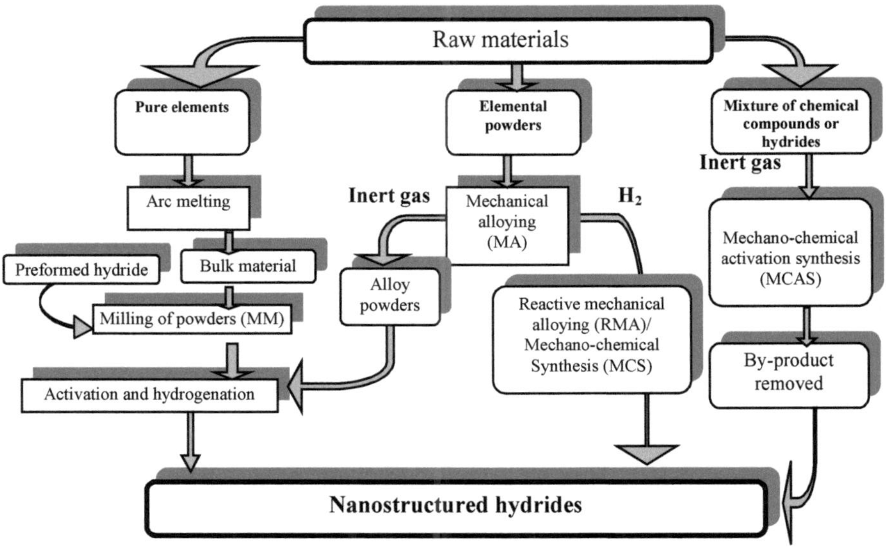

Figure 6.5 Flow chart showing the possible methods of manufacturing nanocrystalline/ nanostructured hydrides (after [36])

compound, which is subsequently pulverized by mechanical milling (MM) and hydrogenated to form the desired intermetallic-based hydride, for example, $TiFeH_{1.7}$ or Mg_2NiH_4.

In the second method, in one route elemental metallic powders of the desired proportion are mixed and then mechanically alloyed (MA) by ball milling under inert gas (*e.g.*, argon) to form intermetallic compounds. Subsequently, those intermetallics powders are hydrogenated to form hydrides. This procedure can be called a "two-step" method. In a "single-step" method the pre-mixed metallic powders are milled under hydrogen atmosphere to directly form an intermetallic hydride. This method is called reactive mechanical alloying (RMA) or mechano-chemical synthesis (MCS). In the specially designed magneto-mill, Uni-Ball Mill 5, in which the movement of steel milling balls is largely controlled by strong external magnets (Section 6.3.1) if hydrogen atmosphere is used, RMA is referred to as controlled reactive mechanical alloying (CRMA), which is based on a mechano-chemical synthesis of the metallic elements and hydrogen into a hydride phase. In CRMA it is possible to study the effect of controlled milling mode on the hydride structure and properties. This type of a magneto-mill will be discussed further later in the chapter.

In the third method, called mechano-chemical activation synthesis (MCAS), a mixture of a chemical compound (*e.g.*, chloride) and a hydride is ball-milled to induce a reaction during milling in which the metal from the constituent hydride forms a compound (*e.g.*, salt) with the non-metal molecules in the chemical compound and the metal from the chemical compound joins the hydrogen group from the constituent hydride and forms a completely new, high capacity hydride. This displacive reaction is also called a metathesis reaction. For example, if a metal

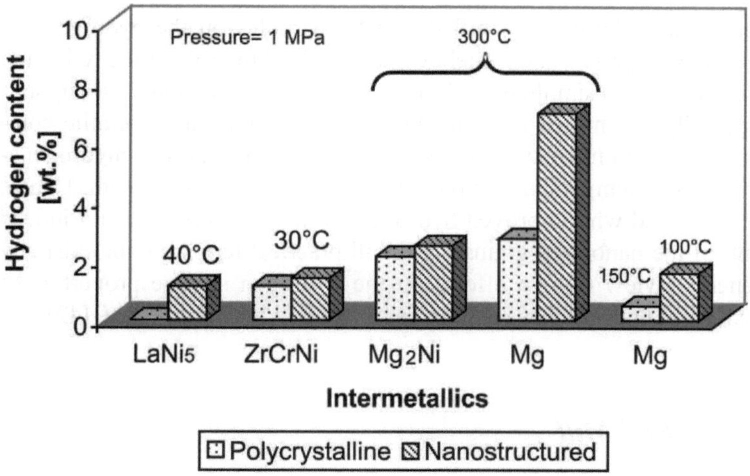

Figure 6.6 Hydrogen content after absorption in polycrystalline and nanostructured intermetallics. For comparison, data for Mg are shown. Absorption temperatures are also shown. Experimental data extracted from [36]

chloride (MCl_n) is used in the reaction with sodium borohydride ($NaBH_4$) then in general terms the metathesis reaction can be written as

$$MCl_n + nNaBH_4 \rightarrow M(BH_4)_n + nNaCl \tag{6.7}$$

where $M(BH_4)_n$ is a newly synthesized complex metal borohydride and NaCl is a salt. The MCAS milling can be done in the Uni-Ball Mill 5 under a controlled mode.

It must be pointed out that in all these three methods the milling leads to a desired synthesis (chemical) reaction and simultaneously induces the formation of nanosized grains inside hydride powder particles and if possible, a substantial reduction in powder particle size. There are no other methods, *e.g.*, chemical, of hydride synthesis that produce so effectively nanocrystalline/nanostructured hydrides.

An interesting question is how does nanostructure affect properties of solid state hydrides? Nanostructure addresses one of the key issues of hydrogen sorption kinetics: the diffusion of hydrogen within the metal leading to the formation of the hydride or back to the metal in a dehydrogenation reaction. Examples in Figure 6.6 show that in practically all intermetallic/metal hydrides, the kinetics of both absorption and desorption can be improved by nanostructuring induced by ball milling.

As reviewed in [36] the hydrides in the early research were not physically ball milled in order to induce nanocrystallinity. In reality, either intermetallic compounds were first synthesized by mechanical alloying or pure metals (*e.g.*, Mg) were simply ball milled under argon and subsequently subjected to hydrogenation at the appropriate temperature and under the required hydrogen pressure. As mentioned earlier in the text such a method is called “a two-step” method. In addition, this early work, especially in the group led by Professor J.O. Ström-Olsen at the McGill University in Montréal, Canada in which a prominent role was played by L. Zaluski and A. Zaluska [37–41], overemphasized the effect of nanograins (crystallites) formed within the heavily milled powder particles on the hydrogen absorption/desorption properties and somehow marginalized the role of the reduction of particle size that occurs simultaneously with the decrease in nanograin (crystallite) size. Especially, for milled Mg, and Mg-based and other intermetallic compounds having nanograin microstructure, subsequent hydrogenation/dehydrogenation cycles at elevated temperatures should give rise to nanograin growth. Hence, it is hard to understand why improved hydrogen storage properties would indeed still be a result of the nanosized grains that in all practical terms do not exist any longer. A general review of size effects on the hydrogen storage properties of nanostructured metal hydrides has recently been reported by Bèrube *et al.* [42].

6.3.1 Magneto Ball Mill

A ball mill apparatus usually consists of a container placed in a rotating or vibrating frame. Different types of ball mills are used to produce mechanically alloyed powders. They differ in their capacity, efficiency of milling, and arrangements for heating and cooling [43, 44]. In conventional ball mills (planetary or vibrational), the trajectories of grinding balls are rather chaotic (Figure 6.7). This creates a continuous and erratic change of various mechanical modes of milling from shearing to impact during the same milling cycle. However, in contrast to the other types of ball mills a magneto mill is characterized by a controlled largely non-chaotic ball movement. In the magneto-mill, Uni-Ball-Mill 5, the trajectories of the balls are controlled by the magnetic field created by extremely strong NdFeB permanent magnets (Figure 6.8). Protective gas such as argon or reactive hydrogen can be admitted through valve to the milling cylinder. The milling mode can be adjusted from shear to impact by changing the angular position of the external magnets, as shown in Figure 6.8. The process variables are not completely independent. For example, the milling mode depends basically on magnet position but also on milling speed and working distance. Moreover, milling time depends on milling mode and ball-to-powder-ratio.

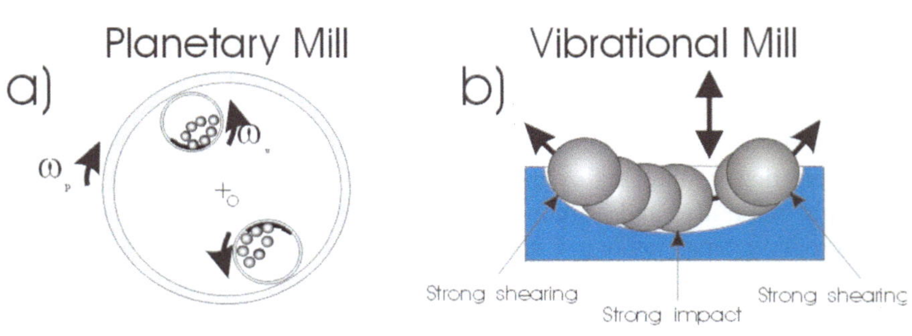

Figure 6.7. Schematic of **a** planetary, and **b** vibrational mill [45]

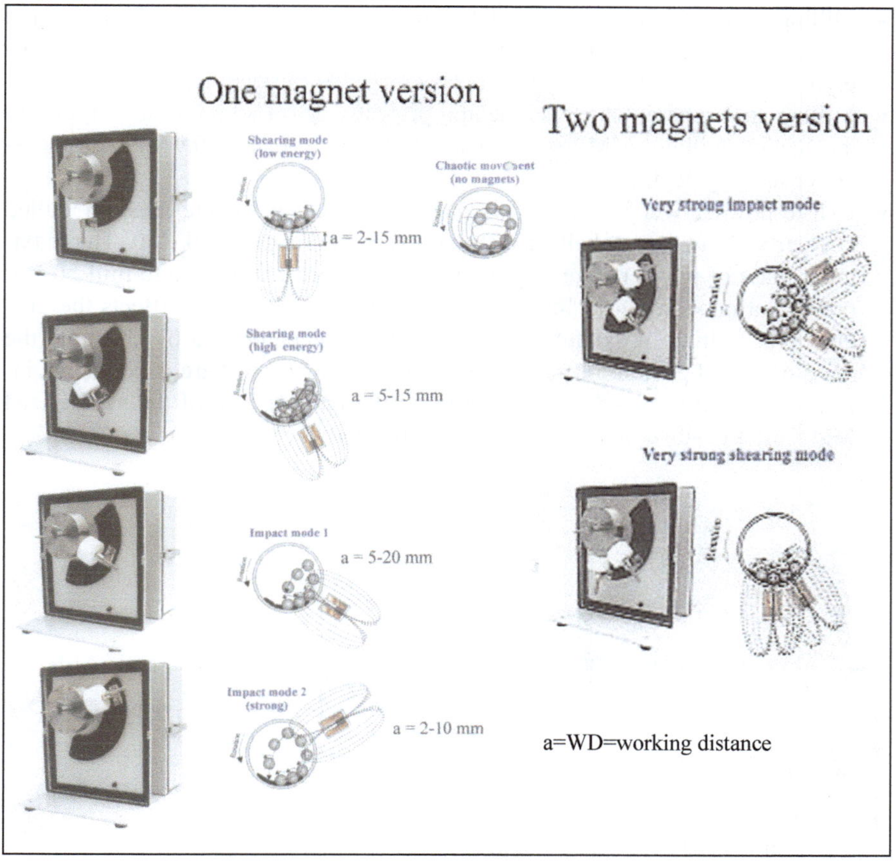

Figure 6.8 Various controlled modes of milling available in the Uni-Ball-Mill 5 [46]

Ball milling in the Uni-Ball-Mill 5 is a complex process that involves optimization of milling parameters to achieve the desired product microstructure and properties. The important parameters are as follows:

- Milling mode, which depends on the magnet position. Changing the magnet position we can change milling mode from low energy shearing, through high energy shearing to impact and strong impact. Two magnet options are also possible (very strong shearing mode and very strong impact mode) (Figure 6.8).
- Number of balls used for milling, which usually varies from two to four balls. A maximum of five balls can by used in one cylinder but the optimal number is four.
- Milling speed, which can be controlled in the range 0–200 rpm and depends on milling mode (more energetic mode usually requires fast rotation).

- Milling time, which is closely related to milling mode, working distance, and type of process (ball milling usually requires a shorter time than mechanical alloying or reactive ball milling).
- Milling atmosphere, which is a neutral protective gas (helium or argon) during mechanical milling (MM) or hydrogen under pressure up to 0.9 MPa during reactive milling (RMA).
- Ball-to-powder weight ratio. This parameter depends on the mass of milled powder and number of balls and usually is in the range of 10–100. The maximum mass of powder in one cylinder is 25 g which allows one to mill 50 g of powder at once in two cylinders. Ball-to-powder weight ratio affects the efficiency of milling or synthesis process and can be controlled by changing the mass of milled powder or the number of balls. However, there is no visible influence of ball-to-powder weight ratio on the particle size of the synthesized MgH_2 after reactive milling for 30 h (Figure 6.9).

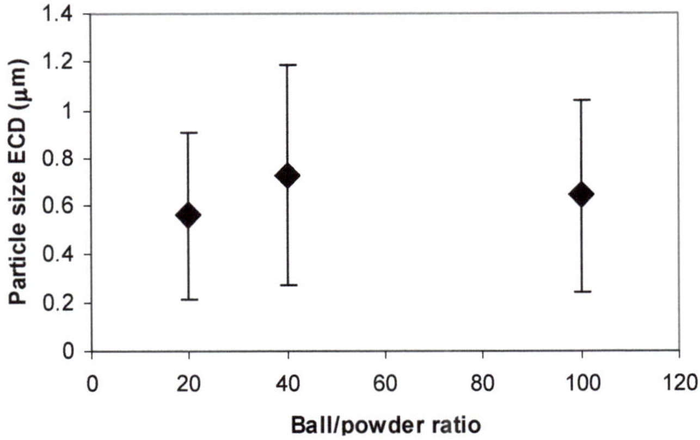

Figure 6.9 Particle size of MgH_2 synthesized by reactive ball milling for 30 h in the Uni-Ball-Mill 5 using various ball-to-powder weight ratios

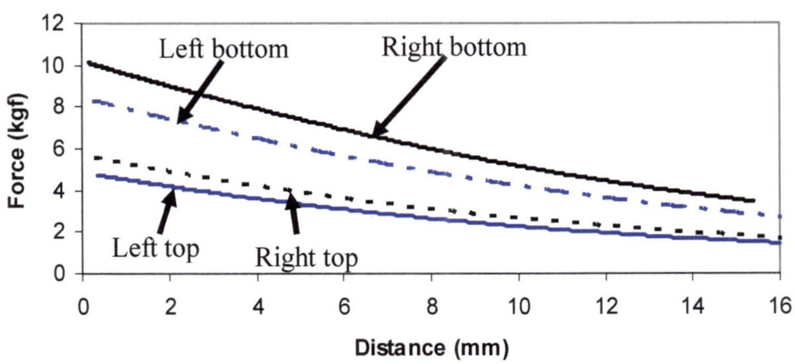

Figure 6.10 Attractive force between 25 mm steel ball and different magnets versus distance from the cylinder (working distance – WD) (Uni-Ball-Mill 5)

- Working distance (WD), which is the distance between magnet (magnets) and cylinder (Figure 6.8). This parameter affects mainly the attractive force between the magnet and balls inside the cylinder (Figure 6.10). It is shown that increasing the WD reduces substantially the attractive force between the magnet and the 25 mm steel ball. The four NdFeB magnets tested in Figure 6.10 vary in attraction force even when produced by the same company. For high energy ball milling, which requires two magnets (Figure 6.8), they are usually applied in tandem such that a stronger magnet is paired with a weaker one on both sides of the Uni-Ball-Mill 5 (left and right) and the weaker magnet is always above (top) the stronger magnet (bottom).

6.3.2 Microstructural Characterization of Ball Milled Hydrides

Two important morphological parameters characterizing ball milled powders are the particle size and grain size of constituent phases residing within the powders. The powder particle size was measured by attaching loose powder to sticky carbon tape and taking pictures in secondary electron (SE) mode with the SEM. The images were then analyzed by image analysis software. The size of the powders was calculated as the particle equivalent circle diameter, ECD = $(4A/\pi)^{1/2}$, where A represents the projected particle area. Usually from 260 to 650 particles were analyzed for each batch.

The crystalline structure of hydride powders is characterized by powder diffraction. The nanograin (crystallite) size of phases residing in the milled powders is calculated from the broadening of their respective X-ray diffraction (XRD) peaks. Since the Bragg peak broadening in an XRD pattern is due to a combination of grain refinement (nanograin/crystallite) and lattice strains, it is customary to use computing techniques by means of which one can separate these two contributions. The separation of crystallite size and strain is obtained from a Cauchy/Gaussian approximation by a linear regression plot according to the following equation [47]:

$$\frac{\delta^2(2\theta)}{\tan^2\theta} = \frac{K\lambda}{L}\left(\frac{\delta(2\theta)}{\tan\theta\,\sin\theta}\right) + 16e^2 \tag{6.8}$$

where the term $K\lambda/L$ is the slope, the parameter L is the mean dimension of the nanograin (crystallite) composing the powder particle, K is a constant ($\approx$1) and e is the so-called "maximum" microstrain (calculated from the intercept), λ is the wavelength and θ is the position of the analyzed peak maximum. The term $\delta(2\theta) = B[1-(b^2/B^2)]$ (rad) is the instrumental broadening-corrected "pure" XRD peak profile breadth [47], where B and b are the breadths in radians of the same Bragg peak from the XRD scans of the experimental and reference powder, respectively. The B value is approximated as the full width at half maximum, FWHM, and calculated by the diffractometer software. The b value is approximated as FWHM from the XRD pattern of a compound LaB_6, the National Institute of Standards and Technology (NIST) standard reference material (SRM) 660 for subtracting the instrumental broadening from the experimental FWHM and finding $\delta(2\theta)$as given above. It must be noted that when FWHMs of the instrumental line profiles are obtained in this manner, the Bragg peaks for the LaB_6 SRM are occasionally at different 2θ angles than those of the analyzed hydride in the milled powders. The interpolated FWHM values between angles for the SRM peaks are found using a calibration curve.

6.4 High Capacity Hydrides

Due to space constraints we will limit the present review to critical discussion of recent advances made in the science and technology of hydrides based on MgH_2, its mixtures with catalytic additives, and finally, (nano)composites of MgH_2 and high capacity hydrides. In particular, the application of nanostructuring by ball milling for the improvement of the storage properties will be assessed. As can be clearly seen in Table 6.2 only high capacity hydrides are important for onboard hydrogen storage for vehicular applications. As pointed out by Read *et al.* [48] depending on the storage material and on the system design, material capacities may need to be a factor of up to two times higher than system capacity targets. If such a rule is to be applied to the high capacity hydrides in Table 6.4, then vis-à-vis the extremely restrictive DOE targets in Table 6.2, most of the high capacity hydrides should not be considered as onboard storage materials for PEMFC vehicles. This of course is very unreasonable and it seems that sooner or later the DOE targets should be relaxed in line with reality. In this review we will only discuss hydrides having a minimum of 5 wt.% H_2 capacity. This is to some extent an arbitrary cut off, but 5 wt.% is still a relatively reasonable capacity. The second parameter of importance is the desorption/absorption temperature. Again, we apply an arbitrary cut off at ~300 °C as a maximum desorption temperature for hydrides to be taken into consideration. Such a temperature range is still reasonable for three reasons. First, there is a recent trend to increase the working temperature of the PEMFC by using an electrolyte membrane of polybenzimidazole (PBI) doped with phosphoric acid [49]. Second, a nanostructured MgH_2 with catalyst with reduced desorption temperature to below 300 °C could be used onboard in a recently proposed two-stage reservoir [50]. Third, there is hope that the 300 °C desorption temperature of simple and complex hydrides could be further reduced by appropriate additives, as will be discussed in this review. Except MgH_2, most simple metal hydrides are eliminated from the present review.

Another very restrictive and debatable DOE requirement for onboard hydrogen storage is the *reversibility* of solid state hydrides (*e.g.*, onboard refueling with the target rate of 1.5 kgH_2/min for 2010, as shown in Table 6.2). However, as pointed out by Sandrock *et al.* [51] it is an immense problem to remove the exothermic heat at that charging rate. For example, if we charged H_2 at 1.5 kg/min into a vehicular storage tank based on $NaAlH_4$ (ΔH=37 kJ/mol H_2), we would have to

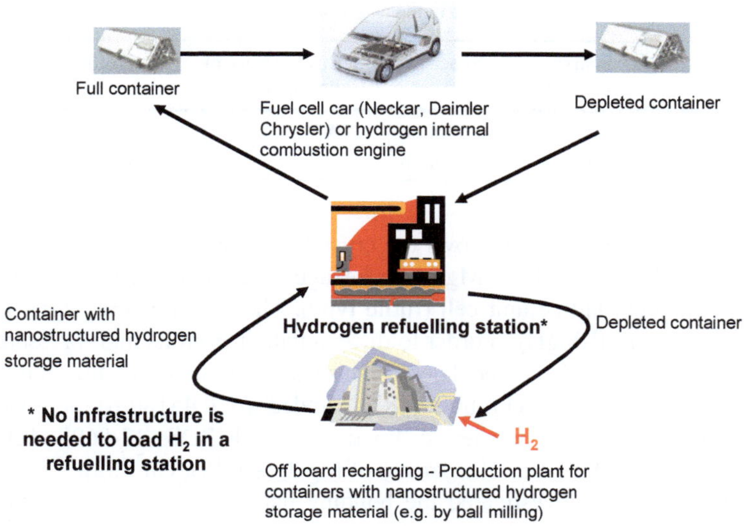

Figure 6.11 Novel refueling/retail station for fuel cell powered vehicles based on the concept of the offboard recharging by the synthesis of nanostructured hydrogen storage materials

remove heat at the rate of 450 kW. This would require very substantial and costly heat-exchangers and in practice would completely eliminate the possibility of quick onboard recharging. Another point of importance is the enormous cost attached to building an entire infrastructure of a hydrogen gas refueling station (steel pipes, steel storage vessels, *etc.*). Last but not least is the hydrogen embrittlement phenomenon mentioned earlier, in which various metals, such as high-strength steel, aluminum, and titanium alloys become brittle and eventually crack under load following exposure to hydrogen. This process could be a problem for hydrogen steel piping and other components of a hydrogen gas refuelling station. In our opinion a much easier solution is offboard recharging in which the depleted solid state hydride container is repleted with nanostructured solid state storage materials manufactured in a dedicated plant, as shown schematically in Figure 6.11. According to this new vision, the refuelling station is just a retail station for hydrogen storage containers where "refuelling" could become eventually a fully atutomated process reduced to a quick replacement of containers. This could also lead to the growth of new businesses for the Hydrogen Economy.

6.4.1 Magnesium Hydride (MgH_2)

6.4.1.1 Crystallographic Characteristics

Magnesium hydride is one of the most researched hydrides. It is potentially very attractive as a storage material because of high capacity ~7.6 wt.%H_2 (Table 6.4)

Table 6.5 Crystallographic data of the phases in the Mg-H system [54]

Phase	Composition at.% H	Pearson symbol	Space group	Strukturbericht designation	Prototype
(Mg)	0 to 11	hP2	$P6_3/mmc$	A3	Mg
MgH_2	66.7	tP6	$P4_2/mnm$	C4	TiO_2 (rutile)

and the low cost of Mg. Table 6.5 shows the crystal structure data of the phases existing in the Mg–H system. Pure Mg has a hexagonal crystal structure and its hydride has a tetragonal lattice unit cell (rutile type). The low-pressure MgH_2 is commonly designated as β-MgH_2 in order to differentiate it from its high-pressure polymorph, which will be discussed later. Precise measurements of the lattice parameters of β-MgH_2 by synchrotron XRD yielded $a = 0.45180(6)$ nm and $c = 0.30211(4)$ nm [52]. The powder diffraction file JCPDS 12-0697 lists $a = 0.4517$ nm and $c = 0.30205$ nm. The density of MgH_2 is 1.45 g/cm^3 [53].

Noritake *et al.* [52, 55] investigated the bonding nature of MgH_2 employing the maximum entropy method and synchrotron radiation powder data. They found that the bonding nature of hydrogen in MgH_2 was quite complex consisting of a mixture of ionic and covalent bonding. As stated by the authors there are weak but significant covalent bonds between Mg and H as well as between H and H. The weak covalency of the Mg–H bond may be advantageous for hydrogenation/dehydrogenation performance. Modeled charge density distribution revealed that the ionic charge of Mg and H can be represented as $Mg^{1.91+}$ and $H^{0.26-}$, respectively. This means that Mg is ionized almost as Mg^{2+}, while hydrogen is very weakly ionized.

Under increasing hydrogen pressure substantial changes occur in the Mg–H system. Bastide *et al.* [56] investigated the behavior of MgH_2 phase under high pressures up to 80 kbar and found that at ambient temperature (20 °C) and 80 kbar pressure the β-MgH_2 phase (designated α-MgH_2 in the original paper [56]) transformed partially into another polymorphic phase, γ-MgH_2, forming a mixture β + γ-MgH_2. XRD studies showed that γ-MgH_2 has an orthorhombic unit cell structure of α-PbO_2 type with the lattice parameters $a = 0.453$ nm, $b = 0.544$ nm and $c = 0.493$ nm (this phase is included in the powder diffraction file JCPDS 35-1184). They also found another metastable phase δ-MgH_2 (designated β-MgH_2 in the original paper [56]), which, upon releasing pressure, transformed into γ-MgH_2. They claimed that above 350 °C γ-MgH_2 transformed into the equilibrium β-MgH_2 phase but this may not be quite correct in view of recent findings.

6.4.1.2 Nanostructuring of MgH_2 by Ball Milling

Ball milling brings about substantial changes in the microstructure of hydride powders. For investigating these effects in our laboratory, we used two commercial MgH_2 hydride powders. The first one was purchased from Degussa-Goldschmidt sold under the trade name Tego Magnan with average purity of 95 %

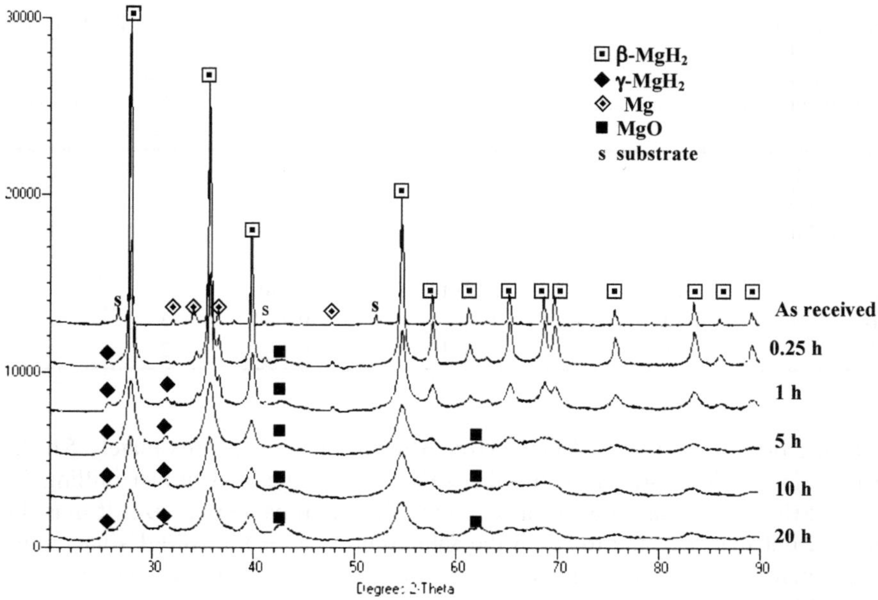

Figure 6.12 Evolution of XRD patterns as a function of ball milling time of ABCR powder under HES57 mode in hydrogen gas atmosphere at ~600 kPa pressure in the magneto-mill Uni-Ball-Mill 5

(remaining Mg), which gives the theoretical purity-corrected hydrogen content ~7.2 wt.%. The second one was purchased from ABCR GmbH&Co.KG, sold under the trade name MG-5026. Its average purity claimed by the supplier is ~98 % (remaining Mg), which gives the theoretical purity-corrected hydrogen content ~7.5 wt.%. For simplicity it will be referred to hereafter as the "ABCR" powder. The as-received Tego Magnan and ABCR powder has an average ECD particle size of 36 μm with standard deviation SD = ±16 μm and 41 μm with SD = ±21 μm, respectively. However, the two powders differ substantially in grain size, which is ~67 nm and ~300 nm for Tego Magnan and ABCR, respectively.

Figure 6.12 shows the evolution of XRD patterns as a function of controlled mechanical milling (CMM) of ABCR powder under high energy shearing mode with two magnets at 5 and 7 o'clock positions (HES57) in the magneto-mill Uni-Ball-Mill 5 (Section 6.3.1). There are three important microstructural changes occurring during milling of MgH_2 that have been discussed in a number of papers [37–41, 57–60].

First, the breadth of the XRD peaks of β-MgH_2 increases with increasing milling time. This is related to the formation of crystallites (nanograins) within the powder particles (Section 6.3.2), which may be accompanied by the introduction of lattice strains. Table 6.6 lists the values of nanograin size and the lattice strain of β-MgH_2 as a function of milling time estimated from the procedure described in Section 6.3.2. As can be seen, the nanograin size is reduced rapidly with milling time such that after 15 min milling it is reduced approximately six-fold compared

Table 6.6 Grain size variations of β-MgH_2 as a function of milling time of ABCR powder under HES57 mode from Figure 6.12

Milling time (h)	Grain size (nm)	Strain	R^2	Number of XRD peaks
0 (as received	299	0	0.9989	4
0 (as received)	303	0	0.9936	6
0.25 (15 min)	52	9.31×10^{-4}	0.9904	4
1	30	3.16×10^{-3}	0.9954	3
5	12	1.17×10^{-3}	0.9901	4
10	10	0	0.9999	3
20	11	5.26×10^{-3}	0.9926	3

with the original as-received grain size. After milling for approximately 5 h the nanograin size is saturated, within the ~ 10–12 nm range, and further milling for 10 to 20 h does not bring about any further change in nanograin size. It must be pointed out that the time to reach the saturation level of the β-MgH_2 nanograin size depends on the milling mode (low-energy shearing (LES), high-energy shearing (HES), or impact) but when the milling process is carried out in the magneto-mill Uni-Ball-Mill 5, saturation time is always a few to tens of hours regardless of the mode. The situation could be slightly different when using a different type of mill, *e.g.*, Spex or Fritsch, but there has been no systematic study of saturation behavior in these mills. We found that within the experimental scatter the grain size saturation level does not depend on the type of commercial MgH_2 powder and mode of milling. The strains of the β-MgH_2 phase in Table 6.6 are minimal (<0.5 %) as opposed to values of over 1 % reported by Huot *et al.* [57]. Taking into account that the β-MgH_2 phase is tetragonal and its atomic bonding is mostly ionic with a little covalency [52, 55], *i.e.*, ceramic-like, one can hardly envision substantial dislocation activities that would lead to dislocation accumulation in the lattice and in turn, strains of such large magnitude (> 1 %). It seems that the strains reported by Huot *et al.* [57] are rather overestimated. For clarification, it is pointed out that the XRD peaks of MgO observed in Figure 6.12 arise due to the exposure of residual Mg present in the as-received MgH_2 powder to air during powder handling for XRD tests, which in a nanostructured form exhibits a very strong affinity to oxygen in air.

Second, even after a short milling time of 15 min (0.25 h), the XRD peaks of an orthorhombic γ-MgH_2 phase appear on the pattern (Figure 6.12). As pointed out by Schulz *et al.* [58, 59] the pressure increase due to the mechanical action of milling balls produces the structural transformation of a tetragonal β-MgH_2 into an orthorhombic γ-MgH_2, which normally would occur under enormous static pressures of around 8 GPa (Section 6.4.1.1). Huot *et al.* [57] estimated that the γ-MgH_2 phase abundance was around 18 % and did not increase with increasing milling time. However, we have found that the volume fraction of the γ phase seems to be slightly higher after milling for 100 h than that after 20 h.

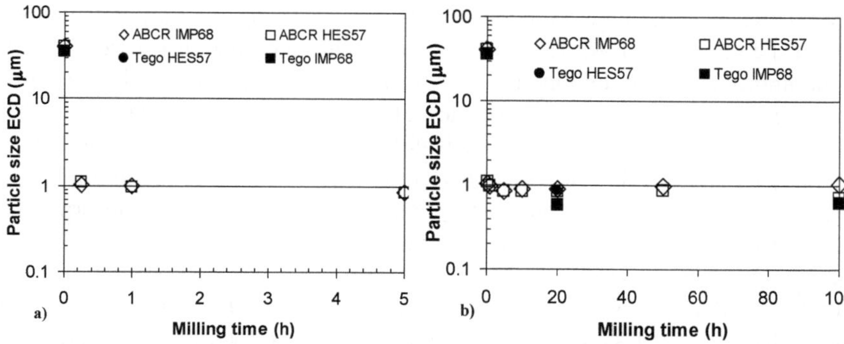

Figure 6.13. Powder particle size versus milling time for two commercial MgH_2 powders, Tego Magnan (Tego) and ABCR, which were milled in the magneto-mill Uni-Ball-Mill 5 under shearing and impact modes: **a** milling up to 5 h; and **b** milling up to 100 h (HES57-high energy shearing with two magnets at 5 and 7 o'clock positions; IMP68-strong impact with two magnets at 6 and 8 o'clock positions)

Third, parallel to the decrease of grain size there is always a decrease of the particle size of milled powders as shown in Figure 6.13. The particle size reduction occurs within a very similar time frame to the reduction in grain size. As can be seen in Figure 6.13a, after only 15 min the particle size is reduced from the initial ~40 µm to ~1 µm. Further prolonged milling for up to 100 h brings about an incremental particle size reduction down to ~0.6 µm (Figure 6.13b). This behavior is of a very general nature and practically, does not depend on the mode of high energy milling and the type of MgH_2 commercial powder, as can be seen in Figure 6.13. Almost identical time frames for grain and particle size variations as a function of milling time, as shown in Table 6.6 and Figure 6.13, makes it rather difficult to identify unambiguously which factor is, indeed, governing hydrogen storage characteristics. This difficulty has led to a common belief that the grain size is mostly responsible for the observed enhancement of hydrogen storage properties, which is not necessarily the case as will be discussed later.

Nanostructuring of magnesium hydride (MgH_2) by high-energy ball milling brings about both beneficial and detrimental effects to its hydrogen desorption characteristics. The major beneficial effect is that the desorption temperature of ball milled, nanostructured MgH_2 hydride is substantially reduced compared with the as-received material as measured by differential scanning calorimetry (DSC). This effect for the ABCR powder is shown in Figure 6.14. The endothermic hydrogen desorption peaks are broad and have characteristic shoulders which suggest overlapping of two peaks. The onset temperature of hydrogen desorption in a DSC curve is designated T_{ON}, the shoulder temperatures are designated LT and the peak maxima are designated HT. All three temperatures decrease systematically with increasing milling time up to 5 h (Figure 6.14a). However, further milling for 10 and 20 h does not reduce desorption temperature any further (Figure 6.14b). According to Gennari *et al.* [60] the low-temperature shoulder/peak is due to the total decomposition of γ-MgH_2 and the partial decomposition of β-MgH_2 whereas the high temperature DSC peak corresponds to the decomposition of the remaining β-MgH_2. The DSC temperature reduction is essentially due to the improvement in the kinetics of desorption which can be clearly seen if one compares the desorption characteristics of the following Tego Magnanhydride powders: (i) non-milled and non-activated in Figure 6.15; (ii) non-milled and activated by short cycling in Figure 6.16 (activation: heating for 15 min to 350 °C under 2.7 MPa H_2 to prevent desorption; three cycles of (a) desorption at 350 °C/0.1 MPa H_2/60 min, (b) annealing under pre-vacuum at 350 °C for 15 min, and (c) absorption at 350 °C/2.7 MPaH_2/30 min; after each desorption at a constant temperature in a volumetric Sieverts-type apparatus under atmospheric pressure of hydrogen the same powder sample was re-absorbed at 350 °C under hydrogen pressure of 2.7–3.5 MPa for 15–30 min and then desorption at a desired temperature); and (iii) ball milled in Figure 6.17 (tested immediately after ball milling).

The non-milled and non-activated Tego Magnan (and the ABCR) powder does not desorb at all below 350 °C (Figure 6.15a). The same non-milled and activated powder in Figure 6.16a shows a trace of desorption at 300 °C and is able to desorb ~5 wt.%H_2 at 325 °C in about 3600 s (1 h). The milled powder in Figure 6.17a desorbs ~4 wt.%H_2 at 300 °C in about 4000 s and ~6.5 wt.%H_2 at 325 °C in about 2000 s. As can be seen in Figure 6.15b, 6.16b and 6.17b, the activation energy values for desorption, E_A, calculated from the Arrhenius plot (Section 6.2.2) are ~120 kJ/mol, ~118 kJ/mol and ~140 kJ/mol, for non-milled and non-activated, non-milled and activated, and milled powder, respectively, with excellent coefficients of fit. It is interesting to note that the non-milled and non-activated, as well as non-milled and activated powders have almost identical values of activation energy and the milled powder with the best desorption kinetics exhibits the highest activation energy value. Tables 6.7, 6.8, and 6.9 show the η values in the JMAK equation (Section 6.2.2). Deleting the highest η value at 300 °C in Table 6.9 gives the activation energy 120 kJ/mol (R^2 = 0.999) for the milled powder, which is almost identical to those for the non-milled, non-activated and activated powders. This shows that the selection of a range of temperatures for calculating the activation energy of desorption can substantially affect the calculated activation energy values.

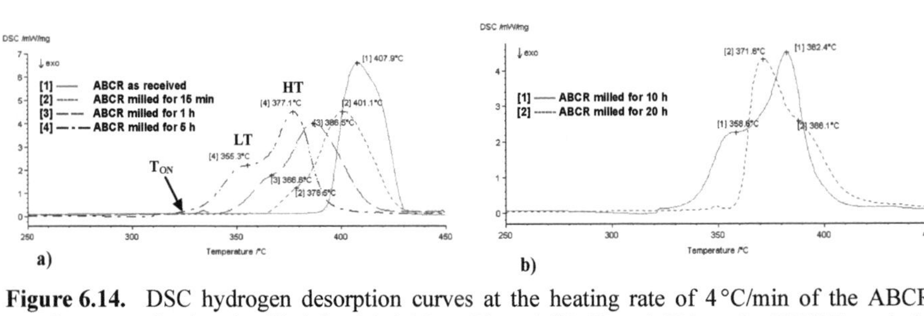

Figure 6.14. DSC hydrogen desorption curves at the heating rate of 4 °C/min of the ABCR powder as received and milled for (**a**) 0.25 to 5 h and (**b**) 10 and 20 h under HES57 mode in hydrogen

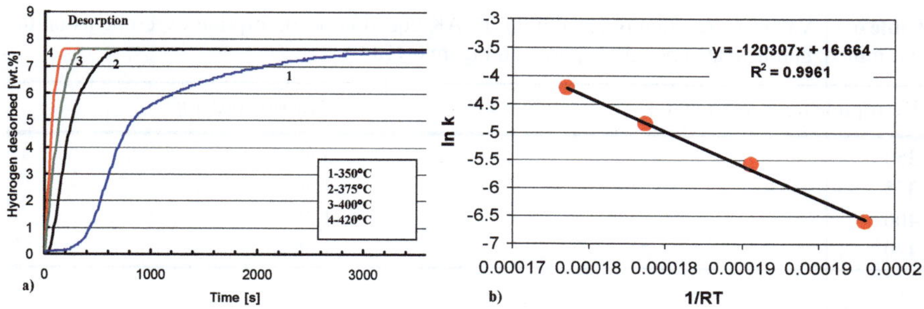

Figure 6.15. **a** Desorption kinetic curves at various temperatures under initial hydrogen pressure of 0.1 MPa of the non-milled, non-activated, commercial MgH_2 powder Tego Magnan. **b** Arrhenius plot of the desorption rate for the estimate of the apparent activation energy, E_A, using kinetics data for four temperatures: 350, 375, 400 and 420 °C (E_A ~ 120 kJ/mol). Coefficient of fit $R^2 = 0.996$

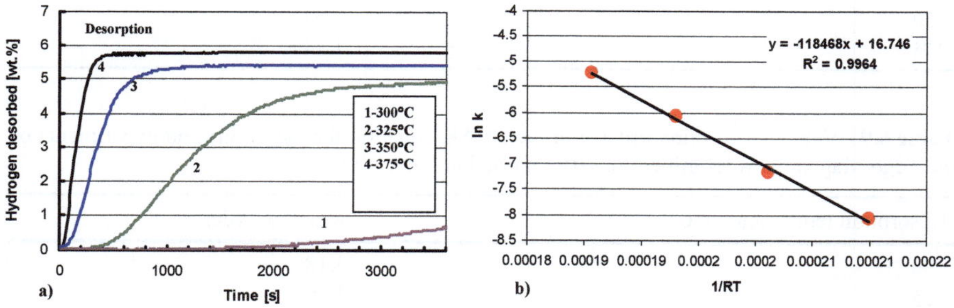

Figure 6.16. **a** Desorption kinetic curves at various temperatures under initial hydrogen pressure of 0.1 MPa of the non-milled, activated commercial MgH_2 powder Tego Magnan. **b** Arrhenius plot of the desorption rate for the estimate of the apparent activation energy, E_A, using kinetics data for four temperatures: 300, 325, 350 and 375 °C (E_A ~ 118 kJ/mol). Coefficient of fit $R^2 = 0.996$

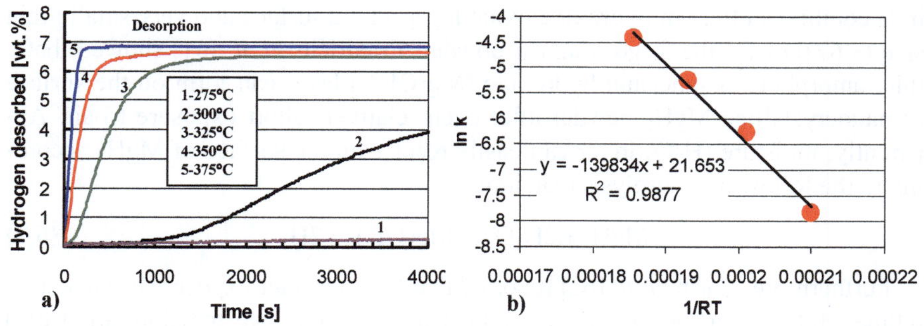

Figure 6.17. **a** Desorption kinetic curves at various temperatures under initial hydrogen pressure of 0.1 MPa of the commercial MgH_2 powder Tego Magnan milled continuously for 20 h under IMP68 mode (in argon). **b** Arrhenius plot of the desorption rate for the estimate of the apparent activation energy, E_A, using kinetics data for four temperatures: 300, 325, 350 and 375 °C (E_A ~ 140 kJ/mol). Coefficient of fit $R^2 = 0.988$. Desorption tests were carried out directly after milling

Table 6.7 Values of reaction order η in the JMAK equation for desorption experiments on the non-milled, non-activated powder reported in Figure 6.15a

Desorption temperature (°C)	Reaction order η
350	3.60
375	1.76
400	1.39
420	1.56

Table 6.8 Values of reaction order η in the JMAK equation for the desorption experiments on the non-milled, activated powder reported in Figure 6.16a

Desorption temperature (°C)	Reaction order η
300	3.52
325	2.44
350	2.04
375	1.84

Table 6.9 Values of reaction order η in the JMAK equation for the desorption experiments on the Tego Magnan powder milled for 20 h reported in Figure 6.17a

Desorption temperature (°C)	Reaction order η
300	3.18
325	1.69
350	1.42
375	1.47

Almost identical values of the activation energy for desorption of the non-milled and non-activated, and non-milled and activated powders can be explained by a combined effect of hydroxide/oxide layer on the surface and very small grain size (~67 nm) for the Tego Magnan powder. Friedrichs *et al.* [61] showed that a thin, amorphous magnesium hydroxide ($Mg(OH)_2$) layer can form on the surface of nanocrystalline MgH_2 powder after even relatively short exposure to air. Apparently, moisture (H_2O) in air can easily react with the surface of MgH_2 according to the following hydrolysis reaction:

$$MgH_2 + 2H_2O \Rightarrow Mg(OH)_2 + 2H_2 \tag{6.9}$$

Furthermore, Varin *et al.* [62] reported that long-term air exposure of nanocrystalline MgH_2 for a few months duration leads to a massive transformation of a large fraction of MgH_2 particles into the crystalline $Mg(OH)_2$ phase. Such a reaction occurs particularly fast on a nanocrystalline MgH_2 synthesized/processed by ball milling.

We can safely assume that the hydroxide/oxide layer is always broken/discontinuous along the intersections of grain boundary planes with the particle surface, which provide excellent paths for hydrogen penetration and diffusion into the bulk. Since Tego Magnan has very small, nearly nanometric, grain size, the layer on its particles will be broken into small pieces and practically fully discontinuous due to the large number of intersections of grain boundary planes with the surface. Activation, in principle, will not change this picture because the layer cannot be broken into even smaller pieces. Since hydrogen can penetrate and diffuse at many locations at the surface with the same ease for non-activated and activated powder, the activation energy of desorption remains at the same, relatively low level before and after thermal activation of the Tego Magnan powder.

For most hydrides the rate of absorption is usually much faster than desorption. One can compare Figure 6.18a showing absorption curves of milled (HES57; 5 h), activated and cycled ABCR powder with Figure 6.16a for a non-milled and activated powder. The former absorbs ~2 wt.%H_2 at the low temperature of 250 °C after about 4000 s. Obviously, no desorption can occur at all at this temperature range for non-milled and activated powder (Figure 6.16a). As shown in Figure 6.19a ball milling additionally improves absorption kinetics even if the powder is activated and cycled after milling. Pre-milled powder absorbs ~3 wt.%H_2 at 200 °C after about 4000 s. At 250 °C absorption rate is quite fast and ~5 wt.%H_2 is absorbed after about 1500 s. The increase of absorption rate by ball milling is indeed substantial. However, pre-milling does not seem to affect measurably the apparent activation energy for absorption. As seen in Figures 6.18b and 6.19b the apparent activation energy for absorption for non-milled and milled ABCR powder is ~65 and 73 kJ/mol, respectively. The small difference between values is within the experimental error. In general, it seems that the apparent activation energy is not sensitive enough to show real benefits of ball milling.

An important question now arises such as what are the factors responsible for improved desorption/absorption properties of ball milled MgH_2? First, we thermally cycled in hydrogen the ball milled ABCR powder according to the following procedure: heating to 325 °C for ~15 min under 3.4 MPa H_2 to prevent desorption and then two desorptions at 325 °C under 0.1 MPa H_2 pressure for ~4700 s with intermediate annealing under pre-vacuum at 350 °C for 15 min and absorption at 350 °C under 3.4 MPa H_2 for 30 min. The microstructure of cycled powder was investigated by XRD. Figures 6.20 and 6.21 show the DSC trace and XRD pattern of the cycled powder, respectively. The DSC curves for cycled samples in Figure 6.20 are smooth and symmetrical without any shoulders as opposed to the DSC curves of the same powder after ball milling (Figure 6.14). However, the desorption peaks are shifted back to higher temperatures while they still show a systematic decreasing trend with increasing milling time. XRD patterns of milled and cycled samples show only peaks of β-MgH_2 in contrast to milled powders which contain both β-MgH_2 and γ-MgH_2 (Figure 6.12). Obviously, thermal cycling completely eliminates γ-MgH_2 from the microstructure of previously milled powder. As reported by Gennari *et al.* [60] the LT shoulder observed in DSC curves of milled MgH_2 powders in Figure 6.14 is most likely due to the hydrogen desorption from the γ-MgH_2 phase. Elimination of γ-MgH_2 after thermal cycling makes DSC curves smooth since they now correspond only to the hydrogen desorption from β-MgH_2.

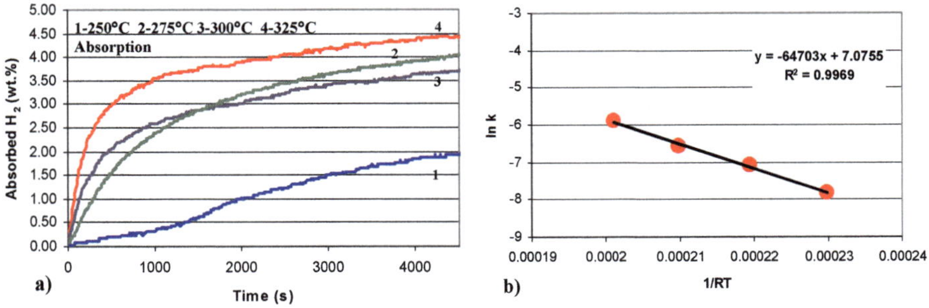

Figure 6.18. **a** Absorption kinetic curves of non-milled, activated and cycled ABCR powder; and **b** estimate of apparent activation energy of absorption from the Arrhenius plot of ln *k* versus 1/*RT* using data for all four temperatures: 250, 275, 300, and 325 °C (E_A ~65 kJ/mol). Coefficient of fit R^2 = 0.997 (Activation: heating 350 °C/3.5 MPa/30–45 min; 9 desorption/absorption cycles: desorption 350 °C/annealing at 350 °C/pre-vacuum/15 min/cooling under pre-vacuum to the required absorption temperature and finally absorption at *T*=325–200 °C (at every 25 °C)/1.2 MPa/4750 s)

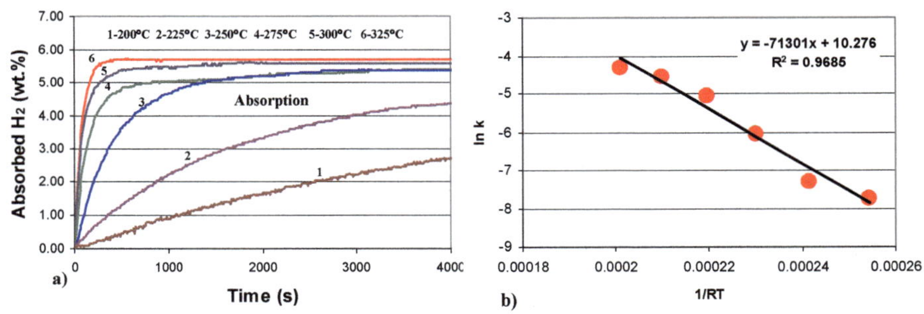

Figure 6.19. **a** Absorption kinetic curves of milled (HES57;5 h), activated and cycled ABCR powder; and **b** estimate of apparent activation energy of absorption from the Arrhenius plot of ln *k* versus 1/*RT* using data for all six temperatures: 200, 225, 250, 275, 300, and 325 °C (E_A ~ 73 kJ/mol). Coefficient of fit R^2 = 0.969

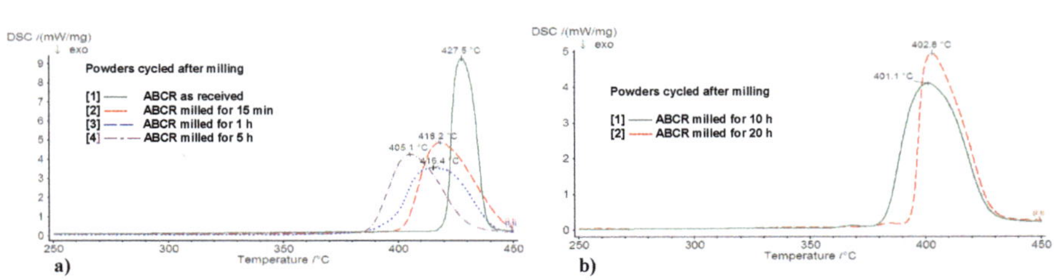

Figure 6.20. DSC traces at the heating rate of 4 °C/min of the ABCR powder from Figures 6.12 and 6.14 after milling for (**a**) 0.25 to 5 h and (**b**) 10 and 20 h, and thermal cycling in hydrogen (after last absorption)

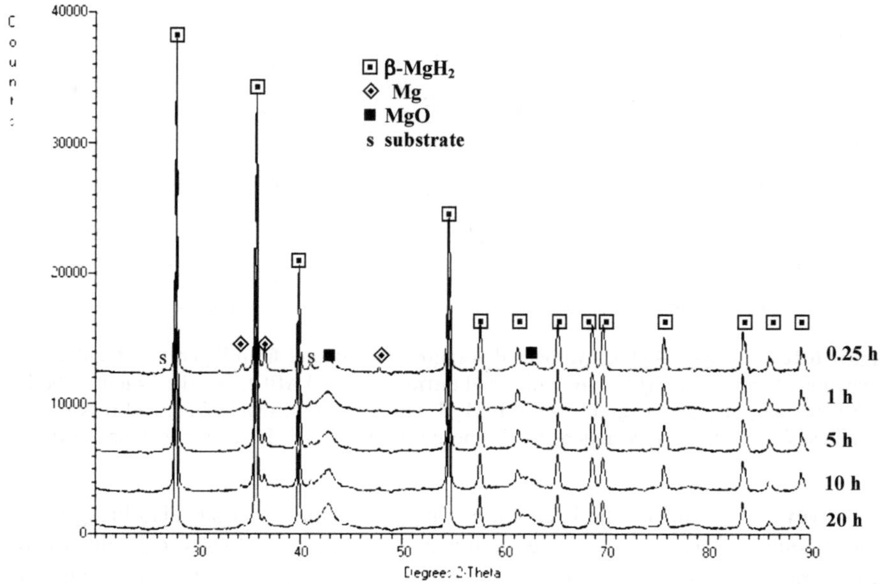

Figure 6.21 XRD patterns of ABCR powder after milling for various times and thermal cycling in hydrogen as described in the text

The DSC onset (T_{ON}) and peak maximum (T_{peak}) temperatures from Figures 6.14 and 6.20 are plotted in Figure 6.22a and b, respectively, as a function of particle size (corresponding grain size of β-MgH_2 is also shown beside each data point). It must be pointed out, that within the experimental scatter the particle sizes of cycled powders are unchanged with respect to the particle sizes of milled powders. This fact has been confirmed by careful measurements. The plotted curves for cycled samples in Figure 6.22a and b, which correspond to the hydrogen desorption from β-MgH_2, are shifted up with respect to the milled samples in which desorption occurs from a phase mixture of β-MgH_2 + γ-MgH_2. The shape of the as milled and cycled curves is very similar, *i.e.*, gradual decrease in the first stage and subsequent fast decrease after reaching a critical value of the particle size (~ 1000 nm). The grain size of the β-MgH_2 phase does not seem to affect the DSC hydrogen desorption temperature in any systematic manner. Accordingly, based on this quantitative evidence it can be concluded that two microstructural factors such as the γ-MgH_2 phase residing within the powder particles and refined powder particle size, acting additively, contribute to a substantial reduction of hydrogen desorption temperature of MgH_2 hydride as observed in DSC.

A substantial apparent reduction of hydrogen desorption temperature with decreasing particle size as obtained due to ball milling is a very beneficial effect although the reduced onset and peak hydrogen desorption temperatures are still slightly higher than 300 °C, *i.e.*, much too high for supplying a PEMFC.

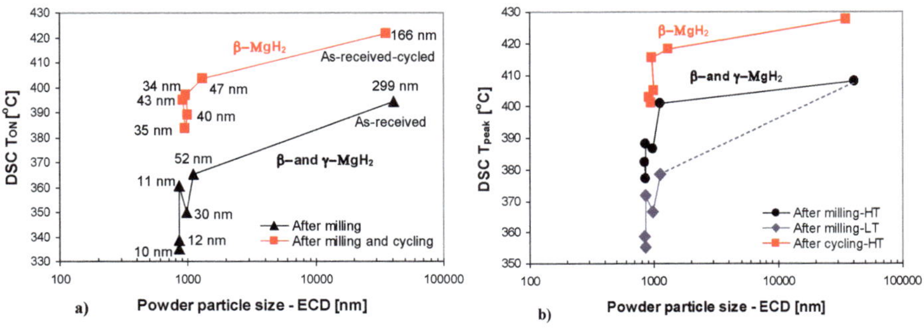

Figure 6.22. Changes of DSC hydrogen desorption temperatures from Figures 6.14 and 6.20 as a function of particle size (ECD) of milled and thermally cycled ABCR powder. Numbers beside each data point indicate the grain size of β-MgH_2. **a** Onset temperature (T_{ON}); and **b** peak temperatures (T_{peak}). Standard deviations for the mean particle size (ECD) are omitted for clarity

However, ball milling and resulting nanostructuring of MgH_2 also bring about a detrimental effect which is a reduction in the hydrogen storage capacity of the milled powders. The purity-corrected hydrogen capacity of the Tego Magnan MgH_2 powder is around 7.2 wt.% at 95 % purity. This capacity is in practical terms obtained by hydrogen desorption from a non-milled and non-activated commercial Tego Magnan powder over the temperature range 350–420 °C (Figure 6.15). However, this capacity can not be obtained after desorption from the milled powder (Figure 6.17) even at temperatures as high as 350–375 °C. We prepared a Tego Magnan sample milled for 100 h under high energy impact mode (two magnets; IMP68) and investigated its microstructure by XRD after desorption at 350, 375 and 400 °C for up to 4000 s under 0.1 MPa hydrogen pressure (atmospheric). As shown in Figure 6.23, after desorption at all three temperatures there are high-intensity peaks of newly-formed Mg and small but sharp peaks of retained MgH_2 discernible on the XRD pattern. Most likely, this is due to the following phenomena. First, the γ-MgH_2 phase is always formed during milling due to the transformation of β-MgH_2 into γ-MgH_2 (Section 6.4.1.1). However, during subsequent high temperature desorption or cycling this orthorhombic hydride phase quickly disappears (Figure 6.21). According to Gennari *et al.* [60] the initial decomposition of γ phase produces synergetic effects during hydrogen desorption that stimulate β-MgH_2 decomposition by creating a volume contraction which, in turn, generates stresses acting on β-MgH_2. Conversely, one may argue that if the γ phase decomposes too quickly then the β phase may become too stable and small amounts of it may persist even up to high desorption temperatures. Second, during desorption of a milled MgH_2 powder there always occurs a simultaneous growth of nanograins of β-MgH_2 (Table 6.10). Since the initial particle size is not changed during desorption, one may hypothesize that the growth of nanograins within the β-MgH_2 particles might somehow decelerate the decomposition of β-MgH_2. However, we have also found that during desorption under primary vacuum β-MgH_2 decomposes completely.

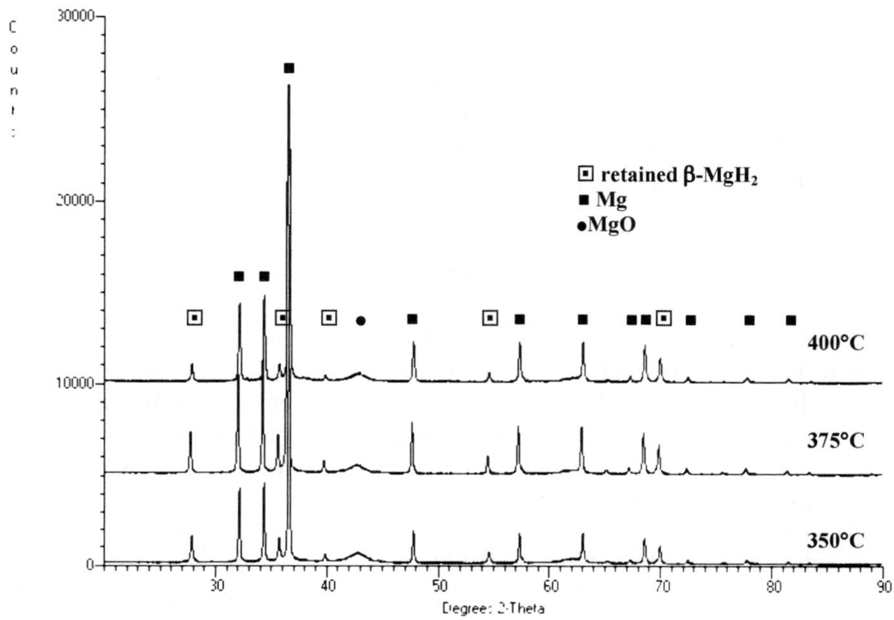

Figure 6.23 XRD patterns of MgH_2 (Tego Magnan) powders milled continuously for 100 h under IMP68 mode and subsequently desorbed directly after milling in a Sieverts-type apparatus under initial hydrogen pressure of 0.1 MPa at various temperatures

We have also found that discontinuous ball milling in which the milling vial is periodically opened during milling apparently degrades the desorption properties of milled MgH_2 powder even further. In comparison with continuously milled powders in Figure 6.17, the kinetics and maximum hydrogen capacities of discontinuously milled powders are much worse (not shown here).

In summary, nanostructuring of a magnesium hydride (MgH_2) by high-energy ball milling brings about both beneficial and detrimental effects to its hydrogen desorption characteristics. An experimentally observed beneficial effect is that the apparent hydrogen desorption temperature, as measured in a DSC test, decreases

Table 6.10 Grain size of β-MgH_2, Mg and retained β-MgH_2 in the Tego Magnan® powder after milling and desorption at various temperatures (calculated from peak breadths in Figure 6.23)

Powder	Grain size (nm)	Strain (%)	R^2	Number of peaks
MgH_2-milled	14	0	0.9964	7
350 °C (Mg)	86	8.2×10^{-4}	0.9838	3
350 °C (MgH_2)	55	0	0.9947	4
375 °C (Mg)	99	6.6×10^{-4}	0.9997	3
375 °C (MgH_2)	80	0	0.9973	5
400 °C (Mg)	78	5.4×10^{-4}	0.9948	6
400 °C (MgH_2)	62	0	0.9948	6

with decreasing particle size of a ball milled hydride powder and the kinetics of desorption improves. There are two factors acting additively that contribute to this behavior: the refinement of the hydride powder particle size and the γ-MgH_2 phase residing within the powder particles which has lower desorption temperature than β-MgH_2. A detrimental effect is a reduction in the hydrogen storage capacity after nanostructuring of MgH_2 by ball milling.

6.4.1.3 Nanostructured MgH_2 with Catalytic Additives

As shown in the preceding section, nanostructuring by ball milling has been unable to reduce the desorption temperature of MgH_2 much below 300 °C as was originally hoped for when ball milling was introduced for nanostructuring of solid state hydrides. Therefore, researchers world-wide started looking for catalytic additives to MgH_2 that, combined with a nanostructure, could further improve the hydrogen storage properties of this particular hydride.

The catalytic additives that have been investigated in MgH_2 can be roughly divided into three important groups: metals/intermetallics, oxides, and chemical compounds, which include hydrides. Their effect on the desorption properties of MgH_2 will be discussed in the following sections. We are not going to discuss absorption since as mentioned before absorption is usually much easier than desorption.

Metal/Intermetallic and Metallic Hydrides Additives

A number of researchers have investigated the effects of various metallic additives on the desorption properties of nanostructured, ball milled MgH_2. Liang *et al.* [63, 64], Dehouche *et al.* [65], Bouaricha *et al.* [66], Huot *et al.* [67], Shang *et al.* [68] and Au [69] investigated the effects of the addition of Ti, V, Nb, Mn, Fe, Al, Cu, La, Ni and Pd to commercial MgH_2 or Mg, which was, after milling, separately hydrogenated. The effects of the addition of intermetallics were investigated by Bobet *et al.* (YNi) [70], Hu *et al.* ($TiMn_{1.5}$ and $Ti_{37.5}V_{25}Cr_{37.5}$) [71, 72], Tran *et al.* (Mischmetal) [73], Skripnyuk *et al.* (Mg_2Ni eutectic) [74], Yonkeu *et al.* ($TiV_{1.1}Mn_{0.9}$) [75], Fu *et al.* ($LaNi_5$) [76] and Sai Raman *et al.* (Ce-free MischmetalNi_5-non-milled; synthesized by encapsulation method) [77]. For some of the metallic additives desorption at a low (200–235 °C) temperature [63, 64, 69] was reported and for some intermetallic additives the desorption temperature was as low as 245 °C (Fu *et al.* ($LaNi_5$) [76]) and 250–270 °C (Hu *et al.* ($TiMn_{1.5}$ and $Ti_{37.5}V_{25}Cr_{37.5}$) [71, 72]).

A limited number of hydrides have also been selected as catalytic additives to MgH_2. Ball milling was used by Zaluska *et al.* [78] to produce a nanostructured mixture of 65 wt.%MgH_2 + 35 wt.%Mg_2NiH_4 which showed low desorption temperature around 240–280 °C. Johnson *et al.* [79] produced a mixture MgH_2 + 0.1 mol.%$LiBH_4$ by annealing in a quartz tube for 12 h but no desorption at temperatures lower than 300 °C was reported.

Some researchers investigated multi-phase additives. Kojima *et al.* [80] ball milled MgH_2 with a nano-Ni/Al_2O_3/C composite catalyst. The mixture decomposed in vacuum at a really low temperature of ~200 °C. A complex composite catalyst BCN/Ni/Pd/SWNT (where BCN-barium-calcium niobium compound and SWNT-single-wall-nanotubes) was also used by Yoo *et al.* [81] for ball milled MgH_2. This desorbed ~3 wt.%H_2 in about 3600 s at 230–250 °C in vacuum.

Instead of ball milling Dufour and Huot [82] used cold rolling to produce Mg–Pd laminates, which after activation desorbed ~6 wt.%H_2 at 300 °C in 1200 s in vacuum. The same authors produced Mg_6Pd compound by cold rolling and ball milling [83]. However, the desorption properties were rather mediocre since the compound desorbed at 350 °C in vacuum (0.01 MPa) only about 2.8 wt.% after 2000 min.

Unfortunately, all the efforts described above have a major flaw. In all of them desorption tests in a volumetric Sieverts-type apparatus were conducted in vacuum. It must be stressed that the results obtained in such a way have essentially a very limited value. The equilibrium temperature at atmospheric pressure of hydrogen (1 atm) for pure MgH_2 is ~280 °C. This simply means that for any hydride, desorption at a temperature lower than its equilibrium temperature at atmospheric pressure of hydrogen is thermodynamically impossible. However, in vacuum the thermodynamic barrier for desorption as given by the Van't Hoff relationship disappears and material can desorb at much lower temperatures, and in addition, as shown by Song *et al.* [84, 85], with decreasing desorption pressure the dehydrogenation rate substantially increases. Obviously, in a practical situation, one can hardly imagine a vacuum pump installed onboard a fuel cell powered vehicle. Even so, the membrane of a PEMFC would soon be contaminated by the oil vapors released from the pump unless it was a dry pump.

A number of researchers tested various supposedly catalytic additives to MgH_2 employing desorption at atmospheric pressure of hydrogen (0.1 MPa). Grigorova *et al.* [86] used a reactive mechanical milling for producing nanostructured MgH_2 with Mg_2Ni and $Mg_2Ni_{1-x}M_x$ (M = Fe and Co) additives. Desorption tests were conducted at 300 °C under 0.15 MPa hydrogen pressure. The mixtures desorbed 5–6 wt.%H_2 within 3000 to 7000 s. No desorption tests below 300 °C were reported. Bobet *et al.* [87] investigated the effect of Co but the material desorbed only at 350 °C. Yu *et al.* [88] also used reactive milling for producing MgH_2 with Ni, Cu and $CrCl_3$ additives. The mixture desorbed a moderate 5 wt.% at 300 °C within 2400 s but no desorption below 300 °C was reported. Li *et al.* [89] combined 20 wt.%Ni and 1 wt.%TiO_2 as a complex additive, which was reactively milled with Mg. Reported desorption at 305° was very slow. Instead of ball milling Løken *et al.* [90] resorted to equal channel angular pressing (ECAP) to process heavily deformed Mg-20 wt.%Ni-8 wt.%Mm (Mischmetal) alloy, which was subsequently activated, hydrogenated and dehydrogenated. No desorption below 325 °C at 0.1 MPa hydrogen was observed. Apparently, this kind of processing is no better or even worse than ball milling. Czujko *et al.* [91] investigated ball milled Mg with 10 wt.%V, Zr and Y additives, which was activated, separately hydrogenated and then desorbed. At 300 °C under 0.1 MPa hydrogen pressure the V-doped material desorbed ~5 wt.%H_2 within 3600 s but that with Zr and Y only 2.8 wt.%H_2.

The most interesting studies are those where desorption tests were conducted at the atmospheric pressure of hydrogen, and the catalyzed MgH_2 was able to desorb a relatively large quantity of hydrogen at a temperature equal to or even lower than the equilibrium temperature at atmospheric pressure of hydrogen, *i.e.*, below ~280 °C. Some interesting results were reported by Vijay *et al.* [92]. They synthesized by reactive mechanical alloying/milling (RMA/RMM) the Mg + 5 wt.%FeTi and 30 wt.%FeTi composites, which were subsequently hydrogenated to form the MgH_2 matrix, and the MgH_2 + 40 wt.%FeTiMn composite by mechanical milling. The latter composite after ball milling could absorb 4 wt.%H_2 at 80 °C and desorb ~3.8 wt.%H_2 at 300 °C within 800 s and ~2 wt.%H_2 at 240 °C within 4200 s under 0.3 MPa of hydrogen pressure. The temperature of 240 °C is much lower than the equilibrium temperature at atmospheric pressure of hydrogen, which may be evidence of the lowering of the enthalpy of MgH_2 by the addition of 40 wt.%FeTiMn. The beneficial mechanism of FeTiMn intermetallic additive was not explained but it may have something to do with the fact that the FeTi intermetallic forms low temperature hydride (Table 6.3) and Mn might somehow additionally improve its performance. A disadvantage of this type of approach is the large quantity of additive intermetallic required. Nevertheless, this composite system requires attention and more research.

Wang *et al.* [93] produced Mg–10.9 wt.%Ce alloy by induction melting which after pulverizing and reactive mechanical milling in hydrogen was mixed with nano-Ni (average particle size ~10 nm) and additionally milled for 50 h under argon. As-cast structure contained Mg matrix with precipitates of $CeMg_{12}$ intermetallic, which, after hydrogenation converted to the $CeH_{2.53}$ hydride. The authors reported that this composite was able to absorb ~2.9 wt.%H_2 at 120 °C within ~1800 s. Under 0.1 MPa of hydrogen pressure the composite desorbed ~2 wt.%H_2 at 180 °C within ~1200 s. The reported results look very interesting although certain discrepancies exist in the presented data. For example, the authors showed a PCT curve for the Mg–Ce/nano-Ni composite which evidently exhibits a pressure of 1 atm hydrogen at 280 °C. With such an equilibrium temperature at 1 atm the composite should not have been able to desorb at 180 °C. Also, they claimed that ball milled MgH_2 began desorbing at 200 °C which is simply very unlikely as can be seen in Figure 6.17. Nevertheless, investigation of this kind of composite should be repeated to check if the reported results are indeed reproducible. The same group led by Wang [94] also investigated hydrogen storage properties of a composite $Mg/Mg_2Ni_{0.8}Cr_{0.2}$ containing nano-TiO_2 (average particle size ~40 nm). The material, containing 20 and 50 wt.% $Mg_2Ni_{0.8}Cr_{0.2}$ (plus TiO_2) desorbed 4.2 and 3.2 wt.%H_2 at 240 °C under 0.1 MPa pressure within 2900 and 900 s, respectively.

Very recently Bystrzycki *et al.* [95] studied the possibility of destabilization of MgH_2 by chemical reaction with Si as proposed by Vajo *et al.* [96]. The commercial MgH_2 and Si powder mixture corresponding to the stoichiometry of Mg_2Si was ball milled to obtain nanocrystalline composite structure. The sluggish destabilization of MgH_2 by solid-state reaction with Si forming the Mg_2Si intermetallic compound was observed at 250 °C. It was confirmed that the Mg_2Si compound was formed after the dehydrogenation of the synthesized MgH_2–Si mixture.

The well-known catalytic additive to MgH_2 is Ni. So far, it has been added as a coarse powder with the size range of a few tens of micrometers. A short review of these early experiments is presented in [97]. More recently, Hanada *et al.* [98] investigated the addition of nano-Ni to ball milled MgH_2. Unfortunately, he used continuous desorption under He gas and the material desorbed ~6 wt.%H_2 at 160 °C in 15,000 s. In our laboratory, we focused our efforts on the catalyzing of MgH_2 with nano-Ni powders produced as experimental batches by Vale Inco in Mississauga, Ontario. The effects of micrometric-size nickel (m-Ni)(commercially produced by Inco Type 255™ Ni) having specific surface area (SSA) = 0.7 m^2/g, submicrometric Ni having SSA = 7.5 m^2/g, and an experimental batch of nano-Ni (n-Ni) having SSA = 30.3 m^2/g on the rate of synthesis of MgH_2 and its hydrogen desorption properties have been reported in [97]. Both micrometric and submicrometric nickels greatly improve the rate of hydrogen absorption during controlled reactive mechanical milling (CRMM) and conversion of Mg into the MgH_2 hydride and subsequently, the hydrogen desorption properties of synthesized catalyzed MgH_2. However, by comparison with the previous two, the n-Ni is the most potent catalyst for the conversion of Mg into MgH_2 during CRMM under hydrogen. Up to ~25 h of CRMM the addition of barely 0.5 wt.% of n-Ni increases about twofold the rate of hydrogen absorption compared with undoped Mg. Furthermore, the addition of 2 wt.% n-Ni results in even faster absorption of hydrogen by Mg, resulting in ~6 wt.%H_2 absorbed after ~15 h of CRMM. The hydrogen desorption kinetics at the technological conditions of 0.1 MPa hydrogen pressure and no initial activation by cycling becomes very fast, which is reflected in the reduction of the activation energy for desorption by ~60 kJ/mol K compared with the reference synthesized MgH_2. The effects of n-Ni and nano-oxide additives (Al_2O_3 and Y_2O_3) on the hydrogen storage properties have been reported in [99]. The addition of both oxides has a limited effect on improving the hydrogen storage properties. In contrast, the addition of specialty Inco m-and n-Ni substantially reduces hydrogen desorption temperatures, which is also accompanied by very fast desorption kinetics under 0.1 MPa H_2 pressure. The activation energy of desorption is also substantially reduced. Prolonged milling for 100 h is detrimental for the hydrogen storage properties of the m-and n-Ni-doped MgH_2. It is to be pointed out that a simple mechanical mixing in a glass vial rotating in a steel milling cylinder of the magneto-mill Uni-Ball-Mill 5 for 1 h without using steel balls of Tego Magnan MgH_2 powder with both m-and n-Ni additives does not affect the desorption properties. Figure 6.24 shows the DSC traces of the MgH_2 as received and mechanically mixed with the Inco m-and n-Ni (SSA = 30.3 m^2/g) in a glass vial. The endothermic hydrogen desorption peaks of MgH_2 with Ni additives are not shifted with respect to the peaks of the as-received powder.

The situation is changed dramatically when the powder containing the Ni additive is ball milled even for a relatively short time. Very recently, we have investi-

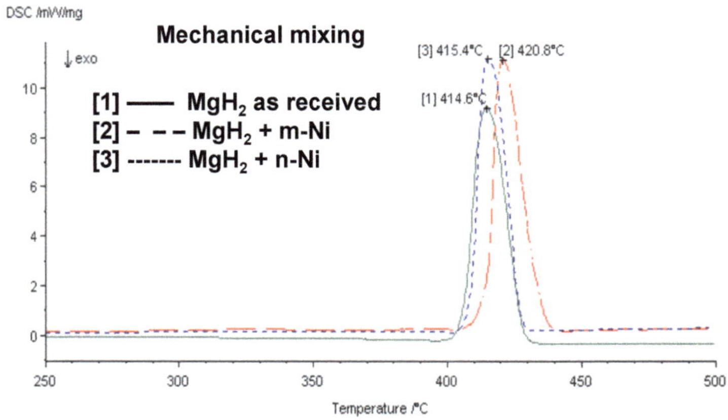

Figure 6.24 DSC traces of MgH_2 doped with m- and n-Ni just mixed for 1 h without milling (no steel balls) (heating rate 4 °C/min; argon flow rate 25 mL/min)

gated the effect of milling time in the magneto-mill Uni-Ball-Mill 5 on the hydrogen storage properties of ABCR MgH_2 powder doped with micrometric and nanometric Ni [100]. Figure 6.25 shows SEM micrographs of micrometric (m-Ni) (Type 255™ produced by Inco) and nanometric Ni (n-Ni) used for this study. Micro-Ni has a very unusual and complex shape. At a relatively low magnification (×5 k), it exhibits a filamentary shape (Figure 6.25a). However, at a magnification ×25 k it is clearly seen that each filament is composed of flower-like corollas (Figure 6.25b), joined closely together, each one resembling a rose. At a very high magnification (×50 k) each rose-like corolla consists of small petal-like features whose thickness is on the order of or less than ~ 100 nm (Figure 6.25c). Although in the macroscale the Type 255™ Ni is micrometric in size the thickness of the petal-like features falls within the nearly-nanometric range. The Inco nano-Ni (n-Ni) has a filamentary shape but resembles a delicate "coral colony" and its dimensions are truly nano, *i.e.*, below 100 nm (Figure 6.25d). Measured mean diameter of a "coral filament" is 42±16 nm.

Figure 6.26 shows the microstructure of the powder after milling for 15 min with the addition of 5 wt.% Inco m-Ni and n-Ni. The distribution of m-Ni is not quite uniform (Figure 6.26a), in contrast to the n-Ni, which shows a very uniform distribution of Ni particulate (bright specks) in the entire mass of the powder (Figure 6.26b). The average ECD particle size of the powder milled with the m-Ni additive is slightly larger than that of the one milled with n-Ni although both are below 1 μm mark. XRD pattern in Figure 6.27a shows that the microstructure of both powders consists of hydride phases β-and γ-MgH_2, Ni and a small amount of retained Mg and MgO, which is formed during the XRD test from the retained Mg. Table 6.11 shows grain size of the phases in the ABCR MgH_2 ball milled with Inco m-and n-Ni additives and in the same powders cycled as will be discussed later. After milling for 15 min using HES57–two magnets mode under 700 kPa hydrogen pressure, the grain size of β-MgH_2 and Mg phase is on the order of ~ 40 nm. Nano-Ni retains its as-received grain size on the order of ~ 20–25 nm.

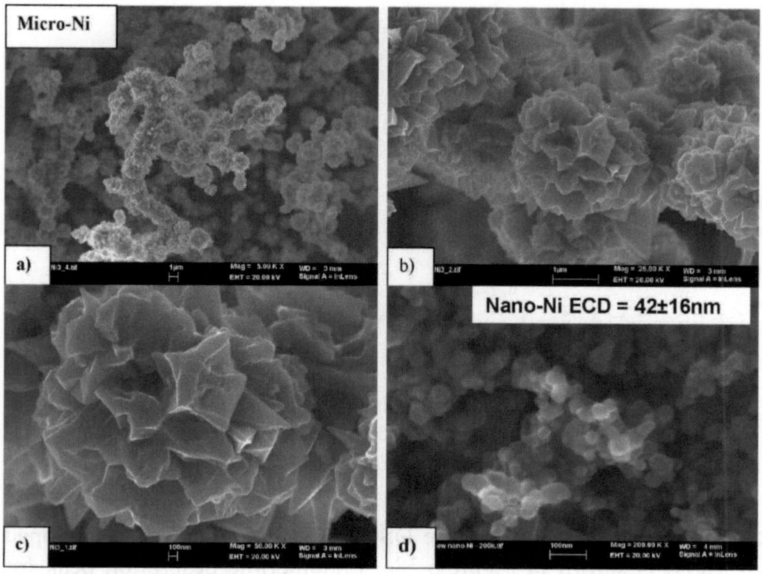

Figure 6.25. SEM micrographs of (**a**,**b**,**c**) micrometric Inco Ni (Type 255™) having SSA = 0.7 m^2/g, taken at various magnifications to reveal its morphology, and (**d**) nano-Ni (n-Ni) having SSA = 14.5 m^2/g

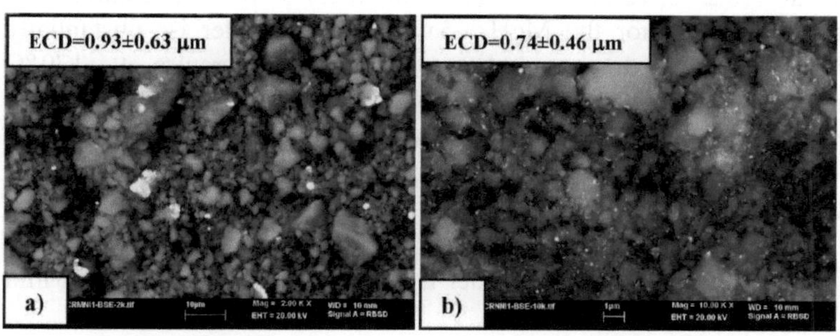

Figure 6.26. SEM micrographs of the ABCR powder after milling for 15 min (HES57–two magnets mode; hydrogen 700 kPa) with the addition of 5 wt.% Inco (**a**) m-Ni and (**b**) n-Ni. Average ECD particle size with standard deviation is showed in the insets

Table 6.11 Grain size of the phases in the milled and cycled powder ABCR MgH_2 containing Inco m- and n-Ni additive

Sample	Grain size of MgH_2 [nm]	Grain size of Ni [nm]	Grain size of Mg [nm]	Grain size of Mg_2NiH_4 [nm]
ABCR + 5 wt.% m-Ni	40.0±2.9	25.5±6.0	41.4±9.6	---
ABCR + 5 wt.% n-Ni	34.9±2.4	17.3±1.3	36.0±4.2	---
ABCR + 5 wt.% n-Ni*	90±10	---	133±17	27±2
ABCR + 5 wt.% n-Ni**	86±9	---	132±19	24±2

* Powder milled and cycled at 300 °C; ** powder milled and cycled at 300 °C + test for E_A.

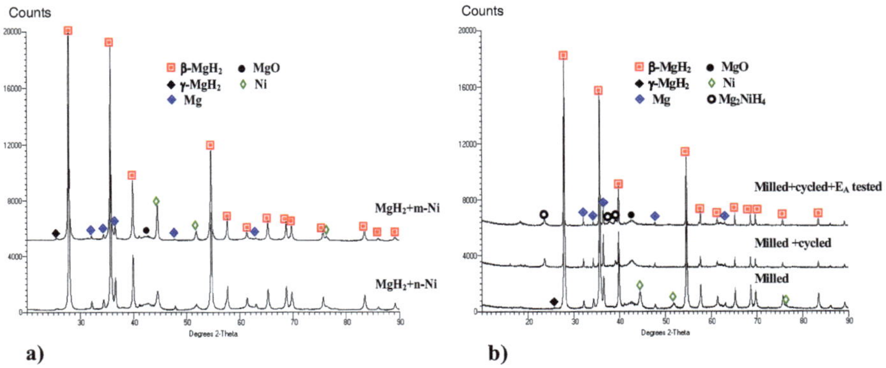

Figure 6.27. XRD patterns of ABCR powder after (**a**) milling for 15 min with the addition of 5 wt.% Inco m-Ni and n-Ni (HES57–two magnets mode; hydrogen 700 kPa), and (**b**) after cycling and testing for activation energy of desorption

The effect of a ball milling time on the DSC desorption behavior of undoped and Ni-doped MgH_2 is shown in Figure 6.28a. Endothermic desorption peak for the MgH_2 + 5 wt.% m-Ni powder milled for 15 min is only modestly shifted to lower temperatures showing the onset temperature at ~350 °C and the peak maximum at 392.8 °C compared with a pure MgH_2 with the onset at ~380 °C and the maximum at 418.2 °C, respectively, also milled for 15 min. In contrast, the hydrogen desorption peak for the MgH_2 + 5 wt.% n-Ni powder is substantially shifted to the lower temperatures and shows the onset temperature at ~170 °C and the peak maximum at 244.5 °C. However, when the MgH_2 + 5 wt.% m-Ni powder is milled for 20 h its desorption properties are much improved such that onset is at ~275 °C and the peak maximum at 302.3 °C. Apparently, longer milling time reduces desorption temperature, most likely due to a better dispersion of Ni particles within the MgH_2 matrix. Nevertheless, the desorption temperature of the 20 h milled MgH_2 + 5 wt.% m-Ni powder is still worse than that of the 15 min

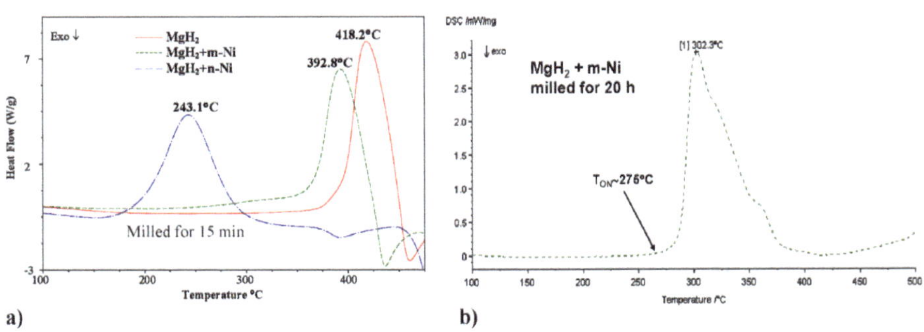

Figure 6.28. **a** DSC traces of ABCR MgH_2 doped with 5 wt.% Inco m-and n-Ni (SSA = 14.5 m^2/g) subjected to 15 min of ball milling (HES57–two magnets mode; hydrogen 700 kPa). **b** DSC desorption peak of Tego Magnan MgH_2 + 5 wt.% Inco m-Ni after milling for 20 h (heating rate 4 °C/min; flow rate of argon 25 mL/min)

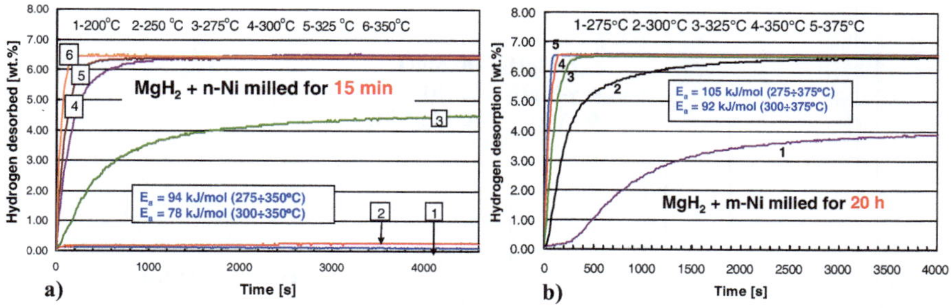

Figure 6.29. Desorption kinetic curves at various temperatures obtained in a Sieverts-type apparatus under 0.1 MPa of hydrogen pressure for (**a**) ABCR MgH_2 + n-Ni (SSA = 14.5 m^2/g) ball milled for 15 min, and (**b**) Tego Magnan MgH_2 + m-Ni ball milled for 20 h

milled MgH_2 + 5 wt.% n-Ni. This clearly shows that n-Ni exhibits superb catalytic properties.

Figure 6.29 shows desorption curves from the ball milled powders. The mixture with n-Ni in Figure 6.29a shows very fast desorption in the range 300–350 °C and moderate desorption at 275 °C. No desorption has occurred at 200 and 250 °C, which indicates that the basic thermodynamic properties of the system (enthalpy) is not changed from that for pure MgH_2. We calculated the apparent activation energy of desorption of the mixture with n-Ni using data points from two ranges of temperatures: 275–350 °C and 300–350 °C and obtained 94 kJ/mol (R^2 = 0.982) and 78 kJ/mol (R^2 = 0.989), respectively. The mixture of Tego Magnan MgH_2 with m-Ni ball milled for 20 h in Figure 6.29b shows slower desorption kinetics than that milled for 15 min, which is also supported by greater apparent activation energies 105 kJ/mol (R^2 = 0.983) and 92 kJ/mol (R^2 = 0.989) for the 275–375 °C and 300–375 °C range, respectively. Once again the n-Ni clearly shows its superb catalytic properties.

The cycling behavior of the MgH_2 mixture with n-Ni was also studied. The cycling process consisted of five desorption/absorption cycles at 300 °C. Desorption was carried out under atmospheric pressure of hydrogen and absorption was realized under 4.0 MPa pressure of hydrogen for 15 min. XRD after cycling showed formation of a small amount of Mg_2NiH_4 hydride (Figure 6.30). A quite profound grain growth of the constituent phases is observed upon cycling, as shown in Table 6.11. Compared with the grain size of the phases directly after milling, the grain size of β-MgH_2 (after the 5th absorption) increases 2–3-fold upon cycling and the grain size of Mg (after the 5th desorption) increases almost 4-fold. Desorption kinetics of the cycled mixture (after the 5th absorption) was again investigated and the pertinent desorption curves are shown in Figure 6.31. In the 275–350 °C range the rate of desorption is slightly slower than that shown in Figure 6.29a but this is not reflected in the apparent activation energy of desorption, E_A, equal to 99 kJ/mol (R^2 = 0.937) and 68 kJ/mol (R^2 = 0.945) for the range 275–350 °C and 300–350 °C, respectively. Most importantly, one can see a substantial loss of final capacity after cycling at each desorption temperature compared with the as-milled material in Figure 6.29a. In general, this effect is

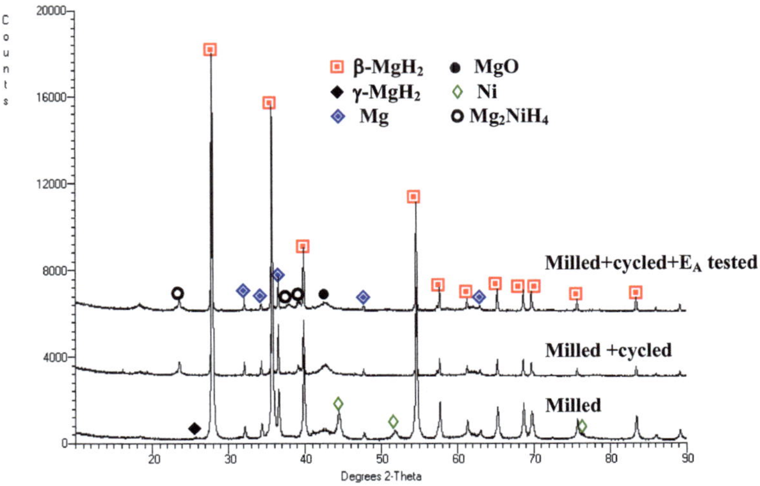

Figure 6.30 XRD pattern of the ABCR MgH_2 + n-Ni (SSA = 14.5 m^2/g) mixture after milling and cycling as well as after testing for activation energy of desorption after cycling

very persistent after cycling of MgH_2 hydride as discussed in Section 6.4.1.2 and obviously, is unrelated to the presence of n-Ni in the mixture with MgH_2.

The combined effect of specific surface area and chemical composition of Inco n-Ni has also been investigated [100]. Table 6.12 lists the SSA, the content of oxygen and carbon, and morphology of various experimental batches of Inco n-Ni that have been studied as catalyst for MgH_2. The SSA covers a wide range from about 4 to 85 m^2/g. The ABCR MgH_2 mixed with 5 wt.% n-Ni having increasing SSA was ball milled under 700 kPa hydrogen for 15 min using HES57 mode (two magnets). For comparison, pure MgH_2 was also milled under the same conditions. Figure 6.32 shows the microstructure of ABCR powder milled with the n-Ni having increasing SSA. It is seen that the spherical Ni particles are distributed in some localized places such as the contact areas between the MgH_2 particles while the filamentary n-Ni is more or less smeared off on the surface of the MgH_2 particles although it does not form any continuous or semi-continuous film. The average ECD particle size is within the range 0.8–1.0 μm.

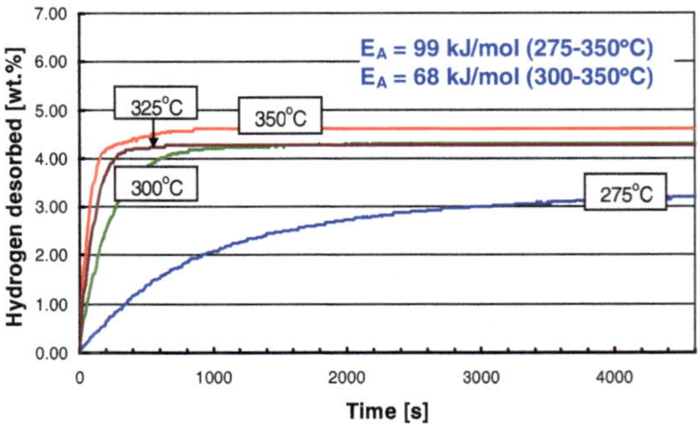

Figure 6.31 Desorption kinetic curves at various temperatures of the cycled ABCR MgH_2 + n-Ni (SSA = 14.5 m^2/g) mixture from which the activation energy E_A is calculated

Table 6.12 Specific surface area, the content of carbon and oxygen and morphology of various experimental batches of Inco n-Ni

Sample	SSA [m^2/g]	Grain size [nm]	C [wt.%]	O [wt.%]	Morphology
NI4	4.02	60	0.53	0.69	Spherical
NI5	6.41	---	2.97	0.051	Spherical
NI6	9.49	---	0.46	0.11	Spherical
NI3	14.50	17	0.34	2.46	Filamentary
NI7	18.82	37	0.61	3.5	Spherical
NI8	28.18	---	2.3	2.08	Filamentary
NI9	60.46	---	1.86	11.4	Filamentary
NI10	84.70	---	0.46	11.7	Filamentary

The XRD traces in Figure 6.33 shows primarily the presence of β-and γ-MgH_2 and some retained Mg. Bragg diffraction peaks of Ni can be recognized in all samples but their intensity gradually decreases with increasing SSA, which indicates that n-Ni having very large SSA is more intimately mixed with MgH_2. Specifically, this is observed for the filamentary n-Ni in Figure 6.33b. This observation correlates well with the SEM micrographs in Figure 6.32. MgO peaks as usual are due to oxidation.

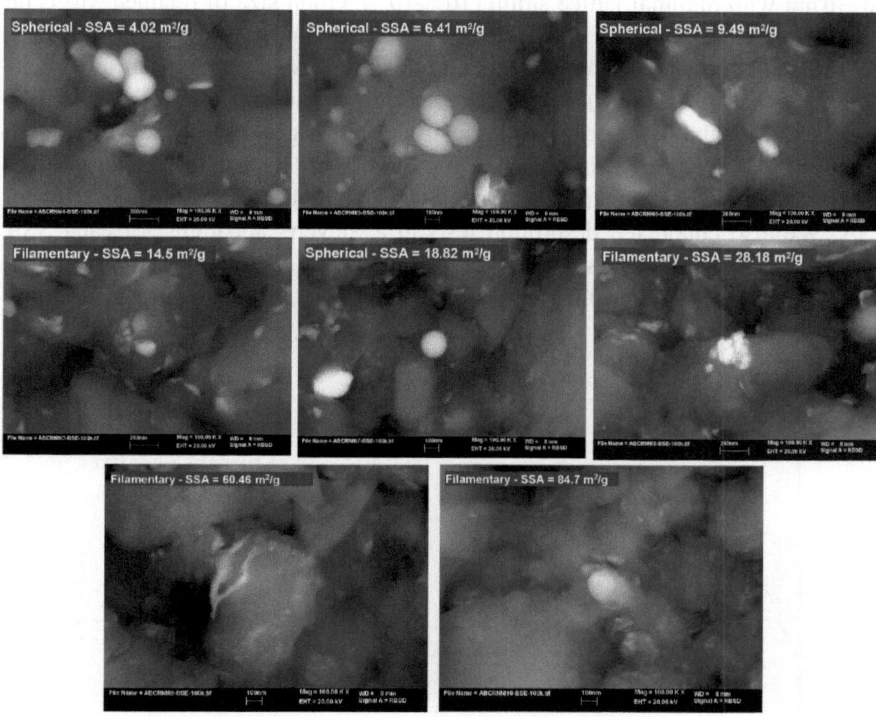

Figure 6.32 Scanning electron micrographs in backscattered electrons (BSE) mode at ×100,000 of the ball milled ABCR MgH_2 + 5 wt.% n-Ni mixtures under 700 kPa hydrogen with varying SSA (see insets in pictures)

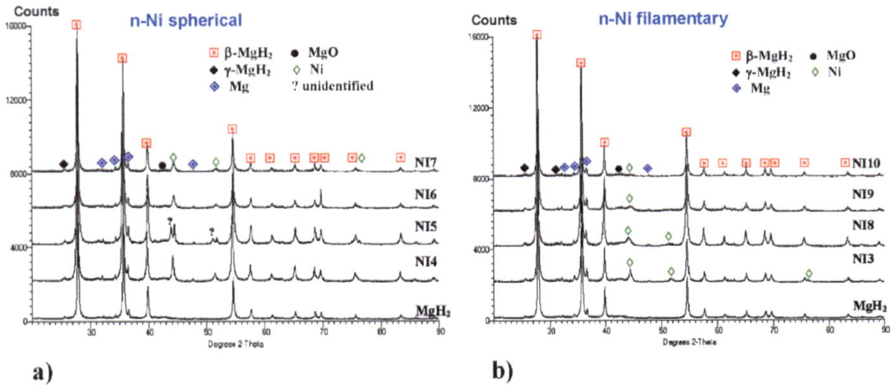

Figure 6.33. XRD patterns of ABCR $MgH_2 + 5$ wt.% n-Ni after milling for 15 min: **a** spherical; and **b** filamentary morphology

of the retained Mg. No formation of Mg_2NiH_4 has occurred upon milling (Figure 6.33) although there is a very high likelihood that it will be formed upon subsequent thermal cycling as evidenced in Figure 6.30. Grain size of β- and γ-MgH_2 calculated from the broadening of XRD peaks (Section 6.3.2) is on the order of 20–30 nm with minimal lattice strain (10^{-3}–10^{-4}). Grain size of retained Mg is on the order of 40 nm.

DSC tests show a substantial reduction of the hydrogen desorption onset (circles) (T_{on}) and peak (T_{peak}) temperatures owing to the catalytic effects of n-Ni compared with the hydrogen desorption from pure MgH_2 also milled for 15 min (Figure 6.34). It is interesting to note that there is no measurable difference between spherical and filamentary n-Ni although there seems to be some effect of SSA. We also conducted desorption tests in a Sieverts apparatus for each SSA and obtained kinetic curves (Figure 6.35) from which the rate constant, k, in the JMAK equation (Section 6.2.2) was calculated. The enhancement of desorption rate by n-Ni is clearly seen. At the temperature 275 °C, which is close to the equi-

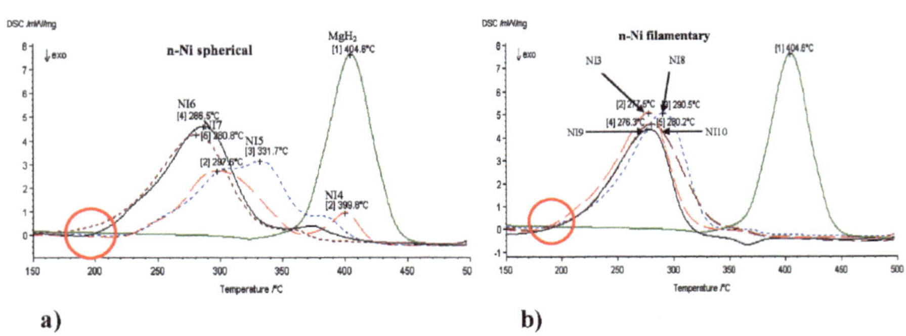

Figure 6.34. DSC traces of ABCR $MgH_2 + 5$ wt.% n-Ni after milling for 15 min: **a** spherical; and **b** filamentary morphology (heating rate 10 °C/min; argon flow rate 50 mL/min). The range of onset temperatures of n-Ni-containing mixtures is encircled

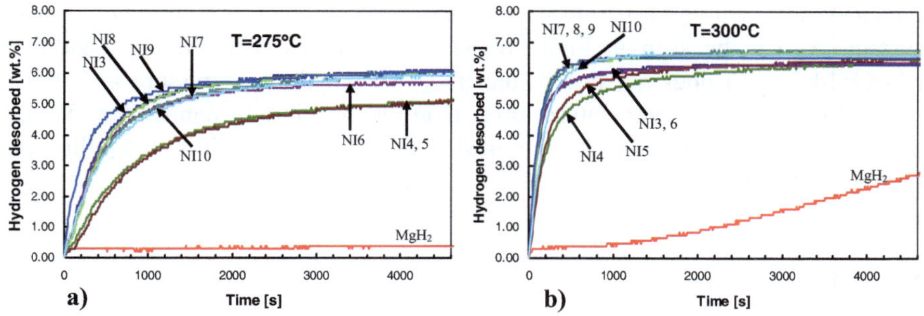

Figure 6.35. Desorption kinetic curves at (**a**) 275 °C and (**b**) 300 °C obtained in a Sieverts-type apparatus under 0.1 MPa of hydrogen pressure of ball milled ABCR $MgH_2 + 5$ wt.% n-Ni having varying SSA

librium at atmospheric pressure (0.1 MPa), all samples desorb from 4 to 5.5 wt.% H_2 within 2000 s.

We plot in Figure 6.36a the onset and peak temperatures from Figure 6.34a and b as a function of SSA for each n-Ni additive. It is seen that a dramatic drop in the hydrogen desorption temperature of MgH_2 occurs only up to a certain value of SSA, approximately 15 m^2/g. Greater values of SSA do not have better effect on the desorption temperature. The plot of k values as a function of SSA is shown in Figure 6.36b for two desorption temperatures 275 and 300 °C. The k dependence on SSA is very similar to that for the desorption temperature in Figure 6.36a, meaning that the k increases up to SSA of ~ 15 m^2/g and then there is no further dependence with increasing SSA. Another important finding, as can be seen in Figure 6.36b, is that there is no apparent effect of carbon (0.34–2.97 wt.%) and oxygen (0.05–11.7 wt.%) content in the n-Ni on the hydrogen storage properties of MgH_2. Finally, it is also obvious that there is no apparent effect of the n-Ni morphology on the hydrogen storage properties.

It is to be concluded that the observed effects of n-Ni on the hydrogen storage behavior of MgH_2 strongly suggest that all the improvement of hydrogen storage properties is owing to the kinetic effect of n-Ni acting as a powerful catalyst rather

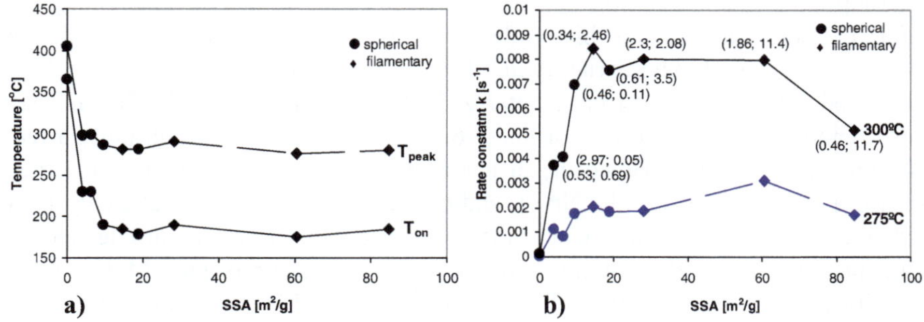

Figure 6.36. **a** DSC peak (T_{peak}) and onset (T_{on}) temperature; and **b** rate constant, k, in the JMAK Equation 6.4 as a function of the SSA of the Inco n-Ni additive

than to a modification of thermodynamic properties (enthalpy) of the MgH_2 + n-Ni system. As can be seen in Figure 6.29a, MgH_2 doped with the n-Ni additive (SSA = 14.5 m^2/g) does not desorb under 0.1 MPa hydrogen pressure at 250 °C, which for MgH_2 is a temperature lower than the equilibrium temperature at that pressure, while it easily desorbs over 4 wt.%H_2 at 275 °C, which is an equilibrium temperature at 0.1 MPa.

Metal Oxide Additives

Metal oxides seem to be one of the most potent catalytic additives to MgH_2. A number of authors investigated only absorption process of MgH_2 catalyzed with a specific metal oxide without an insight into desorption process. Obviously, the absorption kinetics were enhanced as compared to a pure MgH_2. However, since as shown earlier Figures 6.18a and 6.19a for pure MgH_2 the rate of hydrogen absorption is much faster than the rate of desorption, we will not review those papers which were limited to the absorption studies of MgH_2 doped with a specific metal oxide. Therefore, this section will be restricted to review the studies of desorption.

A very unfortunate aspect of the reported research results is that a large number of investigators studied the effect of metal oxides on the desorption properties of MgH_2 in vacuum. As mentioned earlier in this section the results obtained under such a condition have a very limited meaning. Oelerich *et al.* [102, 103] and Jung *et al.* [104] investigated the addition of V_2O_5. In vacuum (0.1 kPa) the doped MgH_2 was able to desorb 2–3 wt.%H_2 at 250 °C while at 300 °C about 6 wt.%H_2 was desorbed within 120–360 s. The addition of Cr_2O_3 to MgH_2 was studied by Jung *et al.* [104], Dehouche *et al.* [105], Barkhordarian *et al.* [106], Bobet *et al.* [107] and Aguey-Zinsou *et al.* [108]. Desorption of 4–7 wt.%H_2 within 300–1000 s was reported at 300 °C while some mixtures desorbed at 250–280 °C in vacuum. Some studies were also made on the effect of oxides such as Fe_2O_3 and Fe_3O_4 [103, 106], Mn_2O_3 [103], MgO [108], Al_2O_3 [104, 109] and La_2O_3 [110] without, however, any noticably better effect on the improvement of MgH_2 desorption in vacuum than that brought about by V_2O_5, Cr_2O_3 or n-Ni for that matter. The most thoroughly researched oxide is Nb_2O_5 which was extensively studied by a group from GKSS Geesthacht, Germany, led by Klassen and Bormann [32, 106, 109, 111–115]. In vacuum this oxide seems to make MgH_2 desorb very fast at 300 °C such that Barkhordarian *et al.* [106, 109, 111, 114] reported 5–7 wt.%H_2 desorbed within 90–250 s from a catalyzed MgH_2 ball milled for 20–200 h. They also reported [106] desorption of about 6.6 wt.%H_2 at 250 °C within quite a reasonable time of 600 s. However, in a recent paper from this group [114] they raised a question whether the Nb_2O_5 additive indeed acts as a typical catalysts or whether the kinetic enhancement is mainly due to their effect on particle size reduction. They found that the desorption kinetics in vacuum of pure MgH_2 milled for 700 h is the same as that of the mixture MgH_2 + 17 wt.% Nb_2O_5 milled for 200 h, where the former is supposed to have nanometric particle size. However, their claimed particle size of pure MgH_2 milled for 700 h being on the order of

0.5–1.5 nm is rather unlikely. Apparently, this interpretation seems to be somehow erroneous. So far, the lowest desorption temperature in vacuum of the ball milled mixture $MgH_2+Nb_2O_5$, equal to 163 and 200 °C, was reported by Hanada *et al.* [112] and Bhat *et al.* [115], respectively. Hanada *et al.* reported ~5 wt.%H_2 desorbed within 6000 s. They also reported that the activation energy for hydrogen desorption of the mixture of MgH_2 + 1 mol.%Nb_2O_5 milled for 20 h was ~71 kJ/mol. This value is quite close to the activation energy which we obtained for MgH_2 milled with Inco n-Ni (Figure 6.29a and 6.31a), which was obtained from the desorption experiments conducted in a Sieverts-type apparatus under atmospheric pressure of hydrogen (0.1 MPa). Apparently, Nb_2O_5 does not seem to be more efficient catalyst than n-Ni.

Great pitfalls of conducting desorption experiments in vacuum are well illustrated by the results obtained by Song *et al.* [84, 85]. In a Sieverts-type apparatus at 300 °C the ball milled mixture of MgH_2 + 10 wt.%Cr_2O_3 desorbed ~0.4, 2.5, and 1.5 under 1, 1.2, and 1.4 bar of hydrogen pressure, respectively. This results are in strong contrast to the amount of 4–7 wt.%H_2 desorbed at 300 °C and even at 250–280 °C reported in [104–108] for the $MgH_2+Cr_2O_3$ mixture when the desorption process was conducted in vacuum. Song *et al.* also studied the desorption process of the mixture Mg–10 wt.%(Fe_2O_3,MnO,Ni) and found that at 320 °C the material desorbed 2.2, 1.2, 0.9, 0.2, and 0 wt.%H_2 under <1, 1.2, 1.4, 1.6, and 1.8 bar, respectively. As mentioned earlier in the text, these experiments clearly show that with decreasing initial hydrogen pressure the thermodynamic barrier for desorption disappears which allows much faster desorption rate at much lower temperatures.

Carbon/Graphite and Carbon Nanotubes Additives

Imamura *et al.* [117] studied absorption of Mg ball milled with graphite and benzene as a milling additive. In a Sieverts-type apparatus the mixture after 20 h milling was able to absorb at 180 °C. This in itself is nothing outstanding because milled, activated and cycled MgH_2 can also absorb at ~200 °C (Figure 6.19). Bouaricha *et al.* [118] also studied absorption of Mg + graphite mixture at 300 °C which showed much better kinetics than that of just milled Mg. A number of researchers studied both absorption and desorption [119–122]. Unfortunately, all desorption studies were conducted in vacuum. However, it must be pointed out that even in vacuum conditions, desorption kinetics was no better than the one obtained with a number of other additives discussed earlier in the text. The lowest desorption temperature applied was 290 °C.

The addition of carbon nanotubes which were either reactively milled under hydrogen mixed with Mg powder [123] or simply mixed with MgH_2 and subsequently milled [124, 125], was investigated. In vacuum desorption at 200 °C gave 3.6 wt.% within 1800 s [123]. Another reference reports ~5 wt.H_2 desorbed at 300 °C within 3600 s and 6 wt.%H_2 at 350 °C in about 300 s from the ball milled mixture of MgH_2 + single-walled nanotubes (SWNTs) tested in vacuum [124]. The results indicate that there is no particular advantage of adding carbon nanotubes to Mg/MgH_2.

There are also a few papers reporting results on the hydrogen desorption of the ball milled mixtures of MgH_2 and carbon/graphite studied by temperature programmed desorption (TPD), DSC and TGA [126–128]. However, they do not report any particularly exciting results that would warrant a more thorough discussion.

In summary, there is no compelling evidence that carbon, graphite and carbon nanotubes can act as potent catalytic additives for the enhancement of absorption/desorption properties of Mg/MgH_2. They are not really better than, for example, some elemental metals such as n-Ni, which can be treated as a catalytic benchmark.

6.4.2 (Nano)composites of Magnesium Hydride (MgH_2) and Complex Hydrides

The general term "complex hydrides" refers to a large group of hydrides which contain a "complex" anion in which hydrogen is covalently bonded to central atoms while the "complex" itself is ionically bonded to a metallic element [130]. One type in this large group is Mg–transition metal (TM) complex hydrides that have $(TMH_x)^-$ complex, for example, $(FeH_6)^{4-}$ and $(NiH_4)^{4-}$ bonded to Mg, which forms Mg_2FeH_6 and Mg_2NiH_4, respectively. Another type of complex hydride consists of Group I and II salts of $(AlH_4)^-$, $(NH_2)^-$ and $(BH_4)^-$, which are known as "alanates", "amides", and "borohydrides", respectively. The latter have recently received considerable attention as potential solid state hydrogen storage materials. Due to a space constraint they will not be discussed in the present review and the reader is referred to a very recent thorough review of these three types of hydrides [129].

Very recently [130] we have formulated a hypothesis that by compositing hydride constituents having high and low temperatures of desorption, the desorption temperature of the constituent having high desorption temperature can be linearly reduced with increasing volume fraction of the low-desorption temperature constituent according to the composite rule-of-mixtures (ROM). According to this hypothesis, the compositing of high decomposition temperature (T_{high}) hydride with a low decomposition temperature (T_{low}) hydride (metal or complex) would reduce the decomposition temperature of the composite hydride mixture according to the well-known ROM:

$$T_{desorption} = T_{high} V_{high} + T_{low} V_{low} \tag{6.10}$$

and after rearranging

$$T_{desorption} = T^0_{high} - b V_{low} \tag{6.11}$$

where $V_{low,\ high}$ is the volume fraction of corresponding hydride and T^0_{high} is the initial temperature at $V_{low} = 0$. Equation 6.11 requires a linear dependence of decomposition temperature with a negative slope b versus the volume fraction of a constituent hydride having lower desorption temperature. Therefore, after compositing of high and low desorption temperature hydrides by ball milling which, in effect, results in the formation of intimately intermixed nanocomposite hydrides, one would expect that the desorption temperature of the hydride that has the higher desorption temperature to decrease linearly with increasing volume fraction of a constituent hydride having much lower desorption temperature in accord with Equation 6.11. This hypothesis has been experimentally verified for the following composite systems: MgH_2–$VH_{0.81}$, MgH_2–$LiAlH_4$, MgH_2–$NaAlH_4$ and $NaBH_4$–MgH_2, where the second hydride in a pair is assumed to have much lower desorption temperature than the first one. The composites have been synthesized by either CRMM or CMM in a magneto-mill, which in effect produced nanocomposites with the nanometric grain sizes of the constituent phases and substantial reduction of particle size.

6.4.2.1 MgH_2-$VH_{0.81}$ Composite System

The original idea of reactive milling under hydrogen a composite of MgH_2 with a varying content of V was to produce $MgH_2 + VH_2$ hydride composite in which VH_2 as having lower desorption temperature could destabilize MgH_2. The following compositions were processed: 20, 50, 66, and 94.6 wt.%, which corresponds to 5.6, 19.1, 31.5, and 80.5 vol.% of V, respectively. Reactive milling was conducted in the magneto-mill Uni-Ball-Mill 5 (Section 6.3.1) under IMP68 mode (high energy impact with two magnets at 6 and 8 o'clock positions in Figure 6.8) for 20 h. Phase analysis by XRD has shown that phases such as β-MgH_2, γ-MgH_2 and $VH_{0.81}$ exist in the nanocomposites. As discussed in Section 6.4.1.2, the metastable γ-MgH_2 hydride is a high pressure polymorphic form of β-MgH_2. For each composition no unreacted V has been observed by XRD, which means that it is fully converted to the $VH_{0.81}$ hydride. The average grain size of the β-MgH_2 and $VH_{0.81}$ constituent after 20 h of ball milling is within 14–16 nm and 14–23 nm, respectively, as calculated from the XRD peak broadening (Section 6.3.2). Apparently, no VH_2 hydride has been formed during reactive milling. The hydride $VH_{0.81}$ exists within the range from $VH_{0.45}$ to $VH_{\sim 0.9}$ [132] and is usually designated as β_2-phase having a body centered tetragonal (BCT) crystallographic structure [131, 132]. The PCT (pressure–composition–temperature) plateau for a mixture of $\beta_2 + \gamma(VH_2)$ is between 3 and 4 atm at 40 °C [131] and it is considered to be stable at room temperature as opposed to $\gamma(VH_2)$, which is unstable at room temperature but shows slow kinetics of desorption [131]. This strongly suggests that VH_2, even if formed, decomposes during high energy ball milling.

Figure 6.37a shows DSC curves for nanocomposites of MgH_2 with varying volume fraction of V after reactive milling for 20 h. For comparison, the DSC curve of a ball milled single-phase MgH_2 is also shown. With increasing content of V (or $VH_{0.81}$) the DSC curves are substantially shifted to lower temperatures and some of them exhibit peak doublets. As reported by Gennari *et al.* [60] the low temperature peak (LT) in a doublet is most likely due to hydrogen desorption from mostly the γ-MgH_2 phase plus some amount of β-MgH_2 and the high temperature (HT) one is due to desorption from the remaining β-MgH_2.

Since V is fully converted to the $VH_{0.81}$ hydride, therefore, the lowering of DSC desorption temperature is most likely due to the presence of $VH_{0.81}$ rather than V. Therefore, if the ROM of Equation 6.11 was obeyed then $VH_{0.81}$ synthesized by reactive milling and composited with MgH_2 should reduce the desorption temperature of the latter. Figure 6.37b shows the plot of the temperature of HT peak in the doublet and a peak maximum for a single peak as a function of V (or $VH_{0.81}$) content. Indeed, it is seen that up to about 20 vol.%V ($VH_{0.81}$) the ROM is obeyed with a very good coefficient of the fit $R^2 = 0.93$. However, at higher volume fractions of V ($VH_{0.81}$) the ROM is no longer obeyed. Representative SEM pictures of the composite powder morphology in the insets in Figure 6.37b show that with increasing vol.% of V the milling becomes much less effective leading to cold welding and formation of large particulates. As already shown in

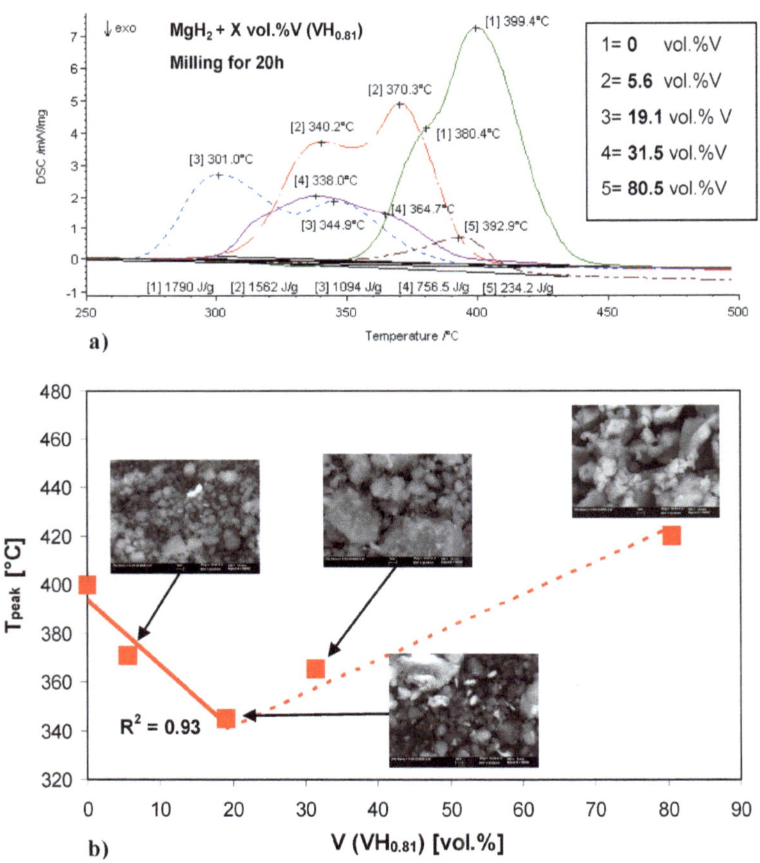

Figure 6.37. **a** DSC traces for MgH_2+X vol%$VH_{0.81}$ (X=5.6, 19.1, 31.5, and 80.5)(heating rate of 10 °C/min under Ar flow). **b** Temperature of higher DSC temperature (HT) peak in the doublet or a peak maximum for a single peak as a function of V (or $VH_{0.81}$) content. Representative SEM pictures showing the morphology of composite powder are shown in the insets (all at the same magnification)

Figure 6.22 the DSC desorption temperature of MgH_2 is substantially reduced only if the particle size becomes smaller than ~1 μm (1000 nm) [133, 134]. Large contents of V in a composite with MgH_2 lead to the formation of a composite particulates with average size of a few tens of micrometers as can be seen in Figure 6.37b, which is much larger than the required ~1 μm, which again increases the desorption temperature of MgH_2.

6.4.2.2 MgH_2-$LiAlH_4$ Composite System

The theoretical storage capacity of $LiAlH_4$ is 10.6 wt.%H_2. It is well established in the literature [135–141] that hydrogen desorbs from a pure, undoped $LiAlH_4$ hydride in a three-step decomposition the first of which goes through the melting of $LiAlH_4$:

$$\text{(R1a) } LiAlH_4(s) \rightarrow LiAlH_4(l) \tag{6.12}$$

$$\text{(R1b) } LiAlH_4(l) \rightarrow 1/3Li_3AlH_6(s) + 2/3Al(s) + H_2(g) \tag{6.13}$$

$$\text{(R2) } 1/3Li_3AlH_6(s) \rightarrow LiH + 1/3Al + 0.5H_2 \tag{6.14}$$

$$\text{(R3) } LiH \rightarrow Li + 0.5H_2 \tag{6.15}$$

where s-solid, l-liquid and g-gas. R1a is endothermic, R1b is exothermic, R2 and R3 are both endothermic reactions. There is some limited experimental evidence that at very low heating rates on the order of 0.5 °C/min the melting (R1a) may not occur [139]. Also, melting may not occur for $LiAlH_4$ doped with $TiCl_4$ [138]. (R1b)–(R3) proceeds with a theoretical hydrogen release of 5.3, 2.6, and 2.6 wt.%, respectively. One must keep in mind that these numbers will be *lower* for purity-corrected capacity. In thermal analysis (R1a,b) occurs around 112–220 °C, (R2) takes place around 127–260 °C, and (R3) occurs at too high temperatures (400–450 °C) to be practical for hydrogen storage applications [138].

Figure 6.38a shows a typical DSC trace from a pure, undoped $LiAlH_4$ (purity 97 %) obtained in our laboratory. It is clearly seen that (R1a, b) is preceded by an endothermic peak centered at 152.5 °C. In the literature this first endotherm is usually assigned to the interaction of $LiAlH_4$ with hydroxyl impurities [138]. An exothermic melting peak of $LiAlH_4$ is centered at 172 °C (R1a) and an endothermic decomposition peak (R1b) is centered at 193.9 °C. It is immediately followed by another endotherm (R2) centered at 242.7 °C. The decomposition of LiH (R3) is endothermic with the maximum at 438.1 °C. It is interesting to note that severe ball milling of $LiAlH_4$ for 20 h does not change in any substantive manner, within the experimental scatter, the positions of the DSC peaks as shown in Figure 6.38b. Hydroxyl reaction peak and (R1b) and (R2) peaks are slightly shifted to lower temperatures. However, more studies are needed to confirm this effect. Ball milling of $LiAlH_4$ doped with $TiCl_4$ may lead to the total transformation/decomposition into Li_3AlH_6 [140, 141]. However, this effect usually occurs at higher milling energy intensity of doped $LiAlH_4$ [138]. Chen *et al.* [136] reported substantial grain size refinement to ~18 nm of $LiAlH_4$ doped with

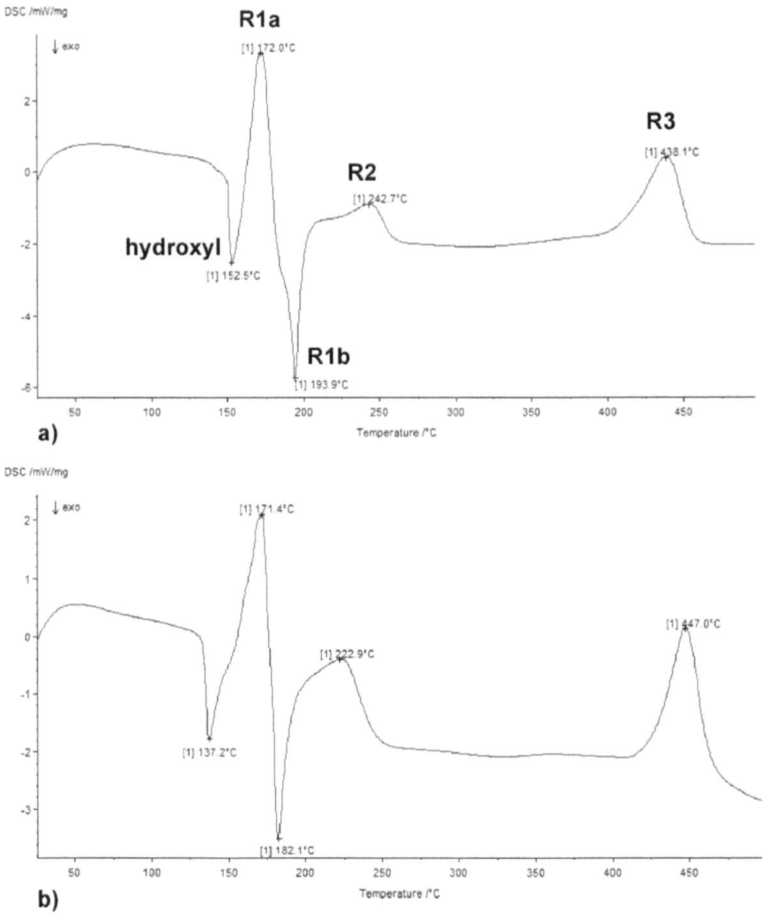

Figure 6.38. **a** DSC trace of as-received, undoped $LiAlH_4$; and **b** the same hydride after milling in the magneto mill Uni-Ball-Mill 5 under HES57 mode (two magnets at 5 and 7 o'clock) for 20 h. Heating rate 10 °C/min at argon flow 50 mL/min

$TiCl_3 \cdot 1/3AlCl_3$ after milling for 0.5–1 h. However, their result is in contradiction to our own average grain size of 58 nm for $LiAlH_4$ after ball milling for 20 h under HES57 (two magnets at 5 and 7 o'clock) mode. In turn, our result agrees well with Andreasen *et al.* [137] who reported a size of ~ 50 nm for $LiAlH_4$ after 6–10 h milling in a Retsch PM 100 planetary mill. They also showed that reaction constant, *k*, in Equations 6.4 and 6.6 increases with decreasing grain size of ball milled $LiAlH_4$ within the range 50–150 nm.

Since major decomposition events of $LiAlH_4$ occur at the low temperature range of 150–250 °C, this complex hydride is an excellent constituent for testing the ROM behavior given by Equation 6.11. Composites MgH_2 + X wt.%$LiAlH_4$ (X = 10, 20, 30, 50, and 70) (converted to vol.% for ROM analysis) were milled for 20 h in the magneto mill Uni-Ball-Mill 5 under HES57 mode (high energy shearing with two

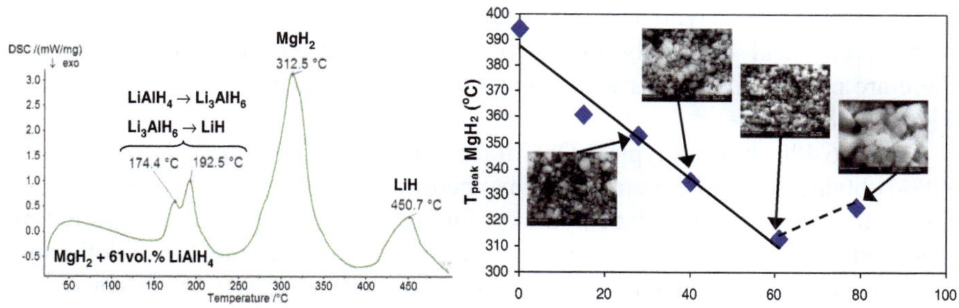

Figure 6.39. **a** Representative DSC curve for $MgH_2 + 61$ vol.%$LiAlH_4$ composite. Heating rate 10 °C/min at argon flow 50 mL/min. **b** Desorption temperature of the MgH_2 constituent in the $MgH_2 + LiAlH_4$ composite as a function of the $LiAlH_4$ content. Representative SEM pictures showing the morphology of composite powders are shown in the insets (all at the same magnification)

magnets at 5 and 7 o'clock; Figure 6.8). Figure 6.39a shows a representative DSC curve for the composition $MgH_2 + 61$ vol.%$LiAlH_4$. In this particular composite $LiAlH_4$ decomposes up to ~220 °C in a three-step process as described above. All other compositions in the MgH_2–$LiAlH_4$ system exhibited very similar DSC traces. The hydrogen desorption peak for MgH_2 centers at 312.5 °C. After 20 h milling the average grain size of the MgH_2 constituent in the composition range up to ~30 vol.%$LiAlH_4$ is ~13 nm but it increases to ~30 nm with a large standard deviation of ±20 nm for higher contents of $LiAlH_4$. Apparently, with increasing content of $LiAlH_4$ the MgH_2 components is being milled less effectively. The grain size of $LiAlH_4$ is ~50 nm. Figure 6.39b shows the ROM for the MgH_2 desorption temperature in the MgH_2–$LiAlH_4$ nanocomposite. It is obeyed up to about 60 vol.%$LiAlH_4$ with an excellent coefficient of fit $R^2 = 0.97$. At higher contents of $LiAlH_4$ in the composite the ROM breaks down, most likely due to very ineffective milling and formation of large cold-welded composite particulates as can be seen in the representative SEM insets in Figure 6.39b. In general, the $LiAlH_4$ constituent in the composite is much more difficult to effectively ball mill than MgH_2 since it behaves as sort of "grease" when present at a relatively large content.

6.4.2.3 MgH_2–$NaAlH_4$ Composite System

The thermal decomposition of $NaAlH_4$ is in essence quite similar to that of $LiAlH_4$ and also occurs in a three-step manner:

$$\text{(R1a) } NaAlH_4(s) \rightarrow NaAlH_4(l) \tag{6.16}$$

$$\text{(R1b) } NaAlH_4(l) \rightarrow 1/3Na_3AlH_6(s) + 2/3Al(s) + H_2(g) \tag{6.17}$$

$$\text{(R2) } 1/3Na_3AlH_6(s) \rightarrow NaH + 1/3Al + 0.5H_2 \tag{6.18}$$

$$\text{(R3) } NaH \rightarrow Na + 0.5H_2 \tag{6.19}$$

(R1b)–(R3) proceeds with a theoretical hydrogen release of 3.7, 1.9, and 1.8 wt.%, respectively (and obviously lower for a purity-corrected capacity). However, there are certain differences with respect to the DSC traces for $LiAlH_4$ shown in Figure 6.38.

Figure 6.40a shows a typical DSC trace from a pure, undoped $NaAlH_4$ (purity 90 %) obtained in our laboratory. The exothermic peak at 175.4 °C is most probably due to the presence of surface hydroxyl impurities in the powder as reported by Andreasen [138] for $LiAlH_4$ and observed in Figure 6.38a. For the endothermic 182.1 °C peak (R1a), the corresponding TGA weight loss (TGA traces not shown here) is almost non-existent. As a result, it is most probably due to the melting of $NaAlH_4$ which is reported by Claudy *et al.* [142] and Gross *et al.* [143] to occur at ~ 180 °C, rather than the first-step decomposition of $NaAlH_4$ claimed by Zaluski *et al.* [144]. The origin of the small peak at 258.6 °C is not clear. TGA shows a very small but recognizable weight loss at this temperature range. Claudy *et al.* [142] reported a phase transition of pseudocubic α-Na_3AlH_6 into face-centered cubic β-Na_3AlH_6 at about 252 °C which is close enough to the peak centered at 258.6 °C.

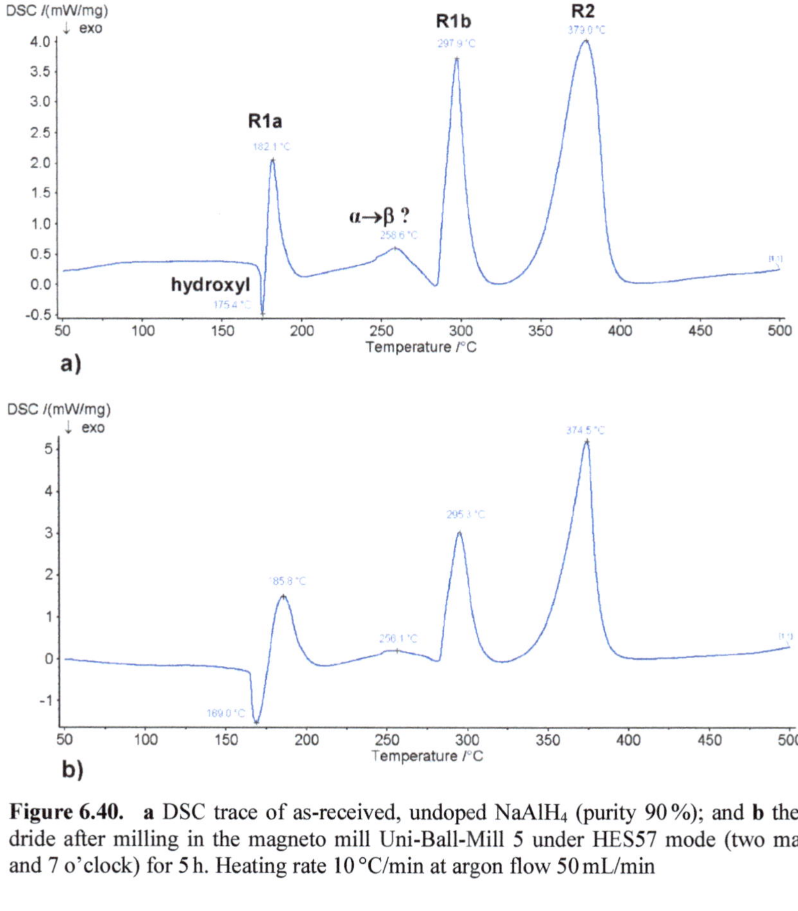

Figure 6.40. **a** DSC trace of as-received, undoped $NaAlH_4$ (purity 90 %); and **b** the same hydride after milling in the magneto mill Uni-Ball-Mill 5 under HES57 mode (two magnets at 5 and 7 o'clock) for 5 h. Heating rate 10 °C/min at argon flow 50 mL/min

However, such an explanation would require the existence of α-Na_3AlH_6 from at least partially decomposed $NaAlH_4$. TGA does not provide any firm evidence of partial decomposition of $NaAlH_4$ in this range of temperatures although it is possible that the decomposition of $NaAlH_4$ starts at this temperature with a simultaneous transformation of α-Na_3AlH_6 into β-Na_3AlH_6. For the peak at 297.9 °C the corresponding TGA shows in the temperature range 250–320 °C a substantial weight loss of ~3.5 wt.%, which is close to the theoretical purity-corrected value of ~3.3 wt.%H_2 of the first-step decomposition of $NaAlH_4$ (R1b). As a result, the peak is most probably attributed to the decomposition of $NaAlH_4$ into Na_3AlH_6, Al and hydrogen (R1b). For the peak at 379 °C, the corresponding TGA weight loss (350–400 °C) is ~1.6 wt.%, which is close to the theoretical value ~1.7 wt.%H_2 of the second-step decomposition of $NaAlH_4$ (R2). As a result, the peak is most probably due to the decomposition of Na_3AlH_6 into NaH, Al and hydrogen (R2). It seems that NaH might decompose at certain temperature higher than 400 °C, which gives a weight drop in the TGA curve. Further XRD study of powders heated to certain DSC peak temperatures is needed to reaffirm the nature of DSC-TGA thermal events in $NaAlH_4$. It is to be pointed out that severe ball milling of $NaAlH_4$ for 5 h does not change, within the experimental scatter, the positions of the DSC peaks as shown in Figure 6.40b. This behavior is exactly the same as observed for $LiAlH_4$ in Figure 6.38b, with the exception that (R2) occurs at much higher peak temperature than that for $LiAlH_4$. In addition, the intensity of the peak assigned to the probable $\alpha \rightarrow \beta$ transformation is substantially reduced by milling.

Composites $MgH_2 + X$ wt.%$NaAlH_4$ (X = 10, 30, 50, and 70) (converted to vol.% for ROM analysis) were milled for 5 h in the magneto mill Uni-Ball-Mill 5 under HES57 mode (two magnets at 5 and 7 o'clock; Figure 6.8). Figure 6.41a shows a representative DSC curve for the composition $MgH_2 + 73$ vol.%$NaAlH_4$, which is compared with the DSC curve of pure ball milled $NaAlH_4$ (from Figure 6.40b). The second stage of decomposition in which Na_3AlH_6 decomposes to NaH, Al and H_2 occurs at ~375 °C. As such, the decomposition temperature of Na_3AlH_6 overlaps with the decomposition temperature of MgH_2 in the MgH_2–$NaAlH_4$ composite. Apparently, because of this peculiar overlap of the desorption temperatures the ROM behavior for the MgH_2 temperature is not obeyed for this composite system as shown in Figure 6.41b. In addition, this composite system exhibits one of the worst milling behaviors in which regardless of the composition, the composite particulate remains relatively coarse as can be seen in the representative SEM insets in Figure 6.41b.

6.4.2.4 $NaBH_4$–MgH_2 Composite System

Figure 6.42 shows a DSC trace of pure $NaBH_4$ obtained in our laboratory. Two endothermic events are clearly observed. The peak centered at 498.1 °C corresponds to the melting of hydride and that at 577.2 °C corresponds to its decomposition according to the reaction [145]:

$$NaBH_4 \rightarrow Na + B + 2H_2 \tag{6.20}$$

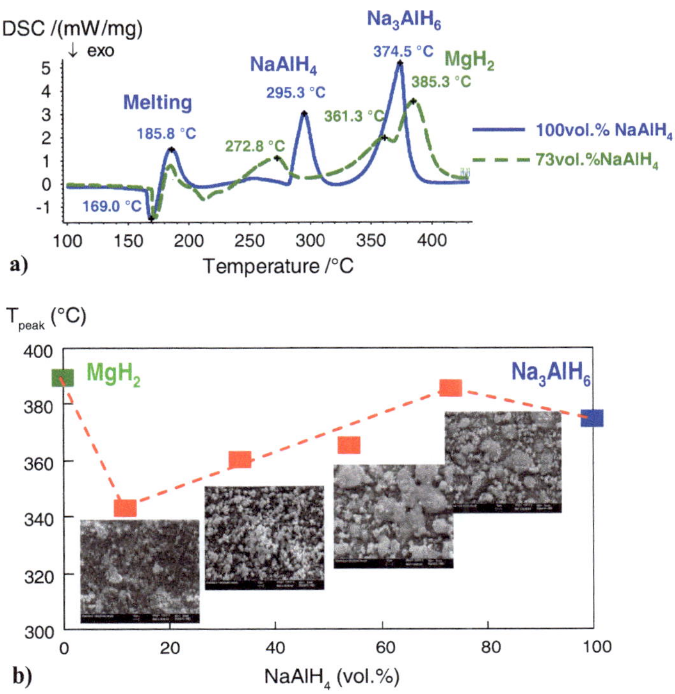

Figure 6.41. **a** Representative DSC curves for $NaAlH_4$ (purity 90 %) and $MgH_2 + 73$ vol.% $NaAlH_4$ composite. Heating rate 10 °C/min at argon flow 50 mL/min. **b** Desorption temperature of the MgH_2 constituent in the $MgH_2 + NaAlH_4$ composite as a function of the $NaAlH_4$ content. Representative SEM pictures showing the morphology of composite powders are shown in the insets (all at the same magnification)

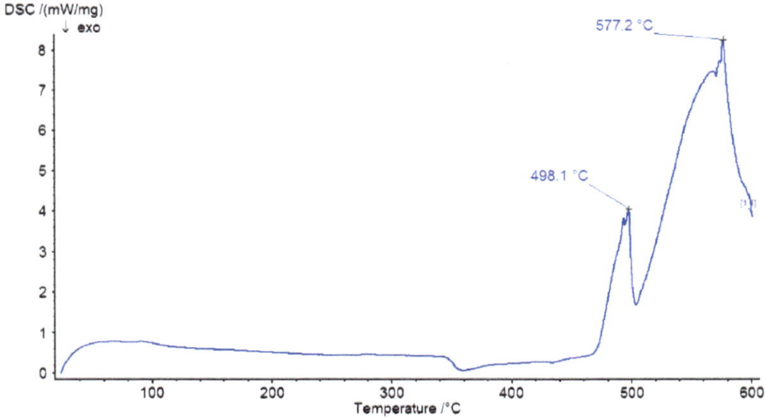

Figure 6.42. DSC trace of as-received $NaBH_4$ (purity 98 %). Heating rate 10 °C/min at argon flow 50 mL/min

In the $NaBH_4$–MgH_2 composite the MgH_2 constituent has much lower desorption temperature than $NaBH_4$ and decomposes first as shown in Figure 6.43a for the $NaBH_4$ + 30 vol.%MgH_2 composition. In the next stage the $NaBH_4$ component undergoes melting and finally decomposes at 500 °C to elemental Na, B, and H_2 as shown by Equation 6.20 [145]. However, a very interesting behavior is observed with increasing concentration of MgH_2 in the composite. Figure 6.43b shows that the DSC melting peak of $NaBH_4$ disappears and only the decomposition peak can be observed. Figure 6.44 shows the ROM plot for the melting and decomposition temperature of $NaBH_4$. The melting temperature shows very weak dependence on the content of MgH_2, which nevertheless can still be described quite well by the linear ROM dependence with $R^2 = 0.85$. In contrast, the decomposition temperature of $NaBH_4$ shows a strong linear dependence on the MgH_2 content, with an excellent fit to the ROM with $R^2 = 0.97$ as required by Equation 6.11.

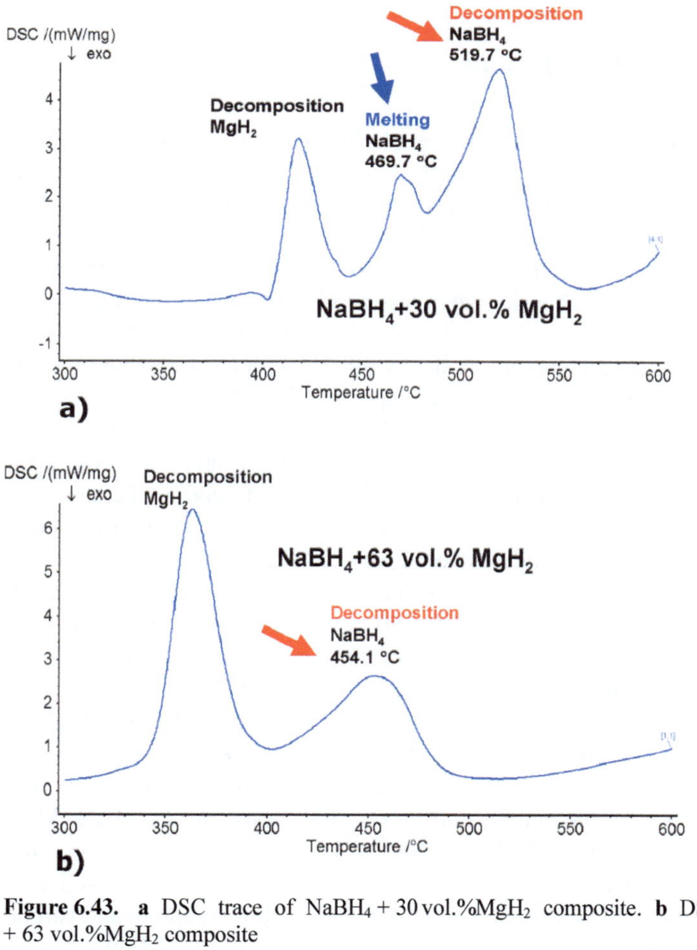

Figure 6.43. **a** DSC trace of $NaBH_4$ + 30 vol.%MgH_2 composite. **b** DSC trace of $NaBH_4$ + 63 vol.%MgH_2 composite

Figure 6.44 Melting and desorption temperature of the $NaBH_4$ constituent in the $NaBH_4+MgH_2$ composite as a function of the MgH_2 content (lower decomposition temperature constituent)

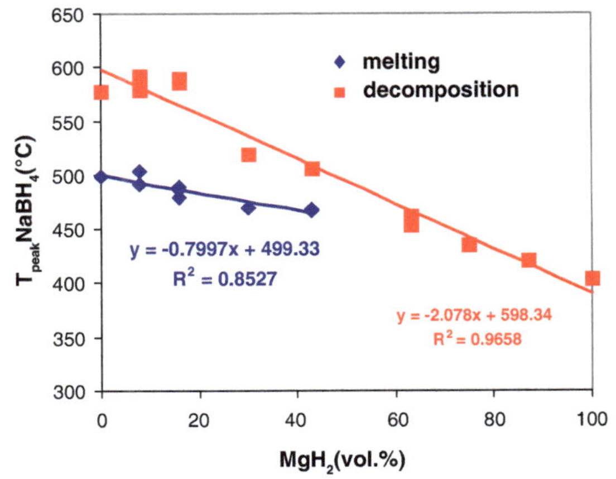

6.5 Summary and Conclusions

So far, MgH_2 remains the only fully reversible hydride for solid state hydrogen storage, which exhibits quite large theoretical capacity of ~7.6 wt.% at a moderately high temperature. Its hydrogen sorption behavior can be improved by ball milling. This improvement is primarily due to the formation of a metastable γ-MgH_2 phase and reduction of the particle size to ~1 μm, which is also accompanied by the formation of nanometric size grains within the powder particles. However, a relatively high enthalpy of formation precludes the decomposition of even ball milled MgH_2 at temperatures much below 300 °C with a reasonable kinetics. A drawback of ball milling is a small loss of capacity of MgH_2.

Catalytic additives improve even further the sorption kinetics of ball milled MgH_2. Nevertheless, a large number of additives to MgH_2 have been tested for desorption under vacuum, which is irrelevant to the commercial application of a PEM fuel cell operating at atmospheric pressure of 0.1 MPa or slightly higher. From all metal/intermetallic additives tested at 0.1 $MPaH_2$ and reported in the literature, such as $Mg_2Ni/Mg_2Ni_{1-x}M_x$ (M = Fe, Co) [86], Co [87], V, Zr and Y [91], FeTi and FeTiMn [92], Ce [93], $Mg_2Ni_{0.8}Cr_{0.2}+TiO_2$ [94] and Inco n-Ni [97–99 and present work], the most potent catalyst appears to be nano Ni. The other two catalysts of interest are FeTiMn [92] and Ce [93] but the tests with them should be repeated to establish if the results claimed in the literature are indeed reproducible. The effects of the oxide additives on the sorption properties of MgH_2 were tested primarily in vacuum, and include the following oxides: V_2O_5 [104], Cr_2O_3 [104–108], Fe_2O_3 and Fe_3O_4 [103, 106], Mn_2O_3 [103], MgO [108], Al_2O_3 [104, 109], La_2O_3 [110], and Nb_2O_5 [106, 109, 111–115]. In vacuum the most potent improvement of sorption properties of MgH_2 was observed with Nb_2O_5 but the tests should be repeated at 0.1 $MPaH_2$. Carbon, graphite and carbon nanotubes additives to MgH_2 were extensively investigated [117–128] but they do not seem to provide better enhancement of sorption properties than metal or oxide additives.

It must be pointed out that overwhelming experimental evidence indicates that catalytic and other additives to MgH_2 do not reduce its enthalpy of formation/decomposition and as such its equilibrium temperature at 0.1 MPa of hydrogen pressure as dictated by the Van't Hoff equation (Equation 6.2). The lowest desorption temperature of MgH_2 doped with n-Ni is still around 275 °C [this work].

For the first time nanocomposites of the systems $NaBH_4$–MgH_2, MgH_2–$VH_{0.81}$, MgH_2–$LiAlH_4$ and MgH_2–$NaAlH_4$ having nanograin sizes of the constituent phases have been successfully synthesized in a very wide range of compositions. In general, the hydrogen desorption temperature of the composite constituent with the higher desorption temperature in the systems such as $NaBH_4$–MgH_2, MgH_2–$VH_{0.81}$, and MgH_2–$LiAlH_4$, substantially decreases linearly with increasing volume fraction of the constituent having lower desorption temperature according to the well-known composite Rule-of-Mixtures. Composite systems of various hydrides show great promise for finding appropriate combinations of hydrides for an efficient onboard storage. This research area is now in its infancy and should be pursued more vigorously. In particular, the enthalpy of various promising hydride composite systems should be thoroughly assessed to established unambiguously if the reduction of the desorption temperature is indeed a thermodynamic effect.

Finally, it must be concluded that at the present moment there is no solid hydride or hydride system that conforms to all the DOE target requirements in Table 6.2. Nevertheless, relaxing, for example, the requirement of onboard hydrogen recharging and moving to off-board recharging would bring a number of solid hydrides and hydride systems to a nearly commercial stage of application.

References

[1] Bockris JOM. Will lack of energy lead to the demise of high-technology countries in this century? Int. J. Hydrogen Energy. 2007; 32:153–158.

[2] Hashimoto K, Habazaki H, Yamasaki M, Meguro S, Sasaki T, Katagiri H, Matsui T, Fujimura K, Izumiya K, Kumagai N, Akiyama E. Advanced materials for global carbon dioxide recycling. Mater. Sci. Eng. A 2001; 304–306:88–96.

[3] Yildiz B, Kazimi MS. Efficiency of hydrogen production systems using alternative nuclear energy technologies. Int. J. Hydrogen Energy. 2006; 31:77–92.

[4] Turner JA, Williams MC, Rajeshwar K. Hydrogen Economy based on renewable energy sources. The Electrochem. Soc. Interface 2004; 13(3):24–30.

[5] Ramani V, Kunz HR, Fenton JM. The polymer electrolyte fuel cell. The Electrochem. Soc. Interface 2004; 13(3):17–45.

[6] Lovins AB. Twenty hydrogen myths. Rocky Mountain Institute, 2005; www. rmi. org.

[7] Scott DS. Smelling Land-The Hydrogen Defense Against Climate Catastrophe. Canadian Hydrogen Association, 2007.

[8] Ewan BCR, Allen RWK. A figure of merit assessment of the routes to hydrogen. Int. J. Hydrogen Energy. 2007; 30:809–819.

[9] Züttel A. Hydrogen storage methods. Naturwissenschaften 2004; 91:157–172.
[10] Sandi G. Hydrogen storage and its limitations. The Electrochem. Soc. Interface 2004; 13(3):40–44.
[11] Satyapal S, Petrovic S, Thomas G, Read C, Ordaz G. The U. S. national hydrogen storage project. Proceedings of the 16th World Hydrogen Energy Conference, 13–16 June, 2006, Lyon, France, pp. 1/10–9/10.
[12] US Department of Energy Basic Research Needs for the Hydrogen Economy, Second Printing, February 2004,, Basic Research Challenges for Hydrogen Storage, pp. 31–51.
[13] Grochala W, Edwards PP. Thermal decomposition of the non-interstitial hydrides for the storage and production of hydrogen. Chem. Rev. 2004; 104:1283–1315.
[14] Sandrock G. A panoramic overview of hydrogen storage alloys from a gas reaction point of view. J. Alloys Compd. 1999; 293–295:877–888.
[15] Schlapbach L, Züttel A. Hydrogen-storage materials for mobile applications. Nature 2001; 414:353–358.
[16] Slattery DK, Hampton MD. Complex hydrides for hydrogen storage. Proceedings of the 2002 USDOE. Hydrogen Program Review NREL/CP-610-32405.
[17] Sandrock G, Bowman RC Jr. Gas-based hydride applications: recent progress and future needs. J. Alloys Compd. 2003; 356–357:794–799.
[18] Ritter JA, Ebner AD, Wang J, Zidan R. Implementing a hydrogen economy. Mater. Today 2003; 6(9):18–23.
[19] Züttel A. Materials for hydrogen storage. Mater. Today 2003; 6(9):24–33.
[20] Schüth F, Bogdanović B, Felderhoff M. Light metal hydrides for hydrogen storage. Chem. Commun. 2004; 2249–2258.
[21] Chandra D, Reilly JJ, Chellappa R. Metal hydrides for vehicular applications: the state of the art. J. Mater. 2006; 58:26–32.
[22] Varin RA, Li S, Wronski Z, Morozova O, Khomenko T. The effect and sequential and continuous high energy impact mode on the mechano-chemical synthesis of nanostructured complex hydride Mg_2FeH_6. J. Alloys Compd. 2005; 390:282–296.
[23] Orimo S, Nakamori Y, Kitahara G, Miwa K, Ohba N, Towata S, Züttel A. Dehydriding and rehydriding reactions of $LiBH_4$. J. Alloys Compd. 2005; 404–406: 427–430.
[24] Eun Jeon, Young Whan Cho. Mechanochemical synthesis and thermal decomposition of zinc borohydride. J. Alloys Compd. 2006; 422:273–275.
[25] Srinivasan S, Stefanakos E, Goswami Y. New transition metal assisted complex borohydrides for hydrogen storage. Proceedings of the 16th World Hydrogen Energy Conference, 13–16 June, 2006, Lyon, France, pp. 1/6–6/6.
[26] Andreasen A. PhD thesis. Risø National Laboratory, Roskilde, Denmark, 2005.
[27] Schwarz RB. Hydrogen storage in magnesium-based alloys. MRS Bull. 1999; 24(11): 40–44.
[28] Bazzanella N, Checchetto R, Miotello A. Appl. Phys. Lett. 2004; 85: 5212–5214.
[29] Kelton KF. Analysis of crystallization kinetics. Mater. Sci. Eng. A. 1997; 226–228: 142–150.
[30] Mintz MH, Zeiri Y. Hydriding kinetics of powders. J. Alloys Compd. 1994; 216: 159–175.
[31] Jensen TR, Andreasen A, Vegge T, Andreasen JW, Ståhl K, Pedersen AS, Nielsen MM, Molenbroek AM, Besebbacher F. Dehydrogenation kinetics of pure and nickel-doped magnesium hydride investigated by in situ time –resolved powder X-ray diffraction. Int. J. Hydrogen Energy. 2006; 31:2052–2062.
[32] Barkhordarian G, Klassen T, Bormann R. Kinetic investigation of the effect of milling time on the hydrogen sorption reaction of magnesium catalyzed with different Nb_2O_5 contents. J. Alloys Compd. 2006; 407:249–255
[33] Suryanarayana C. Nanocrystalline materials. Int. Mater. Rev. 1995; 40:41–64.
[34] Gleiter H. Nanostructured materials: state of the art and perspectives. NanoStruct. Mater. 1995; 6:3–14.

[35] Lu K. Nanocrystalline metals crystallized from amorphous solids nanocrystallization, structure, and properties. Mater. Sci. Eng. R 1996; 16:161–221.
[36] R. A. Varin, T. Czujko. Overview of processing of nanocrystalline hydrogen storage intermetallics by mechanical alloying/milling. Mater. Manuf. Process. 2002; 17:129–156.
[37] Zaluski L, Zaluska A, Tessier P, Ström-Olsen JO, R. Schulz. Hydrogen absorption in nanocrystalline Mg_2Ni formed by mechanical alloying. J. Alloys Compd. 1995; 217: 245–249.
[38] Zaluski L, Zaluska A, Tessier P, Ström-Olsen JO, Schulz R. Catalytic effect of Pd on hydrogen absorption in mechanically alloyed Mg_2Ni, $LaNi_5$ and FeTi. J. Alloys Compd. 1995; 217:295–300.
[39] Zaluski L, Zaluska A, Ström-Olsen JO. Nanocrystalline metal hydrides. J. Alloys Compd. 1997; 253–254:70–79.
[40] Zaluska A, Zaluski L, Ström-Olsen JO. Nanocrystalline magnesium for hydrogen storage. J. Alloys Compd. 1999; 288:217–225.
[41] Zaluska A, Zaluski L, Ström-Olsen JO. Structure, catalysis and atomic reactions on the nano-scale: a systematic approach to metal hydrides for hydrogen storage. Appl. Phys. A 2001; 72:157–165
[42] Bèrube V, Radtke G, Dresselhaus M, Chen G. Size effects on the hydrogen storage properties of nanostructured metal hydrides: A review. Int. J. Energy. Res. 2007; 31:637–663.
[43] Suryanarayana C. Mechanical alloying and milling. New York, Marcel Dekker, 2004.
[44] Lü L, Lai MO. Mechanical Alloying. Boston, Kluwer Academic, 1998.
[45] Calka A, Varin RA. Application of controlled ball milling in materials processing. Proceedings of Conference on. Processing and Fabrication of Advanced Materials IX, ASM International, pp. 263–265, 2001.
[46] Courtesy A. O. C. Scientific Engineering, Australia; Patents: WO9104810, US5383615, CA2066740, EP0494899, and AU643949.
[47] Klug HP, Alexander L. X-ray Diffraction Procedures for Polycrystalline and Amorphous Materials. New York, John Wiley & Sons, pp. 618–708, 1974.
[48] Read C, Thomas G, Ordaz G, Satyapal S. U. S. Department of Energy's system targets for on-board vehicular hydrogen storage. Mater. Matters (Sigma-Aldrich) 2007; 2(2):3–4.
[49] Jensen JO, Li Q, He R, Pan C, Bjerrum NJ. 100–200 °C polymer fuel cells for use with $NaAlH_4$. J. Alloys Compd. 2005; 404–406:653–656.
[50] Maddalena A, Petris M, Palade P, Sartori S, Principi G, Settimo E, Molinas B, Lo Russo S. Study of Mg-based materials to be used in a functional solid state hydrogen reservoir for vehicular applications. Int. J. Hydrogen. Energy. 2006; 31:2097–2103.
[51] Sandrock G, Reilly J, Graetz J, Zhou W-M, Johnson J, Wegrzyn J. Accelerated thermal decomposition of AlH_3 for hydrogen-fueled vehicles. Appl. Phys. A 2005; 80:687–690.
[52] Noritake T, Towata S, Aoki M, Seno Y, Hirose Y, Nishibori E, Takata M, Sakata M. Charge density measurement in MgH_2 by synchrotron X-ray diffraction. J. Alloys Compd. 2003; 356–357:84–86.
[53] CRC, Physical Constants of Inorganic Compounds, CRC Handbook of Chemistry and Physics, 87th On Line Edition, 2006–2007.
[54] H. Okamoto, Desk Handbook-Phase Diagrams for Binary Alloys. Materials Park, OH, ASM International, p. 430, 2000.
[55] Noritake T, Aoki M, Towata S, Seno Y, Hirose Y, Nishibori E, Takata M, Sakata M. Chemical bonding of hydrogen in MgH_2. Appl. Phys. Lett. 2002; 81:2008–2010.
[56] Bastide J. -P, Bonnetot B, Létoffé J. -M, Claudy P. Polymorphisme de l'hydrure de magnesium sous haute pression. Mater. Res. Bull. 1980; 15:1779–1787.
[57] Huot J, Liang G, Boily S, Van Neste A, Schulz R. Structural study and hydrogen sorption kinetics of ball-milled magnesium hydride. J. Alloys Compd. 1999; 293–295:495–500.
[58] Schulz R, Huot J, Liang G, Boily S, Van Neste A. Structure and hydrogen sorption properties of ball milled Mg dihydride. Mater. Sci. Forum 1999; 312–314:615–622.

[59] Schulz R, Huot J, Liang G, Boily S, Lalande G, Denis MC, Dodelet JP. Recent developments in the application of nanocrystalline materials to hydrogen technologies. Mater. Sci. Eng. A 1999; 267:240–245.
[60] Gennari FC, Castro FJ, Urretavizcaya G. Hydrogen desorption behavior from magnesium hydrides synthesized by reactive mechanical alloying. J. Alloys Compd. 2001; 321:46–53.
[61] Friedrichs O, Sánchez-López JC, López-Cartés C, Dornheim M, Klassen T, Bormann R, Fernández A. Chemical and microstructural study of the oxygen passivation behaviour of nanocrystalline Mg and MgH_2. Appl. Surf. Sci. 2006; 252:2334–2345.
[62] Varin RA, Li S, Calka A. Environmental degradation by hydrolysis of nanostructured β-MgH_2 hydride synthesized by controlled reactive mechanical milling. J. Alloys Compd. 2004; 376:222–231.
[63] Liang G, Huot J, Boily S, Van Neste A, Schulz. Catalytic effect of transition metals on hydrogen sorption in nanocrystalline ball milled MgH_2-Tm (Tm = Ti, V, Mn, Fe and Ni) systems. J. Alloys Compd. 1999; 292:247–252.
[64] Liang G, Huot J, Boily S, Van Neste A, Schulz. Hydrogen storage properties of the mechanically milled MgH_2-V nanocomposite. J. Alloys Compd. 1999; 291:295–299.
[65] Dehouche Z, Djaozandry R, Huot J, Boily S, Goyette J, Bose TK, Schulz R. Influence of cycling on the thermodynamic and structure properties of nanocrystalline magnesium based hydride. J. Alloys Compd. 2000; 305:264–271.
[66] Bouaricha S, Huot J, Guay D, Schulz R. Reactivity during cycling of nanocrystalline Mg-based hydrogen storage compounds. Int. J. Hydrogen Energy. 2002; 27:909–913.
[67] Huot J, Pelletier JF, Lurio LB, Sutton M, Schulz R. Investigation of dehydrogenation mechanism of MgH_2-Nb nanocomposites. J. Alloys Compd. 2003; 348:319–324.
[68] Shang CX, Bououdina M, Song Y, Guo ZX. Mechanical alloying and electronic simulations of (MgH_2 + M) systems (M = Al, Ti, Fe, Ni, Cu and Nb) for hydrogen storage. Int. J. Hydrogen Energy. 2004; 29:73–80.
[69] Au M. Hydrogen storage properties of magnesium based nanostructured/amorphous composite materials. Mater. Res. Soc. Symp. Proc. 2004; 801:BB1. 5.1–5.14.
[70] Bobet J-L, Chevalier B, Song MY, Darriet B, Etourneau J. Hydrogen sorption of Mg-based mixtures elaborated by reactive mechanical grinding. J. Alloys Compd. 2002; 336:292–296.
[71] Hu YQ, Zhang HF, Wang AM, Ding BZ, Hu ZQ. Preparation and hydriding/dehydriding properties of mechanically milled Mg-30wt. %$TiMn_{1.5}$ composite. J. Alloys Compd. 2003; 354:296–302.
[72] Hu YQ, Yan C, Zhang HF, Ye L, Hu ZQ. Preparation and hydrogenation characteristics of Mg-30wt. %$Ti_{37.5}V_{25}Cr_{37.5}$ composite. J. Alloys Compd. 2004; 375:265–269.
[73] Tran NE, Lambrakos SG, Imam MA. Analyses of hydrogen sorption kinetics and thermodynamics of magnesium-misch metal alloy. J. Alloys Compd. 2006; 407:240–248.
[74] Skripnyuk V, Buchman E, Rabkin E, Estrin Y, Popov M, Jorgensen S. The effect of equal channel angular pressing on hydrogen storage properties of a eutectic Mg-Ni alloy. J. Alloys Compd. 2007; 436:99–106.
[75] Yonkeu A, Swainson IP, Dufour J, Huot J. Kinetic investigation of the catalytic effect of a body centered cubic-alloy $TiV_{1.1}Mn_{0.9}$ (BCC) on hydriding/dehydriding properties of magnesium. J. Alloys Compd. 2008; 460:559–564.
[76] Fu Y, Groll M, Mertz R, Kulenovic R. Effect of $LaNi_5$ and additional catalysts on hydrogen storage properties of Mg. J. Alloys Compd. 2008; 460:607–613.
[77] Sai Raman SS, Srivastava ON. Hydrogenation behavior of the new composite storage materials Mg-xwt. %$CFMmNi_5$. J. Alloys Compd. 1996; 241:167–174.
[78] Zaluska A, Zaluski L, Ström-Olsen JO. Synergy of hydrogen sorption in ball-milled hydrides of Mg and Mg_2Ni. J. Alloys Compd. 1999; 289:197–206.
[79] Johnson SR, Anderson PA, Edwards PP, Gameson I, Prendergast JW, Al-Mamouri A, Book D, Rex Harris I, Speight JD, Walton A. Chemical activation of MgH_2; a new route to superior hydrogen storage materials. Chem. Commun. 2005; 2823–2825.

[80] Kojima Y, Kawai Y, Haga T. Magnesium-based nano-composite materials for hydrogen storage. J. Alloys Compd. 2006; 424:294–298.
[81] Yoo Y, Tuck M, Kondakindi R, Seo CY, Dehouche Z, Belkacemi K. Enhanced hydrogen reaction kinetics of nanostructured Mg-based composites with nanoparticle metal catalysts dispersed on supports. J. Alloys Compd. 2007; 446–447:84–89.
[82] Dufour J, Huot J. Rapid activation, enhanced hydrogen sorption kinetics and air resistance in laminated Mg-Pd 2. 5 at. %. J. Alloys Compd. 2007; 439:L5–L9.
[83] Dufour J, Huot J. Study of Mg_6Pd alloy synthesized by cold rolling. J. Alloys Compd. 2007; 446–447:147–151.
[84] Song MY, Mumm DR, Kwon SN, Hong SH, Bae JS. Hydrogen-storage properties of Mg-10wt. % (Fe_2O_3, Ni, MnO) alloy prepared by reactive mechanical grinding. J. Alloys Compd. 2006; 416:239–244.
[85] Song MY, Kwon SN, Hong SH, Mumm DR, Bae JS. Improvement of hydrogen-sorption characteristics of Mg by reactive mechanical grinding with Cr_2O_3 prepared by spray conversion. Int. J. Hydrogen Energy. 2006; 31:2284–2291.
[86] Grigorova E, Khristov M, Khrussanova, Bobet J-L, Peshev P. Effect of additives on the hydrogen sorption properties of mechanically alloyed composites based on Mg and Mg_2Ni. Int. J. Hydrogen Energy. 2005; 30:1099–1105.
[87] Bobet J-L, Chevalier B, Darriet B. Effect of reactive mechanical grinding on chemical and hydrogen sorption properties of the Mg + 10wt. %Co mixture. J. Alloys Compd. 2002; 330–332:738–742.
[88] Yu Z, Liu Z, Wang E. Hydrogen storage properties of nanocomposite Mg-Ni-Cu-$CrCl_3$. Mater. Sci. Eng. A 2002; 335:43–48.
[89] Li F, Jiang L, Du J, Wang S, Liu X, Zhang F. Investigations on synthesis and hydrogenation properties of Mg-20wt%Ni-1wt%TiO_2 composite prepared by reactive mechanical alloying. J. Alloys Compd. 2008; 452:421–424.
[90] Løken S, Solberg JK, Maehlen JP, Denys RV, Lototsky MV, Tarasov BP, Yartys VA. Nanostructured Mg-Mm-Ni hydrogen storage alloy: structure–properties relationship. J. Alloys Compd. 2007; 446–447:114–120.
[91] Czujko T, Varin RA, Chiu Ch, Wronski Z. Investigation of the hydrogen desorption properties of Mg + 10wt. %X (X = V,Y,Zr). J. Alloys Compd. 2006; 414:240–247.
[92] Vijay R, Sundaresan R, Maiya MP, Srinivasa Murthy S, Fu Y, Klein H-P, Groll M. Characterisation of Mg-x wt. %FeTi (x = 5–30) and Mg-40wt. %FeTiMn hydrogen absorbing materials prepared by mechanical alloying. J. Alloys Compd. 2004; 384:283–295.
[93] Wang XL, Tu JP, Wang CH, Zhang XB, Chen CP, Zhao XB. Hydrogen storage properties of nanocrystalline Mg-Ce/Ni composite. J. Power Sources 2006; 159:163–166.
[94] Wang XL, Tu JP, Zhang XB, Chen CP, Zhao XB. Hydrogenation properties of $Mg/Mg_2Ni_{0.8}Cr_{0.2}$ composites containing TiO_2 particles. J. Alloys Compd. 2005; 404–406: 529–532.
[95] Bystrzycki J, Polanski M, Plocinski T. Nano-engineering approach to destabilization of magnesium hydride (MgH_2) by solid-state reaction with Si. J. Nanosci. Nanotechnol. 2009; 9(12):3441–3448.
[96] Vajo JJ, Mertens F, Ahn CC, Bowman RC Jr., Fultz B. Altering hydrogen storage properties by hydride destabilization through alloy formation: LiH and MgH_2 destabilized with Si. J. Phys. Chem. 2004; 108:13977–13983.
[97] Varin RA, Czujko T, Wasmund EB, Wronski Z. Catalytic effects of various forms of nickel on the synthesis rate and hydrogen desorption properties of nanocrystalline magnesium hydride (MgH_2) synthesized by controlled reactive mechanical milling (CRMM). J. Alloys Compd. 2007; 432:217–231.
[98] Hanada N, Ichikawa T, Fujii H. Catalytic effect of Ni nano-particle and Nb oxide on H-desorption properties of MgH_2 prepared by ball milling. J. Alloys Compd. 2005; 404–406: 716–719.

[99] Varin RA, Czujko T, Wasmund EB, Wronski ZS. Hydrogen desorption properties of MgH_2 nanocomposites with nano-oxides and Inco micrometric-and nanometric-Ni. J. Alloys Compd. 2007; 446–447:63–66.
[100] Varin RA, Czujko T, Wronski ZS. Nanomaterials for solid state hydrogen storage. New York, Springer Science+Business Media, 2009.
[101] Oelerich W, Klassen T, Bormann R. Comparison of the catalytic effects of V, V_2O_5, VN, and VC on the hydrogen sorption of nanocrystalline Mg. J. Alloys Compd. 2001; 322: L5–L9.
[102] Oelerich W, Klassen T, Bormann R. Metal oxides as catalysts for improved hydrogen sorption in nanocrystalline Mg-based materials. J. Alloys Compd. 2001; 315:237–242.
[103] Jung KS, Lee EY, Lee KS. Catalytic effects of metal oxide on hydrogen absorption of magnesium metal hydride. J. Alloys Compd. 2006; 421:179–184.
[104] Dehouche Z, Klassen T, Oelerich W, Goyette J, Bose TK, Schulz R. Cycling and thermal stability of nanostructured MgH_2-Cr_2O_3 composite for hydrogen storage. J. Alloys Compd. 2002; 347:319–323.
[105] Barkhordarian G, Klassen T, Bormann R. Fast hydrogen sorption kinetics of nanocrystalline Mg using Nb_2O_5 as catalyst. Scripta Mater. 2003; 49:213–217.
[106] Bobet J-L, Castro FJ, Chevalier B. Effects of RMG conditions on the hydrogen sorption properties of $Mg + Cr_2O_3$ mixtures. Scripta Mater. 2005; 52:33–37.
[107] Aguey-Zinsou K-F, Nicolaisen T, Ares Fernandez JR, Klassen T, Bormann R. Effect of nanosized oxides on MgH_2 (de)hydriding kinetics. J. Alloys Compd. 2007; 434–435: 738–742.
[108] Friedrichs O, Klassen T, Sánchez-López JC, Bormann R, Fernández A. Hydrogen sorption improvement of nanocrystalline MgH_2 by Nb_2O_5 nanoparticles. Scripta Mater. 2006; 54:1293–1297.
[109] Gupta R, Agresti F, Lo Russo S, Maddalena A, Palade P, Principi G. Structure and hydrogen storage properties of MgH_2 catalysed with La_2O_3. J. Alloys Compd. 2008; 450: 310–313.
[110] Barkhordarian G, Klassen T, Bormann R. Effect of Nb_2O_5 content on hydrogen reaction kinetics of Mg. J. Alloys Compd. 2004; 364:242–246.
[111] Hanada N, Ichikawa T, Hino S, Fujii H. Remarkable improvement of hydrogen sorption kinetics in magnesium catalyzed with Nb_2O_5. J. Alloys Compd. 2006; 420:46–49.
[112] Hanada N, Ichikawa T, Hino S, Fujii H. Hydrogen absorption kinetics of the catalyzed MgH_2 by niobium oxide. J. Alloys Compd. 2007; 446–447:67–71.
[113] Aguey-Zinsou K-F, Ares Fernandez JR, Klassen T, Bormann R. Effect of Nb_2O_5 on MgH_2 properties during mechanical milling. Int. J. Hydrogen Energy. 2007; 32:2400–2407.
[114] Bhat VV, Rougier A, Aymard L, Nazri GA, Tarascon J-M. High surface area niobium oxides as catalysts for improved hydrogen sorption properties of ball milled MgH_2. J. Alloys Compd. 2008; 460:507–512.
[115] Mandzhukova Ts, Khrussanova M, Grigorova E, Stefanov P, Khristov M, Peshev P. Effect of $NiCo_2O_4$ additives on the hydriding properties of magnesium. J. Alloys Compd. 2008; 457:472–476.
[116] Imamura H, Takesue Y, Akimoto T, Tabata S. Hydrogen-absorbing magnesium composites prepared by mechanical grinding with graphite: effects of additives on composite structures and hydriding properties. J. Alloys Compd. 1999; 293–295:564–568.
[117] Bouaricha S, Dodelet JP, Guay D, Huot J, Schulz R. Activation characteristics of graphite modified hydrogen absorbing materials. J. Alloys Compd. 2001; 325:245–251.
[118] Huot J, Tremblay M-L, Schulz R. Synthesis of nanocrystalline hydrogen storage materials. J. Alloys Compd. 2003; 356–357:603–607.
[119] Bobet J-L, Grigorova E, Khrussanova M, Khristov M, Stefanov P, Peshev P, Radev D. Hydrogen sorption properties of graphite-modified magnesium nanocomposites prepared by ball-milling. J. Alloys Compd. 2004; 366:298–302.

[120] Dal Toè S, Lo Russo S, Maddalena A, Principi G, Saber A, Sartori S, Spataru T. Hydrogen desorption from magnesium hydride-graphite nanocomposites produced by ball milling. Mater. Sci. Eng. B 2004; 108:24–27.
[121] Wu CZ, Wang P, Yao X, Liu C, Chen DM, Lu GQ, Cheng HM. Effect of carbon/noncarbon addition on hydrogen storage behaviors of magnesium hydride. J. Alloys Compd. 2006; 414:259–264.
[122] Chen D, Chen L, Liu S, Ma CX, Chen DM, Wang LB. Microstructure and hydrogen storage property of Mg/MWNTs composites. J. Alloys Compd. 2004; 372:231–237.
[123] Wu CZ, Wang P, Yao X, Liu C, Chen DM, Lu GQ, Cheng HM. Hydrogen storage properties of MgH_2/SWNT composite prepared by ball milling. J. Alloys Compd. 2006; 420:278–282.
[124] Yao X, Wu CZ, Wang H, Cheng HM, Lu GQ. Effects of carbon nanotubes and metal catalysts on hydrogen storage in magnesium nanocomposites. J. Nanosci. Nanotechnol. 2006; 6:494–498.
[125] Imamura H, Tabata S, Shigetomi N, Takesue Y, Sakata Y. Composites for hydrogen storage by mechanical grinding of graphite carbon and magnesium. J. Alloys Compd. 2002; 330–332:579–583.
[126] Imamura H, Kusuhara M, Minami S, Matsumoto M, Masanari K, Sakata Y, Itoh K, Fukunaga T. Carbon nanocomposites synthesized by high-energy mechanical milling of graphite and magnesium for hydrogen storage. Acta Mater. 2003; 51:6407–6414.
[127] Huang ZG, Guo ZP, Calka A, Wexler D, Liu HK. Effect of carbon black, graphite and carbon nanotubes additives on hydrogen storage properties of magnesium. J. Alloys Compd. 2007; 427:94–100.
[128] Orimo S-I, Nakamori Y, Eliseo JR, Züttel A, Jensen CM. Complex hydrides for hydrogen storage. Chem. Rev. 2007; 107:4111–4132.
[129] R. A. Varin, T. Czujko, Ch. Chiu, R. Pulz, Z. Wronski. Synthesis of nanocomposite hydrides for solid state hydrogen storage by controlled mechanical milling techniques. J. Alloys Compd. 2009; 483:252–255.
[130] Reilly JJ, Wiswall Jr RH. The higher hydrides of vanadium and niobium. Inorg. Chem. 1970; 9:1678–1682.
[131] Chandra D, Sharma A, Chellappa R, Cathey WN, Lynch FE, Bowman Jr. RC, Wermer JR, Paglieri SN. Hydriding and structural characteristics of thermally cycled and cold-worked V-0.5 at. %C alloy. J. Alloys Compd. 2008; 452:312–314.
[132] Varin RA, Czujko T, Chiu Ch, Wronski Z. Particle size effects on the desorption properties of nanostructured magnesium dihydride (MgH_2) synthesized by controlled reactive mechanical milling (CRMM). J. Alloys Compd. 2006; 424:356–364.
[133] Varin RA, Czujko T, Wronski Z. Particle size, grain size and γ-MgH_2 effects on the desorption properties of nanocrystalline commercial magnesium hydride processed by controlled mechanical milling. Nanotechnology 2006; 17:3856–3865.
[134] Dilts JA, Ashby EC. A study of thermal decomposition of complex metal hydrides. Inorg. Chem. 1972; 11:1230–1236.
[135] Chen J, Kuriyama N, Xu Q, Takeshita HT, Sakai T. Reversible hydrogen storage via titanium-catalyzed $LiAlH_4$ and Li_3AlH_6. J. Phys. Chem. 2001; 105:11214–11220.
[136] Andreasen A, Vegge T, Pedersen AS. Dehydrogenation kinetics of as-received and ball-milled $LiAlH_4$. J. Solid State Chem. 2005; 178:3672–3678.
[137] Andreasen A. Effect of Ti-doping on the dehydrogenation kinetic parameters of lithium aluminum hydride. J. Alloys Compd. 2006; 419:40–44.
[138] Brinks HW, Hauback BC, Norby P, Fjellvåg H. The decomposition of $LiAlD_4$ studied by in-situ X-ray and neutron diffraction. J. Alloys Compd. 2003; 351:222–227.
[139] Balema VP, Dennis KW, Pecharsky VK. Rapid solid-state transformation of tetrahedral $[AlH_4]^-$ into octahedral $[AlH_6]^{3-}$ in lithium aluminohydride. Chem. Commun. 2000; 1665–1666.

[140] Balema VP, Wiench JW, Dennis KW, Pruski M, Pecharsky VK. Titanium catalyzed solid-state transformations in $LiAlH_4$ during high-energy ball milling. J. Alloys Compd. 2001; 329:108–114.
[141] Claudy P, Bonnetot B, Chahine G, Letoffe JM. Etude du comportement thermique du tetrahydroaluminate de sodium $NaAlH_4$ et de l'hexahydroaluminate de sodium Na_3AlH_6 de 298 a 600 K. Thermochem. Acta 1980; 38:75–88.
[142] Gross KJ, Guthrie S, Takara S, Thomas G. In-situ X-ray diffraction study of the decomposition of $NaAlH_4$. J. Alloys Compd. 2000; 297:270–281.
[143] Zaluski L, Zaluska A, Ström-Olsen JO. Hydrogenation properties of complex alkali metal hydrides fabricated by mechano-chemical synthesis. J. Alloys Compd. 1999; 290:71–78.
[144] Stasinevich DS, Egorenko GA. Thermographic investigation of alkali metal and magnesium tetrahydroborates at pressures up to 10 atm. Russian J. Inorg. Chem. 1968; 13: 341–343.

MIX
Papier aus verantwortungsvollen Quellen
Paper from responsible sources
FSC® C105338

If you have any concerns about our products,
you can contact us on
ProductSafety@springernature.com

In case Publisher is established outside the EU,
the EU authorized representative is:
Springer Nature Customer Service Center GmbH
Europaplatz 3, 69115 Heidelberg, Germany

Printed by Libri Plureos GmbH
in Hamburg, Germany